Molecular Biology Intelligence Unit

Toll and Toll-Like Receptors: An Immunologic Perspective

Tina Rich, Ph.D.
Department of Pathology
University of Cambridge
Cambridge, United Kingdom

Landes Bioscience / Eurekah.com
Georgetown, Texas
U.S.A.

Kluwer Academic / Plenum Publishers
New York, New York
U.S.A.

Toll and Toll-like Receptors: An Immunologic Perspective

Molecular Biology Intelligence Unit

Landes Bioscience / Eurekah.com
Kluwer Academic / Plenum Publishers

Printed in the U.S.A.

Kluwer Academic / Plenum Publishers, 233 Spring Street, New York, New York, U.S.A. 10013
http://www.wkap.nl/

Please address all inquiries to the Publishers:
Landes Bioscience / Eurekah.com, 810 South Church Street, Georgetown, Texas, U.S.A. 78626
Phone: 512.863.7762; Fax: 512.863.0081
http://www.eurekah.com
http://www.landesbioscience.com
ISBN: 0-306-48237-1

Toll and Toll-Like Receptors: An Immunologic Perspective, edited by Tina Rich, Landes / Kluwer dual imprint / Landes series: Medical Intelligence Unit

Library of Congress Cataloging-in-Publication Data

Toll and toll-like receptors : an immunologic perspective / [edited by] Tina Rich.
p. ; cm. -- (Molecular biology intelligence unit)
Includes bibliographical references and index.
ISBN 0-306-48237-1
1. B cells--Receptors. 2. Macrophage--Receptors. 3. Dendritic cells--Receptors. 4. Natural immunity. 5. Cellular immunity. I. Rich, Tina. II. Series: Molecular biology intelligence unit (Unnumbered)
[DNLM: 1. Receptors, Cell Surface--immunology. 2. Membrane Glycoproteins--immunology. QW 504.5 T651 2005]
QR185.8.B15T65 2005
616.07'98--dc22

2004022707

To Alice, Lotte, Liz and Tracy

CONTENTS

EDITOR

Tina Rich, Ph.D.
Department of Pathology
University of Cambridge
Cambridge, United Kingdom
Introduction

CONTRIBUTORS

Sudhir Agrawal
Hybridon Inc.
Cambridge, Massachusetts, U.S.A.
Chapter 9

Shizuo Akira
Department of Host Defense
Research Institute for Microbial Diseases
Osaka University
and
Akira Innate Immunity Project
ERATO
Japan Science and Technology Corporation
Suita, Osaka, Japan
Chapter 6

Elizabeth H. Bassett
University of Brighton
Brighton, East Sussex, United Kingdom
Introduction

Bruce Beutler
Department of Immunology
The Scripps Research Institute
La Jolla, California, U.S.A.
Chapter 8

Zlatko Dembic
Institute for Oral Biology
Faculty of Dentistry
University of Oslo
Blindern, Oslo, Norway
Chapter 1

Xin Du
Department of Immunology
The Scripps Research Institute
La Jolla, California, U.S.A.
Chapter 8

Jonathan J. Ewbank
Centre d'Immunologie de Marseille Luminy
INSERM/CNRS/Université de la Méditerranée
Marseille, Cedex, France
Chapter 7

Monique Gangloff
Department of Biochemistry
University of Cambridge
Cambridge, United Kingdom
Chapter 2

Nicholas J. Gay
Department of Biochemistry
University of Cambridge
Cambridge, United Kingdom
Chapter 2

Philippe Georgel
Department of Immunology
The Scripps Research Institute
La Jolla, California, U.S.A.
Chapter 8

Frank Henkler
Institute of Cell Biology and Immunology
University of Stuttgart
Stuttgart, Germany
Chapter 4

Kasper Hoebe
Department of Immunology
The Scripps Research Institute
La Jolla, California, U.S.A.
Chapter 8

Tsuneyasu Kaisho
Department of Host Defense
Research Institute for Microbial Diseases
Osaka University
Suita, Osaka, Japan
and
RIKEN Research Center for Allergy
and Immunology
Yokohama Kanagawa, Japan
Chapter 6

Ekambar R. Kandimalla
Hybridon Inc.
Cambridge, Massachusetts, U.S.A.
Chapter 9

Phumzile L. Ludidi
Department of Biochemistry
University of Cambridge
Cambridge, United Kingdom
Chapter 2

Luke A.J. O'Neill
Department of Biochemistry
Trinity College Dublin
Dublin, Ireland
Chapter 5

Nathalie Pujol
Centre d'Immunologie de Marseille
Luminy
INSERM/CNRS/Université
de la Méditerranée
Marseille, Cedex, France
Chapter 7

Peter Scheurich
Institute of Cell Biology
and Immunology
University of Stuttgart
Stuttgart, Germany
Chapter 4

Kathy Triantafilou
Institute of Biomedical
and Biomolecular Sciences
University of Portsmouth
School of Biological Sciences
Portsmouth, United Kingdom
Chapter 3

Martha Triantafilou
Institute of Biomedical
and Biomolecular Sciences
University of Portsmouth
School of Biological Sciences
Portsmouth, United Kingdom
Chapter 3

Harald Wajant
Institute of Cell Biology
and Immunology
University of Stuttgart
Stuttgart, Germany
Chapter 4

Claudia Wietek
Department of Biochemistry
Trinity College Dublin
Dublin, Ireland
Chapter 5

Introduction

Toll Receptors and the Renaissance of Innate Immunity

Elizabeth H. Bassett and Tina Rich

Overview

In the last few pages of *Immunology: The Science of Self-Nonself Discrimination* Jan Klein ponders on what he would study if he were to start over in the lab.[1] Dismissing the antibody, MHC, the T-cell and parasitology, he considers instead the phylogeny of immune reactions, particularly in ancient phyla. As for a favored cell he chooses the macrophage. Describing it as a "*Mädchen für alles,*" (all purpose kitchen maid) Klein believed that this immunocyte still had secrets to reveal. Toll-Like Receptor (TLR) biology would prove to be one of these secrets. Analyses of the evolution of these receptors (Tolls and TLRs) have also helped us to rethink immune system phylogeny. In the first part of this chapter the history of the discovery of Toll and TLR biology is described. The evolution of the TLR genes and theories of immune function are covered in later sections.

The remainder of this book presents work from nine groups active in the field. In the first chapter, "The Function of Toll-Like Receptors", Zlatko Dembic sets the stage by introducing us to many of the components of the immune system and their relationships vis à vis Toll receptors. Zlatko finishes his chapter with a discussion about current immune system models and contributes his own 'integrity model'. Work from the laboratory of Nicholas Gay follows this in "Structures and Motifs Involved in Toll Signaling". Having established the structural similarities between *Drosophila* Toll and the IL-1 receptor in the early nineties, his lab now continues to solve structures of Toll signaling components. Ligand binding and receptor clustering is further analyzed by Martha and Kathy Triantafilou in their chapter "'Supramolecular' Activation Clusters in Innate Immunity". Using the immunologic synapse as an exemplar for Toll signaling, the Triantafilous describe how differential clustering introduces specificity into the immune response. Their work demonstrates how heterotypic complexes assemble at the plasma membrane, generating specific immune outputs. The signaling pathways themselves are then covered in three chapters. Harald Wajant and colleagues compare the molecular mechanisms of IL-1R/TLR signaling in their chapter "Interleukin-1 Receptor/Toll-Like Receptor Signaling". Virus mediated signaling is examined by Claudia Wietek and Luke O'Neill in their chapter "Virus Induced Signaling to Initiate the Interferon Mediated Anti-Viral Host Response". A particular emphasis is placed on the interferons and their regulation by IRF transcription factors. Dendritic cells have proven to be pivotal intermediaries between innate and acquired immunity, and for this reason Tsuneyasu Kaisho and Shizuo Akira describe DC activation by Toll-like receptors in their chapter "The Induction of Dendritic Cell Activation and Maturation

Toll and Toll-Like Receptors: An Immunologic Perspective, edited by Tina Rich. ©2005 Eurekah.com and Kluwer Academic / Plenum Publishers.

by Toll-Like Receptor Signaling". In "Pathogen Avoidance using Toll Signaling in C. elegans", Nathalie Pujol and Jonathan Ewbank remind us that different organisms can use Toll signaling to quite different effects. The discovery of novel signaling intermediates are then described by Bruce Beutler and colleagues in their chapter "Forward Genetic Analysis of TLR Pathways: A Shared System for the Detection of Endotoxin and Viral Infection". Turning to therapeutics, Ekambar Kandimalla and Sudhir Agrawal describe recent advances in immune system modulation using synthetic molecules. Their chapter "Agonists of Toll-like Receptor 9: Modulation of Host Immune Responses with Synthetic Oligodeoxynucleotides" describes how antisense therapies with CpG containing oligos have provided valuable retrospective validation of the safety of CpG containing DNAs as therapeutic agents. This chapter also demonstrates the subtleties of medicinal chemistry and how it can be used to tailor the immune profile.

Beginnings

The phenotypes of the first *toll* mutants were described in the mid-eighties.[2-4] At this point Toll activity was rooted firmly in developmental biology, with *toll* understood to be one of twelve maternal effect genes required for dorsal-ventral patterning. This, along with other discoveries in the control of *Drosophila* development led to the award of the 1995 Nobel Prize in Physiology and Medicine to Christiane Nüsslein-Volhard* and Eric Wieschaus** which they shared with Ed Lewis for his work on homeotic mutants. Soon after reports in the literature began to link Toll activity with immunity in the fly.[5] Toll would be cloned by 1988.[6] The timing was propitious. Globally, sequencing projects were being used to compile the genomes of several organisms. All would provide sequence data with which to sift for *toll* homologues. Uniquely, the high throughput computational technologies involved would have a second, less expected impact; to drive molecular biologists away from reductionist biology, back to integrated systems. For immunologists this meant bridging the gap between the two major wings of the immune system. Crucially, the Toll signaling apparatus was to provide a juncture between the systems of innate and adaptive immunity.

The Innate and Adaptive Immune Systems

Innate immunity is the branch of our immune system that governs our immediate and infection-proximal response to pathogens. Innate immune responses underpin and inform the adaptive response, which is unique to the vertebrate phylum. Because of its need to expand rare lymphocyte clones into effectors, the adaptive response takes several days to get under way, during which time the innate immune effectors prime and recruit other immunocytes. Adaptive immunity gives us a memory of past infection, allowing us to mount a stronger, more specific response on re-challenge. The gene rearrangements used to generate B and T cell receptors also ensure great variability in their specificities. On the other hand, allergy, autoimmunity and graft rejection are all properties of the adaptive immune response. Additionally, the adaptive response is unable to determine the nature of the infecting pathogen.[7] This diagnosis is left to innate immunity and without its early intervention we are extraordinarily vulnerable to infection. Patient mortality in the pre-antibiotic era was evidence enough of this.

A Little History

The founder of the *Cellular Theory of Immunology*, in which the innate system is central, was the Russian zoologist Élie (Ilya) Metchnikoff. Despite a turbulent personal life[8] Metchnikoff none-the-less managed to complete his seminal work using the phagocytic cells of starfish

* http://www.nobel.se/medicine/laureates/1995/nusslein-volhard-lecture.html
** http://www.nobel.se/medicine/laureates/1995/wieschaus-lecture.pdf

larvae. By lodging a rose thorn under the skin of a star-fish larva he observed that phagocytic cells would soon surround the intruder. These were clearly part of the body's immunologic early warning system. The cells and mechanisms of this system were soon to be eclipsed by the antibody; rising star of the *Humoral Theory of Immunology*. The humoral-ists took precedence, despite the joint award of the 1908 Nobel Prize to Metchnikoff and Ehrlich, which was early recognition of the possibility for co-operation between the two systems. So well publicized was the acrimony between the two sides of the immunologic divide that Shaw even satirized their arguments in his play, *The Doctor's Dilemma*.

The emerging discipline of biochemistry was soon to provide the tools with which to study the panacea offered by the antibody. The natural history of T and B cells, the cellular source of the adaptive immune paraphernalia, would prove far more ingenious than anyone had imagined. As for their genetics, the gene rearrangements used to build T and B cell receptors were a major innovation, probably conferred by the insertion of a retroposon carrying recombinase-activating genes.[9] This insertion marked the origin of the adaptive immune response, and allowed for a previously unimaginable degree of diversity to be wrung from a relatively shallow gene pool. Against the backdrop of T and B cell biology it wasn't surprising that the effectors of innate immunity were overlooked. Fortunately, key experiments that would eventually lead to the reunion of the two schools of immunology were being carried out. Interestingly, the breakthrough was to come from the work of developmental biologists.

Saturation Screening of the *Drosophila* Genome

The re-emergence of innate immunology came long after a discovery made in the quite different discipline of developmental biology. Christiane Nüsslein-Volhard had made forays into experimental physics, maths, mechanics and chemistry, before embarking on a course of biochemistry at Tübingen. Nüsslein-Volhard (pictured in Fig. 1) was a polymath and, not surprisingly, found the lectures that were toughest to understand the most exciting.*** Nüsslein-Volhard knew that the field of biology was changing and that developmental biology was providing intriguing questions, some of which could be answered using genetics. *Drosophila* seemed the obvious choice of organism and had been a firm favorite with geneticists since Morgan's proof that genes lay on chromosomes.

Having devised her first post-doctoral project, to identify morphogens by mutational analysis, Nüsslein-Volhard started her work at the Biozentrum in Basel. It was here that she met Eric Wieschaus from whom she would learn *Drosophila* embryology. Initial small-scale mutational screens were eventually scaled-up to genome-wide saturation screens. Now that the *Drosophila* genome is complete, the total number of *Drosophila* genes is known to be 13,639; a figure smaller than the earlier estimates of 20,000. Nüsslein-Volhard knew that her task was enormous, having estimated the percentage of lethal mutations, despite working without the benefits of the genome data. Accordingly, she refined her experiments to examine more embryos, more efficiently than previously possible. Treating the eggs with oil to make the normally opaque chorion transparent, and optimizing egg collection, enabled the screens to be completed in weeks rather than years. Initial experiments allowed the team to recover a mutant of *bicaudal*, the gene that had sparked Nüsslein-Volhard's interest in pattern formation. A screen for maternal mutants, using 100 chromosomes, then recovered mutant C79, later called *dorsal*. We now know that Dorsal gradients drive multiple gene activities, controlling gene thresholds that direct the fate of three primary embryonic tissues: the mesoderm, neurogenic ectoderm and dorsal ectoderm.[10]

***http://www.nobel.se/medicine/laureates/1995/nusselein-volhard-autobio.html

Figure 1. Christiane Nüsslein-Volhard, one of the three winners of the 1995 Nobel prize for Physiology and Medicine.

The Toll Receptor in Development

One screen recovered multiple *toll* mutations. The name "*toll*", German slang for "fantastic", was given in reference to the retrieval of both dorsalizing and ventralizing alleles of *toll*. Later work revealed that the Toll protein functioned as a transmembrane signaler.[6] In the embryonic fly, Toll reacts to the end-products of a chain reaction set up in the maternal cells that surround it. In other words, gene and hence protein gradients established in the mother's cells trigger signals, which are then transmitted by Toll into the embryo. These signals sculpt the embryo's ultimate appearance before cell walls are even laid down. At this point the embryo's phenotype is dictated by its mother's *maternal-effect genes*, rather than by its own genes, which are not yet active.

To understand this process we should first consider that the fly egg is an elongated cylinder. The sperm enters at one end, fuses with the egg, and a single nucleus results (Fig. 2). Rather than successive cleavage events, the nucleus alone divides. Within a couple of hours several thousand nuclei (~6000) are generated. Only then do walls enclose each nucleus, shutting each off from its neighbors. After 24 hours the larva is fully segmented and gastrulation is complete.[#] The larva consists of an anterior feeding structure, three thoracic segments, and eight abdominal structures. Cytoplasm at the anterior end of the egg translates maternally derived *bicoid* mRNA soon after the egg is laid. Bicoid diffuses along the egg, establishing an anterior posterior gradient, which controls the boundaries of the head and thorax. A second gradient, established at ninety degrees to Bicoid is set up by Dorsal, demarcating the dorsal ventral axis. A high concentration of Dorsal protein directs the synthesis of muscle (Fig. 2). As its concentration dips, nerve cells form, and at lower concentrations still, surface or ectodermal cells result.[10]

[#]Excellent on-line descriptions of these processes can be obtained at http://flybase.bio.indiana.edu/ and http://flymove.uni-muenster.de/.

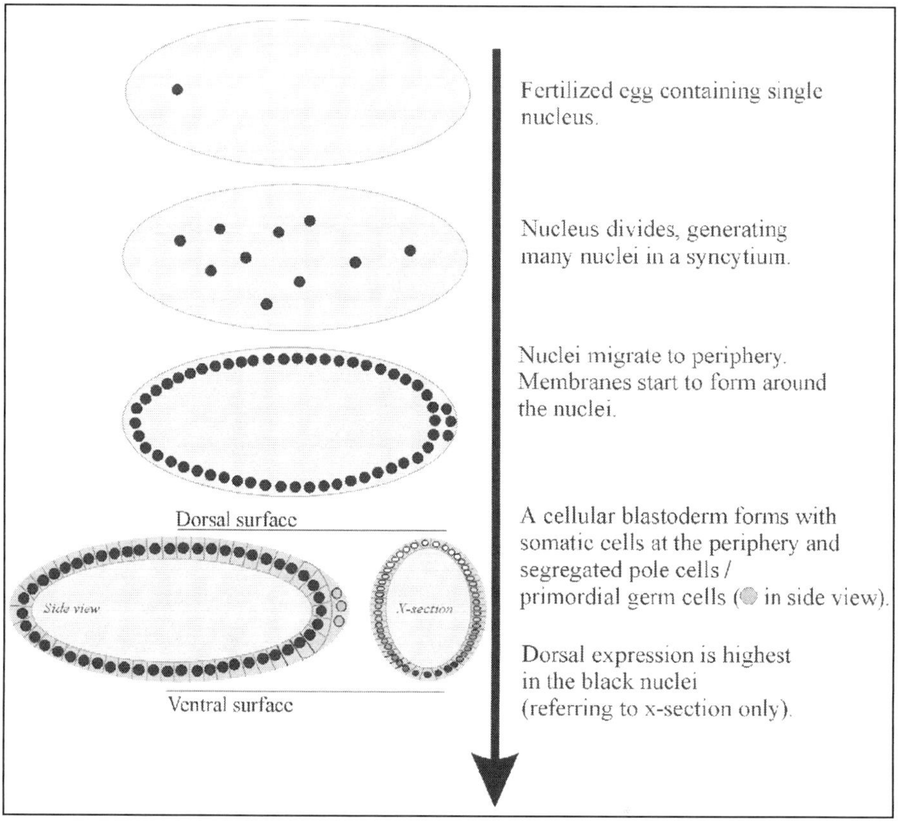

Figure 2. Dorsal Ventral patterning in *Drosophila*. Some of the early steps of embryo development are outlined. The Dorsal nuclear gradient is shown by the black to grey transition in the nuclei (X-section only). After fertilization, Nudel is produced by maternal follicle cells and activates a proteolytic cascade. This cascade results in the processing of the inactive form of Spätzle which then activates Toll. The end-result of Toll activation is the translocation of Dorsal from the cytoplasm to nucleus. This is most prevalent in the ventral regions where Toll signaling is strongest. In the lateral regions, the lower concentration of Spätzle leads to a lower density of activated Toll receptors and a diminished translocation of Dorsal.

The Implication of Toll in Fly Immunity

Toll is evidently a maternal effect gene, at least in the developing fly. But what role, if any, does it play in the adult? Not long after the *toll* and human *IL-1R* genes were cloned researchers began to note the striking similarities between the two receptors.[11,12] Both express a novel domain at their C-termini, which came to be called the Toll-Interleukin-1 receptor domain or TIR. Crucially, the TIR domains provided the signature sequences with which to search for related genes. Toll and IL-1R could also activate NFκB through signals transmitted by homologous kinases (Pelle and IRAK).[4] The finding that the promoter regions of *Drosophila* antimicrobial peptides seemed to carry consensus NF-κB-like binding sites was sufficient to implicate Toll in immunity.[13] The tendency of Toll null flies to succumb to fungal infections cemented the link. Later work showed that Toll signaling regulates the expression of numerous anti-microbial peptides which are synthesized in the fly fat body, equivalent to the human liver.[14,15] A family of *Drosophila toll* genes (*toll, 18-wheeler and toll-3 to 9*) has now been described, although most appear to support developmental rather than immune functions.[16] The

Figure 3. Charlie Janeway Jr, founder of the "Pattern Recognition Hypothesis".

signaling pathways that lead from the Toll receptor to the fly versions of NFκB, and a second pathway (Imd) that detects Gram-negative bacteria independently of Toll, are documented in later chapters. The question of how insects might detect infection is less well understood, which may reflect the paucity of microbes available to researchers with which to infect flies. Most studies rely on direct injection of microbes into the body cavity, a technique which often results in a skewed immune response.[17]

Given that insect larvae often thrive in rotting vegetable matter, and that the adults often carry infectious pathogens, their expression of Toll and Imd immune mechanisms is not surprising. Clearly, insects can count on robust immune protection. In the mid nineties, when the fly laboratories were describing these novel immune mechanisms, immunologists began to question whether the same family of proteins could operate in mammalian cells.

A Problem with Vaccination

Much of the early work with mammalian Toll-Like Receptor genes came from the laboratory of Charles Janeway Jr (pictured in Fig. 3). Janeway had long puzzled over how T and B cells could be alerted to the presence of pathogen in the lag phase between the onset of infection and the expansion of antigen-specific effectors. He came up with the idea that pathogens could be recognized by conserved pathogen specific motifs. Crucially, Janeway realized that these pattern recognition receptors might bridge the gap between innate and adaptive immunity. Janeway's *Pattern Recognition Hypothesis* was published as a chapter in the *Cold Spring Harbor Symposia* of 1989.[18] Could the same phenomenon also explain the need for adjuvants, often containing bacteria or parts thereof, in vaccines? Janeway dubbed the need for adjuvant "Immunology's dirty little secret".* Simply put, vaccinations often work only when emulsified in oils and concoctions of other substances, usually bacterial. These substances are loosely defined as enhancers of the immune response, or adjuvants. Could adjuvants act by opening the dialogue between the innate and adaptive immune response? The description of Hoffmann's fungus infested Toll null flies, led Janeway, now teamed up with Ruslan Medzhitov, to look for mammalian versions of Toll. These could be the pattern recognition receptors that Janeway was

* http://www.the-scientist.com/yr2003/nov/opinion_031103.html

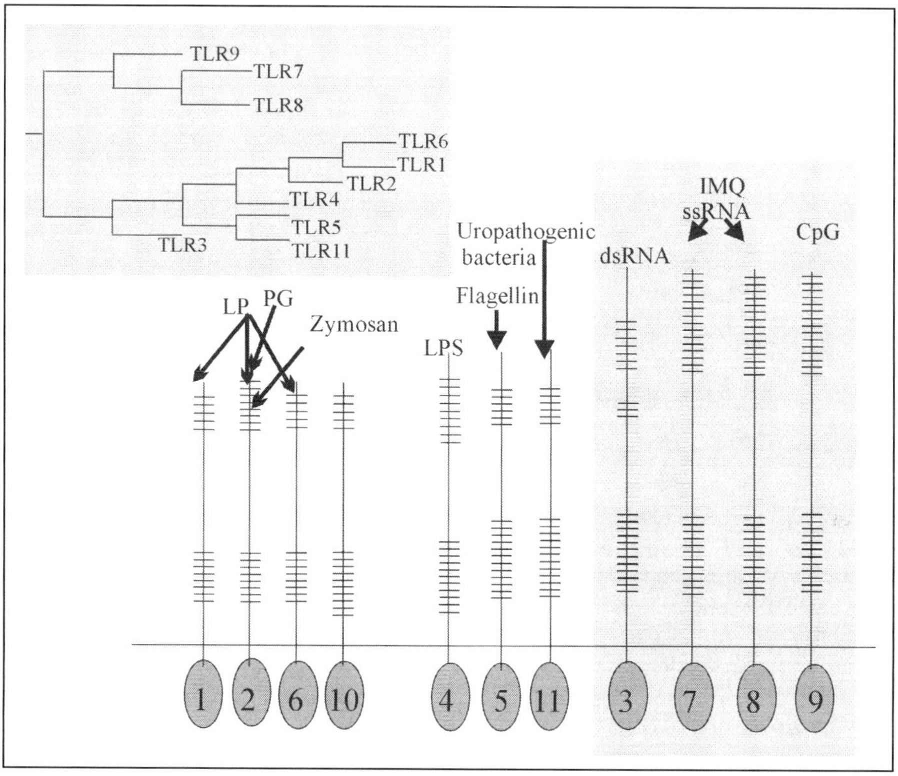

Figure 4. The TLRs and some of their ligands. Horizontal dashes denote leucine-rich patches. Ovals indicate TIR domains. This schematic illustrates only a few of the ligands that are recognized by TLRs. LP denotes lipoprotein; PG is peptidoglycan; LPS denotes lipopolysaccharide, IMQ indicates Imiquimod and CpG indicates unmethylated CpG DNA. The phylogenetic analysis of the TLR family (inset) is re-drawn from ref. 20 and shows the relationship of TLR11 to other family members. Murine TLR11 appears to recognize uropathogenic bacteria. Our expression of a truncated version of TLR11 could make us susceptible to infection from these same bacteria. Note that TLRs 3, 7, 8 and 9 recognize nucleic acid ligands.

looking for. Eleven mammalian Toll-Like Receptors (TLRs), all responding to natural microbial, self and synthetic ligands have now been cloned (Fig. 4).[19] More receptors and their ligands remain to be discovered. Clearly mammalian TLRs comprise a web of homeostatic surveillance with nodes from which the adaptive system can 'listen-in'. Most recently TLR11, was described in March 2004[20] and two more, TLRs 12 and 13 are rumored to be in the pipeline. Novel TLR ligands have also recently been identified. TLRs 7 and 8 have long been known to respond to the anti-viral drug Imiquimod, which has led many to speculate that these receptors play some role in viral recognition. Murine TLR7 and human TLR8 have now been shown to recognize ssRNA, both of viral and endogenous origin.[21,22]

The Question of Escape Mutants

We now know that the TLRs recognize molecules that are displayed on microbes (Fig. 4). The beauty of TLR recognition is that their ligands are invariant. Although pathogens constantly mutate many of their antigenic structures, their TLR ligands appear to be immutable. Pathogens cannot avoid expressing these structures and escape mutants are generally less virulent, viable, or both. Consequently, it is held that pathogens cannot avoid detection, but is this really the case? Plant recognition of bacterial flagellin, the principle component of bacterial 'propellers', involves the FLS2 protein (see "Plant Innate Immunity" section).[23] Like Toll, FLS2 expresses extra-cellular leucine repeats, but has an intracellular kinase cassette rather than a TIR domain. The crucial point here is that FLS2 can discern the flagellae of pathogenic versus symbiont bacteria, the differences being manifested by sequence changes in the Flg22 molecule.[23,24] What prevents the generation of escape mutants in this scenario? Closer to home, how do humans discriminate pathogenic microbes from normal gut flora? In the case of plants it is not yet clear how escape mutants are avoided. In humans the answer seems to be to compartmentalize the flagellin signal. In order to tolerate the large commensal bacterial load of the gut, humans rely on two innovations. First, the epithelial cell is polarized. Its apical face is exposed to the lumen and this surface is devoid of TLR5. The basolateral surface is, however, decorated with TLR5 and the exposure of flagellin to this surface triggers proinflammatory gene expression.[25,26] This exposure necessitates the translocation of flagellin into the cells, a feat which is easily achieved by certain flagellated pathogenic bacteria. Recent work has shown that the sequence recognized by TLR5 is uniquely expressed on monomeric flagellin, not the polymeric version of the filament.[27] The TLR5 recognition site is also fairly permissive for mutations and spans a large number of residues. Mutation of any of these residues can impact bacterial motility. As a result, TLR5 can recognize most flagella and those that are missed are likely to be expressed by defunct non-motile bacteria. However, the first cases of escape mutants are now being reported in the literature. Not surprisingly, *Helicobacter pylori*, the gram negative bacterium linked with peptic ulcers, is leading the way. *H. pylori* neither releases monomeric flagellin nor expresses a flagellin which is overtly pro-inflammatory.[28]

The Evolution of Toll Receptors

The discovery of a set of immune effectors conserved in mammals and flies paved the way for comparative genomic investigations of TLR evolution. These studies provide a snapshot of the evolution of gene organization, although in the future, the *toll* and *tlr* orthologs and paralogs that we detect may be lost or mutated beyond recognition. Genome searches have uncovered TIR domains across the animal and plant kingdoms. These analyses have revealed the selective pressure to evolve new TLR affinities, and provide an indication of the burden of immune recognition carried by TLR signaling.

The ectodomains of Toll and TLRs show considerable variation compared to the intracellular signaling apparatus. This is not surprising given the evolutionary pressure placed on ligand recognition domains. By comparison, the cytoplasmic TIR domains show higher sequence conservation, reflecting their role as structures to which the intracellular signaling apparatus is tethered. The sequence identity between animal and plant TIRs is often less than 20%[29-31] with the majority of conserved residues clustered at the five alpha helical and beta sheet motifs.[30] The finding that TIR domains are involved in both plant and animal immunity suggests that this structural element was expressed by the last common ancestor of plants and animals. The later adoption of the Toll signaling apparatus for developmental patterning in *Drosophila* will have necessitated changes at either end of the signal transduction pathway. These modifications would have merged a pre-existing proteolytic cascade that controls development with the Toll signaling apparatus, and altered the transcription factor output.[29]

In view of the low identities between TIRs, optimized consensus sequences (hidden Markov models) can be constructed and used to mine databases for novel TIR sequences.[29] 'SMART' (Simple Modular Architecture Research Tool) searches have even identified TIR homologies in prokaryotes, although these genes may have been acquired by horizontal transfer and used to negate Toll and TLR signaling in a similar fashion to the poxvirus (see chapter by C. Wietek and L. O'Neill).[32]

It is a paradox that Toll receptors are classified as part of the primitive innate immune response, when one considers that this system is exploited by insects, possibly the most successful organisms on earth. Insects first arose in the early/mid Devonian period (417-354mya). Their evolution, closely tied with the emergence of flowering plants (144-65mya), generated three-quarters of all living and fossil organisms. The completion of the *Drosophila* genome provided the first opportunity to examine the immune gene organization of a fly. The *Drosophila melanogaster* genome comprises 180 Mb of sequence organized on 3 autosomes plus the X and Y chromosomes. The immune response genes are littered throughout the genome, although several small gene clusters are also evident.[33] These encode antimicrobial peptides, pattern recognition proteins, and the Rel proteins, for example Dif and Dorsal. As described earlier, Dorsal regulates patterning in the fly embryo. As for immune system function, both Dorsal and Dif trans-activate *drosomycin* in larvae, whereas Dif alone activates *drosomycin* in the adult fat body. The *dorsal* and *dif* genes appear to have arisen from a duplication event in an ancestral gene. A second cluster contains *necrotic*, along with two related serpin genes, although only necrotic has a role in immunity. Three gene clusters related by function stand out; *spätzle, toll* and *pelle; cactus, dorsal* and *dif; attacin, drosocin* and *metchnikowin*. This clustering may well favor the rapid evolution of variants by unequal crossing-over events.[33] Indeed clustering of immune function genes appears to be synonymous with accelerated evolutionary potential. Multiple *Drosophila* toll genes have now been cloned, although analysis of their function implicates most in development rather than in immunity.

Therapeutic Targets in the Mosquito Genome

The sequencing of a second dipteran genome, the mosquito (*Anopheles gambiae*), made possible the comparison of its genomic organization with *Drosophila*, with whom it shared a common ancestor approximately 250 million years ago.[34] *A.gambiae* is a vector for the protozoan parasites that cause malaria; a disease that infects 500 million people annually, of whom more than a million die. Consequently, a major impetus in genomic analysis is the desire to identify novel drug targets that may disturb the balance between vector and parasite.[35] The aim is to break the lifecycle of the parasite and disturb both its sexual maturation and transmission. The *A.gambiae* genome encodes 10 Tolls, four of which are orthologs of *Drosophila Toll-6,7,8* and 9.[34-36] A second group of *Anopheles* genes, TOLL1A,1B,5A and 5B, appear to be recently duplicated orthologs of *Drosophila* Toll-1 and 5.[36] Evidence for immune function is provided by one of *Anopheles gambiae's* ten Toll receptors which is highly expressed in the gut and may be involved in anti-*Plasmodium* responses.[37] Much of the intracellular signaling pathway is intact. The mosquito genes *myd, tube, pll1, cact,* and *rel1*, are clear orthologs of *myd, tube, pelle, cactus,* and *dorsal.* A *dif* equivalent is missing however, which is interesting given the *dorsal / dif* cluster in *Drosophila.* Six genes also show similarity to *spätzle.*[36]

C. elegans *Tol-1 Signaling*

C. elegans expresses a single Toll receptor gene *tol-1* and several elements of the toll-signaling pathway.[38] Pellino, a *C. elegans* version of the *Drosophila* Pelle associated protein, has also been described.[39] More recently, a second TIR domain containing protein called TIR-1, homologous to the vertebrate SARM protein has also been described.[40] Interestingly, *C. elegans* does not appear to use Tol-1 signaling to combat infection. A combination of TGF-β, DAF-2 and p38 signals are instead used to control the stress and immune response.[41,42] Curiously, Tol-1 signals still play a role in pathogen detection, but in this case controlling the pathogen avoidance reaction of *C. elegans.* Nathalie Pujol and Jonathan Ewbank describe this phenomenon in detail in their chapter. Before we conclude that Tol-1 signals play a different and rather unique function in *C. elegans* we should consider a point raised in a recent review. Schulenburg and colleagues pointed out that much of the *C. elegans* research has been conducted with a single strain. Under laboratory conditions there is little positive selection for the energetically costly immune system, much of which might be jettisoned in favor of reproductive vigour.[42] Arguably, the analysis of natural strains of *C. elegans* will provide a more realistic picture of the role of Tol-1 in immunity.[43]

Viral Subversion

Elements of the Toll signaling pathway have also been found in viruses. Poxviruses are formidable genetic thieves. They steal host immune genes, then alter them, often to convert them into dominant negative mutants.[44] Consequently, an analysis of the Pox genome gives us a good idea as to which host biologic systems are targeted. Vaccinia virus encodes two proteins (A46R and A52R) that interfere with IL-1 and TLR signaling. Luke O'Neill and Claudia Wietek describe these proteins in more detail in their chapter. An insect-infecting Entomopoxvirus has also been found to have acquired a copy of the *pellino* gene.[39]

Teleosts and Their Tolls—Ever Expanding Repertoires

How many *toll* genes do fish have? This is an interesting question as the emergence of jawed fish is presumed to mark the beginning of adaptive immune function. The zebrafish genome (*Danio rerio*) contains 19 putative TLRs (DareTLRs), including orthologs of mammalian TLRs 2-5, 7-9 and a group of fish specific TLRs.[45] The zebrafish genome contains one receptor that appears to be the ancestor of TLRs 1,6 and 10. Unlike the pufferfish (*Fugu rubripes*) the zebrafish also expresses two TLR4 orthologs. All zebrafish TLRs express the classic LRR-TM-TIR domain organisation.[45] Clearly the large number of DareTLRs constitutes a more complex innate immune system than was previously thought, with considerable pressure for the amplification of the TLR repertoire.

Plant Innate Immunity

Consideration of a plant's requirement vis à vis innate immunity draws some interesting conclusions. Plants encounter just as many pathogens as do vertebrates, yet they must do so without an adaptive immune response, specialized cell types (e.g., macrophages, DCs) or a circulatory system. Plants are sessile and cannot avoid a predator, some of which kill their host to extract nutrients (necrotrophs), whilst others require living host tissue to complete their life cycles (biotrophs). To counter these threats, every plant cell must be capable of mounting a cell-autonomous response to infection. Additionally, plants have evolved mechanisms to induce heightened systemic immunity, using salicylic acid, jasmonic acid and ethylene.[46] Although systemic immunity, like adaptive immunity, is durable, it lacks specificity.

The localized resistance response is driven by the 'gene for gene' system in which the activities of pathogenic avirulence genes (Avr) are countered by cultivar specific plant resistance genes (R genes).[30,47] These R genes are bred into crops to improve their resistance. Importantly, the robustness of anti-pathogen responses is determined by the expression of pathogen recognition proteins, and not by any deficiency in the response that these pathogens provoke. The response to different pathogenic insults is also often identical. A rapid oxidative burst that induces apoptotic death is common, the so-called hypersensitive response. Plant defense mechanisms also include the production of lytic enzymes (chitinases, proteases), anti-microbial proteins (defensins), and anti-microbial metabolites (phytoalexins).

The localization of R proteins gives a good indication of where they encounter pathogens. For example, the intracellular *N* R-gene product recognizes the helicase domain of TMV replicase, whilst extra-cellular pathogens are recognized by extra-cellular or TM ligated R proteins. R proteins can even shuttle between cellular compartments. The bacterial type III secretion system, which can best be described as a molecular syringe, has also been implicated in the delivery of pathogenic effectors.[30]

A second non-cultivar specific wing of plant innate immunity also exists. This system appears to use receptors to recognize general elicitors (PAMPs) such as bacterial flagellin, polypeptides, glycoproteins, lipids and even host cell fragments released by pathogen damage.[23] The available evidence suggests that the PAMPs themselves are not subject to frequent mutation, and are largely immutable. This system equates to the TLR-PAMP system of vertebrates although true homologues of TLRs are not found in plants. Plant TIR domains are found in R genes, however.

There are at least 135 TIR domain-containing proteins in the *Arabidopsis* genome, probably many more.[48] In dicots, the TIR domain is conjoined to a number of different domains, including the nucleotide binding site domain (NB) and leucine-rich repeats (LRRs). In plants the most common TIR arrangement is TIR-NB-LRR. This domain organization is found in the prototypic plant defence protein, the tobacco *N* gene product, which confers resistance to tobacco mosaic virus (TMV). Interestingly, TIR-NB-LRR proteins are absent from one of the two subdivisions of angiosperms (flowering plants). Angiosperms arose 130 million years ago and are subdivided into two monophyletic groups, the monocots and the eudicots. The TIR-NB-LRR family of R genes appears to be absent from monocots (cereals), but is found in Gymnosperm databases (e.g., *Pinus Taeda*) and, presumably, in the Angiosperm progenitor plant. We can only conclude that TIR-NB-LRR genes were found in the progenitors of grasses and were amplified in dicots, but were lost or have evolved to an unrecognizable degree in cereals.[30,31] As SMART searches of current databases reveal TIR homologies in prokaryotes, the former conclusion is more likely. Variant TIR domain containing genes have been identified in rice, although these have diverged considerably from the NB-LRR family of genes. One proposal is that cereal genomes have replaced TIR domains with coiled coil domains, another common protein-protein interaction domain. There is also considerable variation in the ability of monocot and dicot plants to recognize non-self. For example the flg22 epitope on flagella, which was described earlier, is not recognized by rice.

The number of R gene sequences in *Arabidopsis* reveals that they may comprise as much as 1% of the total genome, indicating that R genes are used as a pool from which to mold new anti-pathogen specificities. Certainly the clustering of R genes and the high mutation rates in duplicated alleles suggest that selection is driven most rapidly in clusters of related paralogs. However, the number of R proteins is still smaller than the number of potential pathogens. An interesting explanation for the success of the R gene system despite this shortfall uses the olfactory system as a model. In higher vertebrates only 500-1000 odor receptors are necessary for a complete sense of smell. One model that explains this feat relies on a reduced affinity between odorant and olfactory receptor. Similarly, a lower affinity between Avr and R proteins may explain how R proteins of the same order of magnitude can recognize so many pathogens.[49]

R Protein Signaling

Experiments to elucidate how R proteins and their effectors interact have met with limited success. This is due to several factors. R proteins are not abundant and tend to be targeted for destruction after stimulation. Nor are leucine-rich repeat containing proteins amenable targets for two-hybrid assay. Lastly, if the 'olfactory' model is correct, then effector-R protein affinities may not be particularly high. However, yeast two-hybrid has been used to show that a single amino acid substitution in an LRR weakens the interaction of a susceptible strain with its pathogen effector. Hypermutation at LRRs is consistent with these findings. The consensus opinion seems to be that R proteins relay their signals through a complex network of additive and interconnected pathways. How R and Avr proteins interact remains largely unsolved.

Multiple models have been proposed, of which the most influential is the "Guard" hypothesis. This model proposes that the pathogen effector targets a plant protein, presumably to promote disease. A host guard protein monitors the status of the "guardee", and can activate defense responses if a pathogen disrupts its charge.[50]

Self Recognition by Toll-Like Receptors

Thus far we have described the recognition of foreign molecules by TLRs. However, TLRs have also been implicated in the recognition of host self proteins. This aspect of TLR action can be used to test some of the current theorems that surround a central issue of how the immune system works. These theorems are described by Zlatko Dembic in his chapter, "The Function of Toll-Like Receptors", in which he also explains his integrity hypothesis.

Self Protein Signaling via Toll Receptors

Self proteins are usually closeted from the immune system in healthy cells but can be released under certain conditions, of which one is necrotic cell death. Normal, physiologic cell death or apoptosis entails the ordered collapse and removal of a cell. This is an energetic process which drives cell condensation, without membrane rupture, and ultimately generates a number of shrunken-membrane bound apoptotic bodies.[51] These can be quickly recognized and phagacytosed by macrophages and dendritic cells, a process that shields host proteins from the immune system. In contrast, necrotic death and the accompanying rupture of cellular compartments can result in the release of self proteins. Some of the self proteins to which the immune system responds are heat shock proteins, amongst the oldest and most abundant proteins in the cell. Heat shock proteins (HSPs) have been shown to interact with and to activate TLRs, although some of these data have been discredited as artefactual.[52-55]

Heat shock proteins protect the cell machinery from stresses, such as oxidative damage and temperature fluctuations, both hyper and hypothermic. Their abundance makes them an attractive flag for cellular stress. Hsp molecules (gp96) that carry antigenic peptides can also stimulate both wings of the immune system simultaneously.[56] Interestingly, Hsps (gp96, Hsp90) have also been implicated in the chaperoning of Toll like receptors and R proteins in man and plants respectively.[57] One fascinating hypothesis is that these chaperones facilitate phenotypic change by easing the folding and transport of emergent R proteins with variant LRR domains.[58] Immune complexes that accrue during autoimmune responses,[59] the extra domain A of fibronectin and endogenous ssRNAs[21,22] have all been found to act as self-ligands for TLRs. Indeed TLR recognition of self proteins may constitute an important part of the machinery of the Danger Hypothesis.

The Danger Hypothesis

In 1994 Polly Matzinger published her first version of the Danger model.[60-62] She claimed that the immune system responds to dangerous substances, regardless of whether these are foreign. By implication an immune response need not be mounted against a bacterium that is not damaging the host. This model counters the 'dualism' of the self-non-self theory, and instead considers the self and 'other' as parts of an integrated system. The Danger hypothesis takes a holistic approach to the immune response and appears to account for some of the anomalies of the self non-self model of discrimination. Matzinger felt that the textbook explanation of T cell education was unrealistic and that a more plausible model would describe immune reactions tailored to need rather than strict self non-self designations. In defending her hypothesis Matzinger drew on a number of scenarios in which the self non-self hypothesis falls short. The oft cited example is that of tolerance of post puberty proteins. If all thymocytes are educated such that overt self reactors are deleted, what about the 'self' that was absent when the thymocytes went to school? Proteins that are synthesized after puberty fall into this category. Simply put, how can the self non-self model account for the fact that the adult body does not mount an immune response to the lactating breast, when it had no opportunity to designate milk proteins as self?

Tissue trauma is another obvious danger signal. Matzinger proposes that the intensity of the danger signal evoked by damaged tissue explains the success of unmatched kidney transplantation from live donors over matched donations from cadavers. Likewise, the success of liver transplants may be due to their innate regenerative capacity after the initial response to surgical trauma.

There are many more immune model theories. Most acknowledge that the complexity of the immune system is not well served by the self non-self model. Some work uses computer models to simulate the immune system.[63-66] With fuzzy logic, an extension of conventional (Boolean) logic, the 'partial truths' of immunity can be accounted for.

Beyond Self Non-Self: A Critique from the Humanities

Wider searches of the literature on TLR biology and the Danger hypothesis reveal considerable interest from social scientists. An awareness of the range of models currently utilized in Immunologic research should prompt any researcher to question the relationship between these models, the language in which they are expressed, and the direction that experimental work is taking. Social Scientists have invested considerable academic effort in critiquing how scientific language shapes the experimental landscape. Some of this work applies directly to *toll*, at least in its guise as a maternal effect gene. Evelyn Fox Keller, a social scientist who holds a PhD in Physics, argued that the dominance of the discourse of genetics over that of embryology during the first half of the Twentieth century rendered the discovery of maternal effect genes all but impossible.[67] She postulates that as the field of genetics came to eclipse that of embryology, the cytoplasm was demoted, and described as a substance that existed merely to nurture genes. The language not only described known phenomena, it also restricted what experimental work was carried out. According to Fox Keller, the discourse that prioritized gene action and depicted the cytoplasm as by-product "… not only enabled geneticists to get on with their work without worrying about what they did not know; it framed their questions and guided their choices, both of experiments worth doing and of organism worth studying".[68]

The *toll* gene was first identified as a "Maternal Effect" gene by Christiane Nüsslein-Volhard, whose work on *Drosophila* was described earlier in this chapter. Nüsslein-Volhard's work marked a sharp change in direction from the accepted view of cytoplasm as neutral substrate, as it by assigned it a role in the development of the egg. According to Fox Keller, the crucial development in experimental enquiry that Nüsslein-Volhard's work represented was dependent upon a previous shift in scientific discourse; a change in the ways in which the phenomena under study were described and discussed. Fox Keller argues that this change in discourse was a more decisive factor in the discovery of Maternal Effect genes than was the availability of new technology and experimental techniques. In fact, the experimental hardware necessary for Nusslein-Volhard's work had been available to scientists in the 1930s.[69]

The discourse of Immunology has also been subject to critique in the social sciences. Much of the discussion focuses on the use of metaphors used to describe this complex field, and questions how these metaphors may have shaped researchers' lines of enquiry and interpretation of findings. The dominant immunologic model of distinction between self and non-self, with its emphasis on the body's ability to recognize and destroy alien elements, uses metaphors that were developed after the Second World War. The model was developed largely through the work of Frank Macfarlane Burnet and Peter Medawar. It was Medawar's work on skin grafts on an injured Air Force pilot in 1943 that prompted him to consider graft rejection as an immunological reaction. Burnet went on to develop his "self-marker" hypothesis in 1949. Medawar and Burnet were jointly awarded a Nobel Prize in Medicine in 1960. Scientist turned social scientist, Lisa Weasel, is one researcher who has written about the deployment of militaristic language in immunology.[70] The relationship between immunologic theory and language runs both ways, with immunologic metaphor now deployed by the military in descriptions of their strategies.[71] This flow of influence between scientific theory and language and discourse is not limited to the field of immune recognition, and has also been analyzed in reference to cancer and AIDs.[72]

To acknowledge that the self non-self theory was shaped partly by the social and political world outside the research lab is not to question the rigor of the experimental work or the insight of the researchers involved. Rather, recognition of the role played by the discourse through which a theory is developed and discussed allows for acknowledgement of some of the ways in which this language shapes scientific enquiry. Metaphors, often grounded in familiar physical and social experiences, are used to describe complex scientific phenomenon that we are unable to observe directly.* Theories are dependent on language, and metaphors play a crucial role in the development and explication of scientific theory. Until recently however, there has been little acknowledgement in the sciences of the influence of metaphors and scientific discourse over the shaping of scientific enquiry. Research from the social sciences can offer some insights into this and should, perhaps, prompt scientists to take greater responsibility for the language that they use to describe their theories. Arguably, we can never devise theories outside of language, but we can take a step back and question how our use of language shapes theory. If this sounds nebulous to a bench scientist, then look again at the term 'pattern recognition'. TLRs recognize molecules, not patterns, even if the molecular surfaces are often conserved; a point that was also recently made by Beutler.[73] We should be careful not to critique the errors of one generation of immunologic metaphor only to replace them with those of our own making.

To the Clinic

At the end of the day, the discovery of TLR biology must be translated to the clinic if it is to be of benefit. One of the most obvious advances resulted from the seminal work of Bruce Beutler in elucidating how bacterial LPS is recognized.[74] Failure to recognize the LPS of Gram negative bacteria leads to uncontrolled infection and septic shock. Septic shock is a major killer and very few interventions can protect the patient. Through the work of Beutler and others we are now beginning to understand some of the signaling processes that operate in these scenarios. Genetic polymorphisms at TLR loci have been found to correlate with an individual's risk of septic shock. Creating a genetic profile of the susceptibility of patients to septic shock will clearly be of use. However, one of the most profound changes that can be brought through increased knowledge of TLR signaling will be in the field of vaccination. This is a huge field, but the work presented by Ekambar Kandimalla and Sudhir Agrawal in their chapter "Agonists of Toll-like receptor 9: Modulation of Host Immune Responses with Synthetic Oligodeoxynucleotides" succinctly describes the advances that have been made in this area. The sophistication of their nucleic acid vaccines underscores this. Clearly more TLRs will be discovered and more microbial and perhaps self ligands found, providing a clearer picture of the surveillance offered by TLRs. The notion of self protein recognition by TLRs remains controversial. Future work in this area will reveal just how discerning the immune response can be.

* http://www.the-scientist.com/yr2003/nov/opinion_031103.html

References

1. Klein J. Immunology. The Science of Self-Nonself Discrimination. 1st ed. New York: John Wiley & Sons, 1982.
2. Anderson KV, Nüsslein-Volhard C. Information for the dorsal-ventral pattern of the Drosophila embryo is stored as maternal mRNA. Nature 1984; 311:223-227.
3. Christiane Nüsslein-Volhard. The identification of genes controlling development in flies and fishes. Nobel Lecture 1995; Physiology or Medicine: 285-306.
4. Belvin MP, Anderson KV. A conserved signaling pathway: the Drosophila toll-dorsal pathway. Ann Rev Cell Dev Biol 1996; 12:393-416.
5. Lemaitre B, Nicolas E, Michaut L et al. The dorsoventral regulatory gene cassette spatzle/Toll/ cactus controls the potent antifungal response in Drosophila adults. Cell 1996; 86:973-983.
6. Hashimoto C, Hudson KL, Anderson KV. The Toll gene of Drosophila, required for dorsal-ventral embryonic polarity, appears to encode a transmembrane protein. Cell 1988; 52:269-279.
7. Janeway CA, Medzhitov R. Innate immune recognition. Annu Rev Immunol 2002; 20:197-216.
8. Desowitz R. The thorn in the starfish. The immune system and how it works. 1st ed. New York: W.W. Norton & Company, 1987.
9. Agrawal A, Eastman QM, Schatz DG. Transposition mediated by RAG1 and RAG2 and its implications for the evolution of the immune system. Nature 1998; 394:744-751.
10. Stathopoulos A, Levine M. Dorsal gradient networks in the Drosophila embryo. Dev Biol 2002; 246:57-67.
11. Sims JE, March CJ, Cosman D et al. cDNA expression cloning of the IL-1 receptor, a member of the immunoglobulin superfamily. Science 1988; 241: 585- 589.
12. Gay NJ and Keith FJ. Drosophila Toll and IL-1 receptor. Nature 1991; 351:355-356.
13. Engstrom Y, Kadalayil L, Sun SC et al. kappa B-like motifs regulate the induction of immune genes in Drosophila. J Mol Biol 1993; 232:327-333.
14. Hoffmann JA. The immune response of Drosophila. Nature 2003; 426:33-8.
15. Anderson KV. Toll signaling pathways in the innate immune response. Curr Opin Immunol 2000; 12:13-19.
16. Imler JL, Hoffmann JA. Toll receptors in Drosophila: a family of molecules regulating development and immunity. Curr Top Microbiol Immunol 2002; 270:63-79
17. Khush RS, Lemaitre B. Genes that fight infection. What the Drosophila genome says about animal immunity. TIG 2000; 16:442-448.
18. Janeway CA, Jr. Approaching the asymptote? Evolution and revolution in immunology. Cold Spring Harb Symp Quant Biol 1989; 54:1-13.
19. Barton GM, Medzhitov R. Toll-like receptors and their ligands. Curr Top Microbiol Immunol 2002; 270:81-92.
20. Zhang D, Zhang G, Hayden MS et al. A toll-like receptor that prevents infection by uropathogenic bacteria. Science 2004; 303:1522-1526.
21. Heil F, Hemmi H, Hochrein H. Species-specific recognition of single-stranded RNA via toll-like receptor 7 and 8. Science 2004; 303:1526-1529.
22. Diebold SS, Kaisho T, Hemmi H. Innate antiviral responses by means of TLR7-mediated recognition of single-stranded RNA. Science 2004; 303:1529-1531.
23. Gómez-Gómez L, Boller T. Flagellin perception: A paradigm for innate immunity. Trends Plant Sci 2002; 7:251-256.
24. Felix G, Duran JD, Volko S et al. Plants have a sensitive perception system for the most conserved domain of bacterial flagellin. Plant J 1999: 18;265-276.
25. Gewirtz AT, Navas TA, Lyons S. Cutting edge: bacterial flagellin activates basolaterally expressed tlr5 to induce epithelial proinflammatory gene expression. J Immunol 2001; 167:1882-1885.
26. Gewirtz AT. Intestinal epithelial toll-like receptors: to protect. And serve? Curr Pharm Des 2003; 9:1-5.
27. Smith KD, Andersen-Nissen E, Hayashi F et al. Toll-like receptor 5 recognizes a conserved site on flagellin required for protofilament formation and bacterial motility. Nat Immunol 2003; 4:1247-1253.

28. Gewirtz AT, Yu Y, Krishna US et al. Helicobacter pylori Flagellin Evades Toll-Like Receptor 5-Mediated Innate Immunity. Jr J Infect Dis 2004; 189:1914-1920.
29. Beutler B, Rehli M. Evolution of the TIR, tolls and TLRs: functional inferences from computational biology. Curr Top Microbiol Immunol 2002; 270:1-21.
30. Fluhr R, Kaplan-Levy RN. Plant disease resistance: commonality and novelty in multicellular innate immunity. Curr Top Microbiol Immunol 2002; 270:23-46.
31. Robert Fluhr. Sentinels of Disease. Plant Resistance Genes Plant Physiol 2001; 127:1367-1374.
32. Turner JD. A bioinformatic approach to the identification of bacterial proteins interacting with Toll-interleukin 1 receptor-resistance (TIR) homology domains. FEMS Immunol Med Microbiol 2003; 37:13-21.
33. Khush RS, Lemaitre B. Genes that fight infection: What the Drosophila genome says about animal immunity. Trends Genet 2000; 16:442-449.
34. Christophides GK. Immunity-related genes and gene families in Anopheles gambiae. Science 2002; 298:159-165.
35. Blandin S, Levashina EA. Mosquito immune responses against malaria parasites. Curr Opin Immunol 2004; 16:16-20.
36. Christophides GK, Vlachou D, Kafatos FC. Comparative and functional genomics of the innate immune system in the malaria vector Anopheles gambiae. Immunol Rev 2004; 198:127-148.
37. Luna C, Wang X, Huang Y. Characterization of four Toll related genes during development and immune responses in Anopheles gambiae. Insect Biochem Mol Biol 2002; 32:1171-1179.
38. Tan M-W, Aususbel F. Caenorhabditis elegans: A model genetic host to study pseudomonas aeruginosa. Curr Opin Microbiol 2000; 3:29-34.
39. Rich T, Allen R, Lucas A-M et al. Pellino-related sequences from Caenorhabditis elegans and Homo sapiens. Immunogenetics 2000; 52:145-134.
40. Couillault C, Pujol N, Reboul J et al. TLR-independent control of innate immunity in Caenorhabditis elegans by the TIR domain adaptor protein TIR-1, an ortholog of human SARM. Nat Immunol 2004; 5:488-494.
41. Millet CM and Ewbank JJ. Immunity in Caenorhabditis elegans. Curr Opin Immunol 2004; 16:4-9.
42. Schulenburg H, Kurz CL, Ewbank JJ. Evolution of the innate immune system: the worm perspective. Immunol Rev 2004: 198;36-58.
43. Schulenburg H, Müller S. Natural variation in the response of Caenorhabditis elegans towards Bacillus thuringiensis. Parasitology 2004; 128:433-443.
44. Alcami A, Koszinowski UH. Viral mechanisms of immune evasion. Trends Microbiol 2000; 8:410-418.
45. Jault C, Pichon L, Chluba J. Toll-like receptor gene family and TIR-domain adapters in Danio rerio. Mol Immunol 2004; 40:759-771.
46. Feys BJ, Parker JE. Interplay of signaling pathways in plant disease resistance. Trends Genet 2000; 16:449-455.
47. Cohn J, Sessa G, Martin GB. Innate immunity in plants. Curr Opin Immunol 2001; 13:55-62.
48. Jebanathirajah JA, Peri S, Pandey A. Toll and interleukin-1 receptor (TIR) domain-containing proteins in plants: a genomic perspective. Trends in Plant Sciences 2002; 7:388-391.
49. Lancet D, Sadovsky E, Seidemann E. Probability model for molecular recognition in biological receptor repertoires: significance to the olfactory system. PNAS 1993; 90:3715-3719.
50. Jones DA, Takemoto D. Plant innate immunity-direct and indirect recognition of general and specific pathogen-associated molecules. Curr Opin Immunol 2004; 16:48-62.
51. Kerr JFR, Wyllie AH, Currie AR. Apoptosis: A basic biological phenomenon with wide-ranging implications in tissue kinetics. Br J Cancer 1972; 26:239-257.
52. Vabulas RM, Ahmad-Nejad P, Ghose S et al. Hsp70 as endogenous stimulus of the Toll/Interleukin-1 receptor signal pathway. J Biol Chem 2002; 277:15107-15112.
53. Vabulas RM, Ahmad-Nejad P, da Costa C et al. Endocytosed HSP60s use Toll-like receptor 2 (TLR2) and TLR4 to activate the Toll/Interleukin-1 receptor signaling pathway in innate immune cells. J Biol Chem 2001; 276:31332-31339.
54. Vabulas RM, Braedel S, Hilf N et al. The endoplasmic reticulum-resident heat shock protein Gp96 activates dendritic cells via the Toll-like receptor 2/4 pathway. J Biol Chem 2002; 277: 20847-20853.

55. Gao B, Tsan MF. Endotoxin contamination in recombinant human heat shock protein 70 (Hsp70) preparation is responsible for the induction of tumor necrosis factor alpha release by murine macrophages. J Biol Chem 2003; 278:174-179.
56. Vabulas RM, Wagner H, Schild H. Heat shock proteins as ligands of toll-like receptors. Curr Top Microbiol Immunol 2002; 270:169-84.
57. Randow F, Seed B. Endoplasmic reticulum chaperone gp96 is required for inate immunity but not cell viability. Nat Cell Biol 2001; 3:891-896.
58. Rutherford Sl, Lindquist S. Hsp90 as a capacitor for morphological evolution. Nature 1998, 396:336-342.
59. Leadbetter EA, Rifkin IR, Hohlbaum AM et al. Chromatin-IgG complexes activate B cells by dual engagement of IgM and Toll-like receptors. Nature 2002; 416:603-607.
60. Matzinger P. Tolerance, danger, and the extended family. Annu Rev Immunol 1994; 12:991-1045.
61. Matzinger P. An innate sense of danger. Semin Immunol 1998; 10:399-415.
62. Matzinger P. The danger model: A renewed sense of self. Science 2002; 296:301-305
63. Leng Q, Bentwich Z. Beyond self and nonself: fuzzy recognition of the immune system. Scand J Immunol 2002; 56:224.
64. Efroni S, Cohen IR. Simplicity belies a complex system: A response to the minimal model of immunity of Langman and Cohn. Cell Immunol 2002; 216:23-30.
65. Langman RE, Cohn M. If the immune system is large, random, and somatically generated, then... Cell Immunol 2002; 216:15-22.
66. Cohn M, Langman RE, Mata JJ. A computerized model for the self-non-self discrimination at the level of the T(h) (Th-genesis). I. The origin of "primer" effective T(h)cells. Int Immunol 2002; 14:1105-1112.
67. Keller EF. Drosophila embryos as transitional objects: The work of Donald Poulson and Christiane Nüsslein-Volhard. History and Sociology of the Physical Sciences 1996; 26:313-346.
68. Keller EF. What impact, if any, has feminism had on science? J Biosci 2004; 29:7-13
69. Ashburner M. "Epilogue" to The Development of Drosophila melanogaster. Bate M, Arias AM, eds. Cold Spring Harbor, NY, Cold Spring Harbor Laboratory Press, 1993.
70. Weasel L. Dismantling the self/other divide in science: Towards a feminist theory of the immune system. Hypatia 2001; 16:27-44.
71. Haraway D. The biopolitics of postmodern bodies: Determination of self in immune system discourse. Differences 1989; 1:3-43.
72. Sontag S. Illness as metaphor and AIDs and its metaphors. Picador USA, 2001.
73. Beutler B. Not "molecular patterns" but molecules. Immunity 2003; 19:155-156.
74. Beutler B, Rietschel ET. Innate immune sensing and its roots: the story of endotoxin. Nat Rev Immunol 2003; 3:169-176.

Chapter 1

The Function of Toll-Like Receptors

Zlatko Dembic

Introduction

The Toll family of receptors comprises numerous related proteins implicated in the development and defense of plants and animals. Toll was first discovered in *Drosophila melanogaster* as a gene that controlled the dorsal-ventral axis of the developing embryo. Elements of its molecular structure; the extracellular leucine-rich repeat domain (LRR), short cysteine rich patches, a transmembrane portion, and an intracellular domain homologous to that of the human interleukin-1 receptor (IL-1R) are discussed in detail in other chapters. Here we are principally concerned with the role of the Toll like receptors (TLRs) and their signaling pathways in the immune system.

TLRs are found in *Arabidopsis* as intracellular proteins, whereas the Toll proteins of *Drosophila,* Tol in *Caenorhabditis elegans,* and the mammalian TLRs are all transmembrane proteins. The human TLRs comprise a family of ten related proteins (Fig. 1). Across mammalian species the number of TLRs differs, as does their expression in different cell types and their transcriptional regulation in activated cells. Ligands (some synthetic) for all but the tenth TLR have been identified, and their number is rapidly growing, due in no small part to the frenetic research activity in this area.

TLR ligands are varied, comprising bacterial cell wall components, bacterial genomic DNA, fungal, parasitic and viral products and synthetic analogs of natural products. Interestingly, TLRs can also bind autologous (self) molecules such as heat shock proteins (HSPs), intercellular matrix products, and mammalian genomic DNA. In general terms, the ligands for mammalian TLRs are either products of microbial origin that have an unusual molecular motif (pattern) or can be derived from the host species itself. A closer look reveals that host-derived ligands are usually shielded or concealed from the immune system and their emergence, for example after tissue trauma, signals that intervention by the immune system is required.

The presence of LRR modules in plant and animal proteins suggests an evolutionarily conserved role as a molecular pattern recognition receptor. Additionally, the developmental functions of TLRs in the fruit fly and nematode point to another role for TLRs in higher vertebrates, which could be to sense tissue integrity. This role could have arisen after the developmental functions that we ordinarily attribute to TLRs. Additionally; we can envisage other functions for TLRs if we consider that their activation can give rise to ten potentially divergent signals. These signals, modulated by their intensity, the cell type (including differentiation stage) of their derivation, and cellular microenvironment, may synergize or compete with one another to generate distinct TLR signals. Thus, the action of B and T cells could depend on the type of TLR signal generated by antigen presenting cells.

Toll and Toll-Like Receptors: An Immunologic Perspective, edited by Tina Rich. ©2005 Eurekah.com and Kluwer Academic / Plenum Publishers.

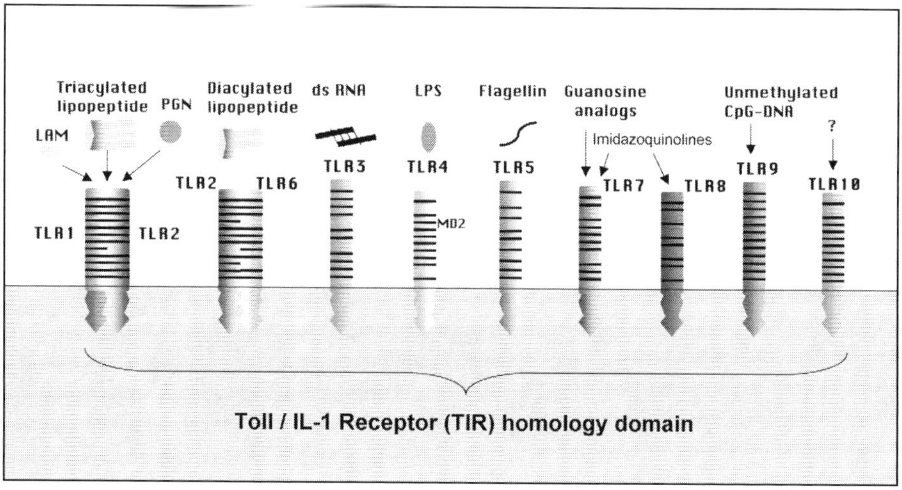

Figure 1. Human TLRs.*

TLRs help the immune system to fight the dangerous, protect the useful and neglect the vast majority of harmless microorganisms that colonize our bodies. These three functions could be generalized in a statement not dissimilar to "immunity maintains the integrity of tissues". This statement can be taken as a start point for the 'Integrity Hypothesis', which proposes three functions for TLRs. The first is to detect unusual molecular patterns. The second is to sense the extent of tissue damage, and the third is to determine the class of immune response. Specialized cells of central immunity such as dendritic cells and T and B cells are principle players in integrating these TLR signals into a specific immune response.

Clearly a better understanding of the function of TLRs in higher vertebrates is crucial for biomedical research as it will allow us to improve health care by refining the therapeutic regimes with which we treat disease.

Developmental Functions

Drosophila

The *Drosophila* genome contains a total of nine Toll-like genes. Toll-1 and its signaling pathway were identified by the genetic analyses of mutants which led to the discovery of maternal signals that are important for the dorsal-ventral patterning (Fig. 2) of the embryo (for a review see ref. 1). The expression levels of the other Toll proteins also change throughout development, suggesting that they may all play a role in embryonic development.

Dorsal-Ventral Polarization of the Embryo

Toll-1 mRNA is maternally transmitted and distributes evenly within the early embryo. Toll-1 is a transmembrane protein and its action begins at stage 5 of embryonic development. Prior to this stage, which is also known as the cell formation stage, the embryo is a sac containing a large number of nuclei. Cell membranes then begin to form around these nuclei, and Toll-1 inserts into the membranes to face out towards the perivitelline space (Fig. 2A). Subsequently, a

* Since going to press, TLR11 was identified, which recognizes a ligand expressed by uropathogenic bacteria (Zhang D et al. Science 2004; 303:1522-1526).

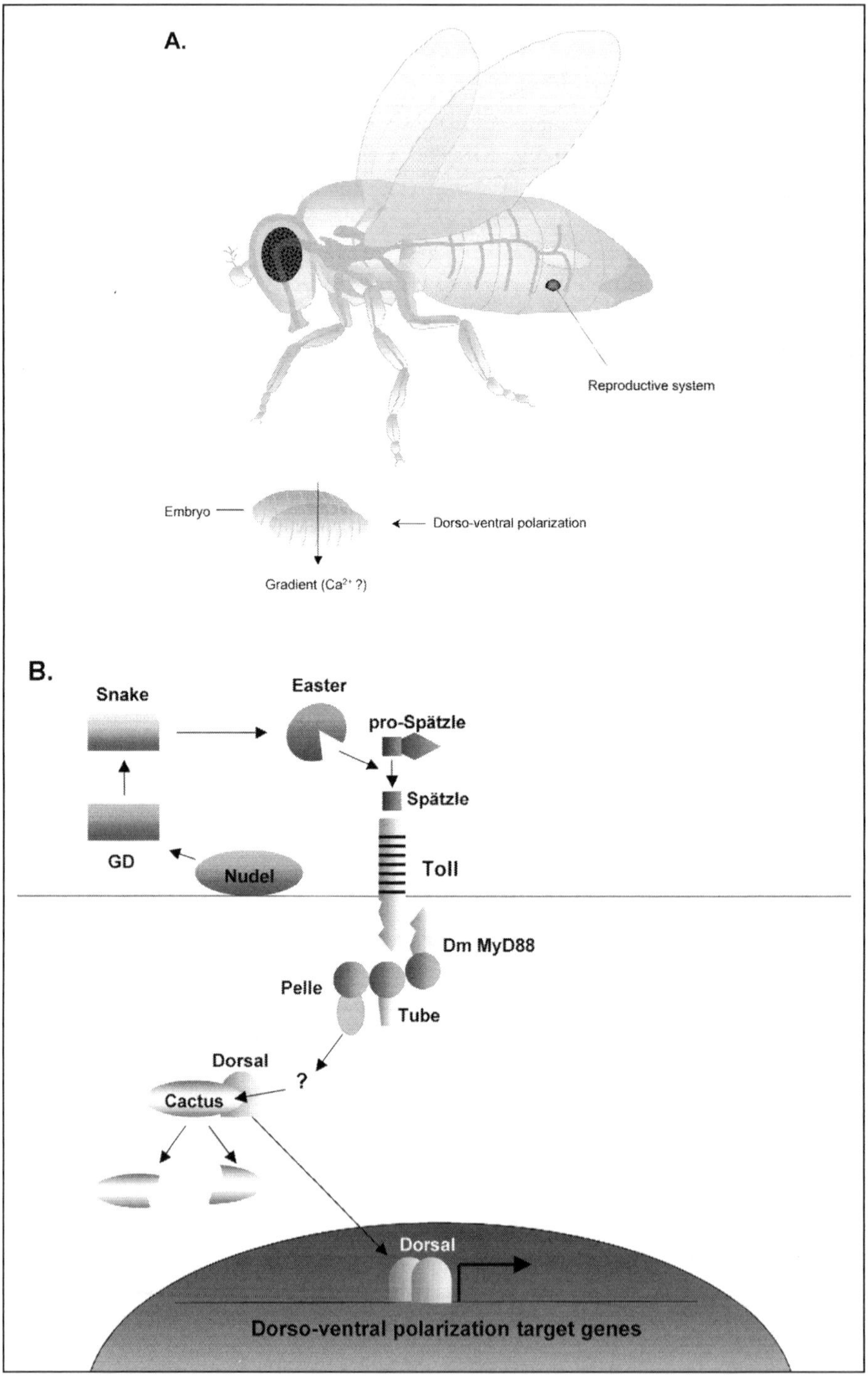

Figure 2. A) *Drosophila* development. B) Toll pathway in *Drosophila* development.

proteolytic cascade is initiated by a positional signal in the ventral region of the embryo (Fig. 2B) that generates Spätzle, which binds Toll-1. Toll-1 then recruits the *Drosophila* death domain homologue of Myd88, then Pelle and Tube. Pelle, a serine/threonine kinase, phosphorylates itself as well as Tube and Toll-1. Downstream signals, possibly mediated by an unknown kinase, eventually result in the cleavage of Cactus, (equivalent to the degradation of I-κB in the I-κB/NF-κB complex) and the liberation of the transcription factor Dorsal. Dorsal translocates to the nucleus to drive the expression of genes responsible for dorsal-ventral polarization and upregulates ventral genes (such as Twist). At the same time Dorsal acts to inhibit the '*dorsal*' genes in a ventral patch of cells in the embryo. This action is also responsible for the formation of a dorsal-ventral calcium gradient at stage 5, approximately 2.5 hours after fertilization.

Muscle Development

Toll-1 is preferentially expressed in some muscle cells and it has been suggested that early development and muscle fiber growth is dependent on Toll action. The insertion of certain muscle cells into epidermal "muscle attachment cells" has also been shown to be dependent on Toll expression.

Motoneuron Growth Cone Guidance

Motoneuron growth cone guidance, also called "axonal pathfinding" is a process whereby growing neurons penetrate tissues until they form a synapse with specific muscles. This guidance is dependent on Toll expression, and specifically on Toll's modulation of the glycosylation pattern of the cells in which it is expressed. Interestingly this action provokes Toll null cells to alter their glycosylation pattern, providing a beacon with which to guide the axon to the muscle cell. It is plausible that the growth-cone receptors can recognize glycosylation patterns and use them to steer axonal growth. The role of Toll expression in other organs, including the salivary glands, pharynx, esophagus, gut, and malpighian tubules of *Drosophila* is unknown.

Caenorhabditis elegans

Components of the *Drosophila* Toll like pathway were also found in the half-millimeter long nematode called *Caenorhabditis elegans* and their developmental roles were examined in deletion mutants of *tol-1, trf-1, pik-1,* and *ikb-1;* the *C. elegans* homologues of *Drosophila toll-1, dtraf, pelle,* and the *cactus* genes respectively. These analyses showed that only tol-1 is essential for *C. elegans* development.

Others

Thus far, the role of TLRs in embryonic development ends with *C. elegans.* It is tempting to speculate that a developmental role for TLRs may exist in higher animals. However, all the murine TLR gene knockouts have been found to develop normally and their sole defects are immune related. It therefore seems unlikely that TLRs will be found to play a role in embryonic development in mammals.

It is worth noting that the somatic cells of mammalian hematopoietic lineages develop or mature upon binding TLR ligands. These include cells of myeloid and lymphoid lineage, including monocytes, precursors of dendritic cells (DCs) and B cells. Though I consider these to be developmental functions, they will be discussed later as they also dovetail with the defense functions of these cells. Similarly, osteoblasts can be induced by TLRs 4 and 9 to exhibit osteoclastogenic activity. TLRs are expressed in many somatic tissues of higher animals, but their functional significance in these tissues remains largely unknown.

Human TLR1 is ubiquitously expressed whereas TLR2 is expressed in the brain, heart, muscle and lungs. TLR3 is, in addition, expressed in the placenta and pancreas. TLR4 is abundant in the placenta, on cardiomyocytes, endothelial and smooth muscle cells and hematopoetic cells. The distribution of TLR5 is similar to TLR4, with the inclusion of its ovarian expression. TLR6 is expressed in the thymus, spleen, ovaries and lungs. However, TLRs 7 and 8 show a more restricted expression profile. TLR7 is found solely on plasmacytoid dendritic cells (pDCs) whereas TLR8 is found on monocyte derived dendritic cells. TLRs 9 and 10 are expressed on dendritic cells, and interestingly, are abundant on activated B cells.

It is possible that the primitive developmental role of TLRs has evolved in higher vertebrates into specific roles for somatic tissues, perhaps controlling their regeneration, or as physiologic mediators of homeostasis (in adults). Drawing attention to these possibilities allows us to understand how TLRs could sense tissue integrity.

The Role in Defense

Plants

The first plant protein to be discovered with any similarity to Toll was the tobacco *N* gene product that confers resistance to the tobacco mosaic virus. However, the similarity to Toll did not extend beyond its LRR and TIR modules. Further, in contrast to Toll proteins, the *N* gene product lacks a transmembrane region and is therefore intracellular. However, the *N* gene product was found to express another module called the nucleotide binding site, which is present in most other plant Toll-like proteins. In the *Arabidopsis* genome, the first plant genome to be completely sequenced, similarity searches have identified an unusually large number of proteins with modular compositions similar to the *N* gene product. A hallmark of most of these plant Toll-like proteins is that they are intracellular and it has been suggested that they serve in defense as pathogen sensors.

Interestingly, another possible pathogen sensor in *Arabidopsis* is the FLS2 protein, a transmembrane receptor, which has extracellular LRRs and a cytoplasmic serine/threonine kinase domain. Analysis of deletion mutants of FLS2 showed that it could be involved in the recognition of flagellin, the principle protein of bacterial flagella. Flagellin retards the growth of wild type plants, which may serve as a defense response as well as an infection avoidance reaction. This response may be analogous to the aversion responses seen in *C. elegans* (see below).

Drosophila

Humoral

Infection with the fungus *Beauveria* provokes Toll-1 receptor signaling which mediates the induction of the antifungal agent drosomycin. There is no direct binding of the fungus to the Toll-1 receptor. Rather, the extracellular Toll-1 cascade, shown in Figure 3, is activated by the endogenous ligand Spätzle. Free Spätzle is normally unavailable, and exists as part of a larger precursor. The serine protease Persephone (whose action is tightly regulated), cleaves the Spätzle precursor into two unequal parts, one of which, Spätzle, binds to Toll-1. Persephone's action can be blocked by a protein called Necrotic. However, Persephone is rendered constitutively inactive by an unknown protease inhibitor. Blockade of this inhibitor (by *Beauveria*) triggers a cascade of events that leads to the anti-fungal response. Downstream of Toll-1, most of the factors involved in developmental signaling are also involved in the immune response, though with some important differences (for a review see ref. 2).

Thus, in *Drosophila* adults, the Toll-1 pathway generates a quite different output. Namely, it regulates the transcription of genes that encode antimicrobial peptides, such as drosomycin and defensin, as well as regulating the genes that alter hemocyte density. Figure 3 illustrates the crucial differences between Toll-1's developmental and defense signaling pathways. In immunity the degradation of Cactus releases, instead of Dorsal, the Dorsal related immunity factor (Dif), which then translocates to the nucleus where it acts in tandem with Relish, a nuclear factor derived from a different signaling pathway.

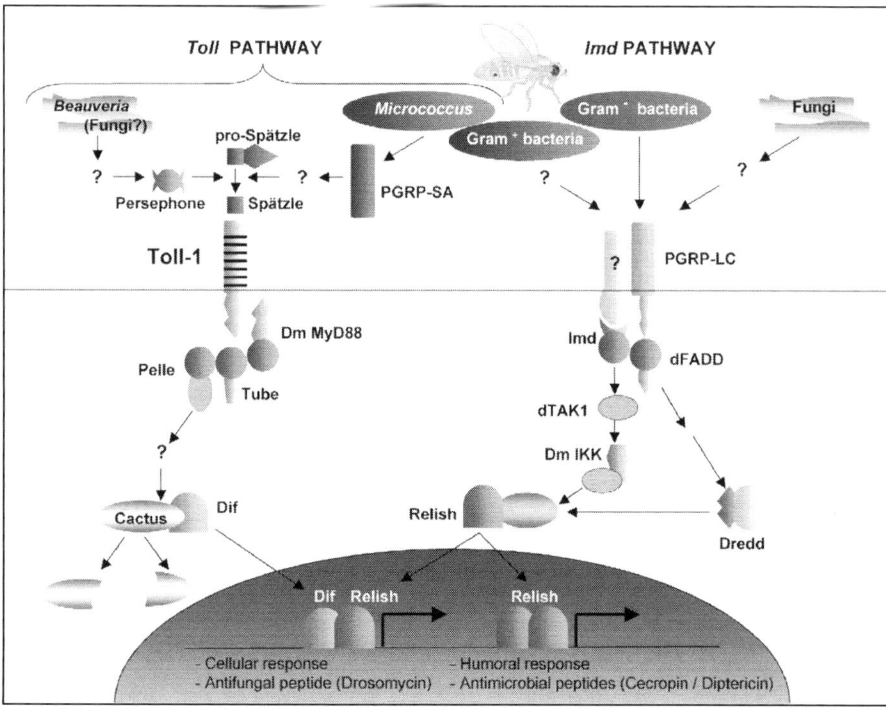

Figure 3. Humoral and cellular immunity in *Drosophila.*

A Gram positive organism *Micrococcus* has been shown to activate the Toll-1 pathway, and drives the expression of the antifungal peptides drosomycin and defensin. *Micrococcus* binds extracellularly to a pattern recognition receptor called the peptidoglycan recognition protein (PGRP-SA), which is encoded by *semmelweis.* PGRP-SA freely circulates in the hemolymph and, using a domain that can bind to the peptidoglycan moiety of Gram positive bacteria, probably activates a serine protease that cleaves pre-Spätzle. This event triggers the intracellular cascade that leads to the successful defense against *Micrococcus* (Fig. 3) (for a review see ref. 3).

Another defense pathway is the immune deficiency (Imd) pathway, named after a particular (*imd*) mutant. Interestingly, a related pattern recognition receptor to that involved in Toll-1 signal transduction may play an important role in this pathway. However, the two pattern recognition molecules (PGRP-LC, and -LE) involved, are transmembrane proteins, probably located in specialized cells. The Imd product is crucial to intracellular signaling, but does not interact with a pattern recognition receptor. Instead Imd signals downstream via a two-pronged pathway that involves the IKK kinase in one arm, and Dredd in the other. Both pathways cleave Relish, to release its nuclear binding portion that then translocates to the nucleus to form homodimers or heterodimers with Dif. These in turn act on various gene targets that are important in humoral defense. Figure 3 illustrates the main defense pathways of *Drosophila.* It appears that the Imd pathway is more potent with respect to humoral defense than the Toll-1 pathway as it controls the expression of a larger number of antimicrobials including cecropin and diptericin. These confer protection against a wide range of pathogenic microorganisms including fungi, Gram negative and Gram positive bacteria. The Gram negative diaminopimelic acid type peptidoglycan is the most potent inducer of the Imd pathway, whilst the Toll pathway is predominantly activated by the Gram positive lysine type of peptidoglycan. Thus, the ability of *Drosophila* to discriminate between Gram positive and Gram negative bacteria may rely on its recognition of specific forms of peptidoglycan.

Cellular

The lymph glands of adult *Drosophila* produce hemocytes that circulate throughout the body. These ensure host defense by encapsulating foreign bodies and phagocytosing smaller toxic objects. Encapsulation and subsequent melanization inactivates the intruder, which remains trapped in the body. Several genes that regulate cellular immunity also control the proliferation and differentiation of the hematopoetic lineage. The Polycomb group of genes regulates lymph gland cell proliferation and hemocyte numbers in the body. Mutant larvae show increased growth of the gland and excessive numbers of hemocytes that occasionally go on to invade tissue as pseudotumors. Mutants with enhanced Toll signaling show a similar phenotype, whereas those with reduced Toll signaling have fewer hemocytes (for a review see ref. 4). In conclusion, the Toll pathway regulates cellular and some humoral aspects of immunity in *Drosophila*.

C. elegans

Caenorhabditis elegans expresses structural homologues of several components of the *Drosophila* Toll pathway. The *tol-1* gene, homologous to *toll*, is required not only for nematode development, but also to avoid the pathogenic microbe *Serratia marcescens*. Analyses of deletion mutants of the *C. elegans* Tol pathway showed that none, save *tol-1*, were important for immunity. Interestingly, *tol-1* deletion mutants lost their *S. marcescens* avoidance behavior though other chemosensory behaviors remained intact. Further analysis revealed that the Tol-1 protein is required for sensing *S. marcescens* by a neural system. Tol-1 is located on the sensory tip of an axon located near the oral orifice and microbial contact generates signals that then induce the evasion response (Fig. 4).[5] This phenomenon is discussed in more detail by Nathalie Pujol and Jonathan Ewbank in their chapter.

The defensive role of the neural system has not been studied sufficiently at the molecular level in higher vertebrates, due to the obvious complexity of the neural networks involved. Nevertheless, our ancestors might have expressed pattern recognition receptors in sensory neurons, similar to those of *C. elegans*. Perhaps taste or olfactory organelles evolved from these ancestors. It is also possible that the detection of dangerous molecular patterns on macromolecular objects using neural tissue was deselected during evolution, as the immune system developed ever more elaborate defense mechanisms.

Mammals

TLRs, 1, 2, 4 and 6, expressed at the surface of many hematopoetic cells, have been shown to bind distinct bacterial cell wall components and viral products. TLR2 associates predominantly with TLRs 1 and 6, but in some cases also with TLR4 (Fig. 5A). The ligands for TLR1, 2 and 6 includes peptidoglycan (PGN) from Gram positive bacteria (e.g., *Staphylococcus aureus* and *Streptococcus pneumoniae*), lipoproteins of spirochetes and mycobacteria, yeast and mycoplasma. The TLR1/2 heterodimer binds triacylated lipopeptides, whereas the TLR2/6 combination is specific for diacylated lipopeptides. The TLR1/2 heterodimer is also implicated in binding products derived from *M. leprae* and *M. tuberculosis*. TLR2 can bind human cytomegalovirus (HuCMV) products as well as measles virus. TLR2 has also been reported to bind bacterial lipopolysaccharides (LPS) (for example, from *Leptospira interrogans* and *Porphyromonas gingivalis*) as well as mycobacterial lipoarabinomannan (LAM). These reports implicated a second molecule, MD-2, which was found associated with TLR4. Consequently, TLR2 may associate with other TLRs in these complexes. On the other hand, TLR4 can homodimerize to bind the LPS of Gram negative bacteria (*E. coli* or *Neisseria meningitides*) in association with

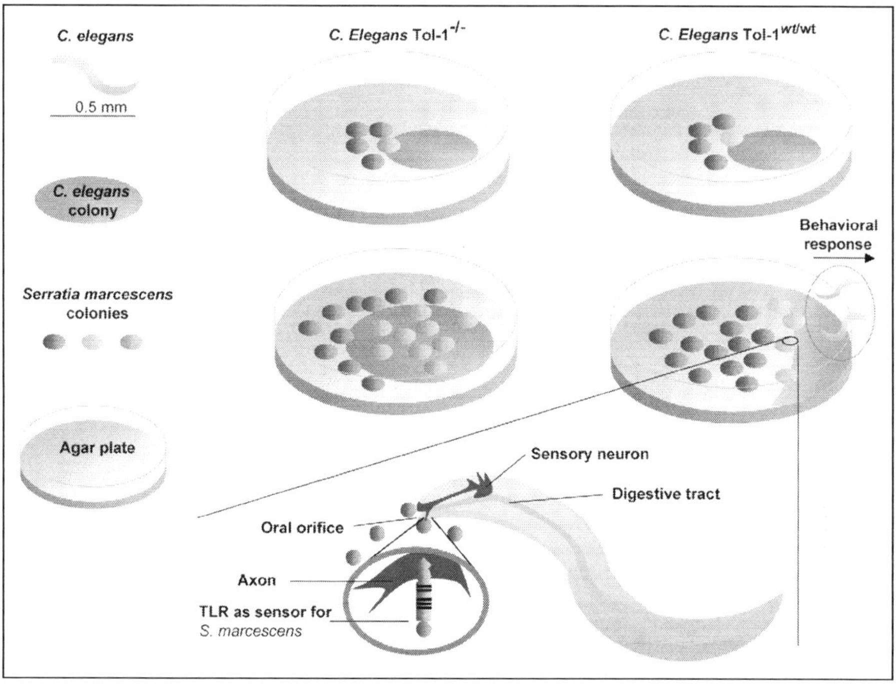

Figure 4. Defensive role of Tol-1 in *Caenorhabditis elegans.*

the MD-2 protein, bacterial lipoteichoic acid (LTA), and bacterial heat shock proteins such as Chlamydial HSP60, and *Toxoplasma gondii* HSP70. Viral products that bind TLR4 have been described in mice infected with respiratory syncytial virus and murine mammary tumor virus, a murine retrovirus. TLR3 binds double stranded RNA, a product of some viral infections, and the synthetic ligand, poly Inosine:poly Cytidine. TLR5 binds flagellin, a bacterial locomotory organelle, and TLR7 binds synthetic nucleoside analogs and, with TLR8, binds imidazoquinolines. TLR9 binds bacterial or viral genomic DNA that contains unmethylated stretches of CpG nucleotides (or synthetic oligonucleotides that contain CpG motifs). No ligand has yet been identified for TLR10, which was identified in humans, but not in mice.

The chromosomal locations for these genes in humans and mice have been elucidated (except for TLR10), and their function in various infectious disease models extensively studied in mice deficient for various TLRs (for a review see ref. 6). Interestingly, TLRs 2 and 4 bind not only bacterial, but also host self molecules, for example the human heat shock proteins (HSP60 and HSP70), and mouse products found in the intercellular space, including fibrinogen, hyaluronan, and heparan sulfate. TLR4 was implicated in binding the extra domain A of mouse fibronectin. Similarly, in low but still notable quantities, mammalian CpG DNA can be found after necrotic cell death, which may explain the capacity for TLR9 to bind double stranded mouse genomic DNA (Fig. 5B).

TLR ligands can instigate intracellular signals similar to those of the IL-1R signaling pathway, but recently differences have begun to emerge. TLR functions have been assessed by studying murine gene deletions in vivo and ex vivo, and in vitro, using *tlr* null cell lines (e.g., HEK293) that are then transfected with the relevant *tlrs*. The TLR signaling pathway includes the activation of a series of adapter molecules: MyD88 and TRAF; protein kinases such as IRAK, the inhibitor of κ-B kinase (IKK), cJun N-terminal kinase (JNK), MAPK, and

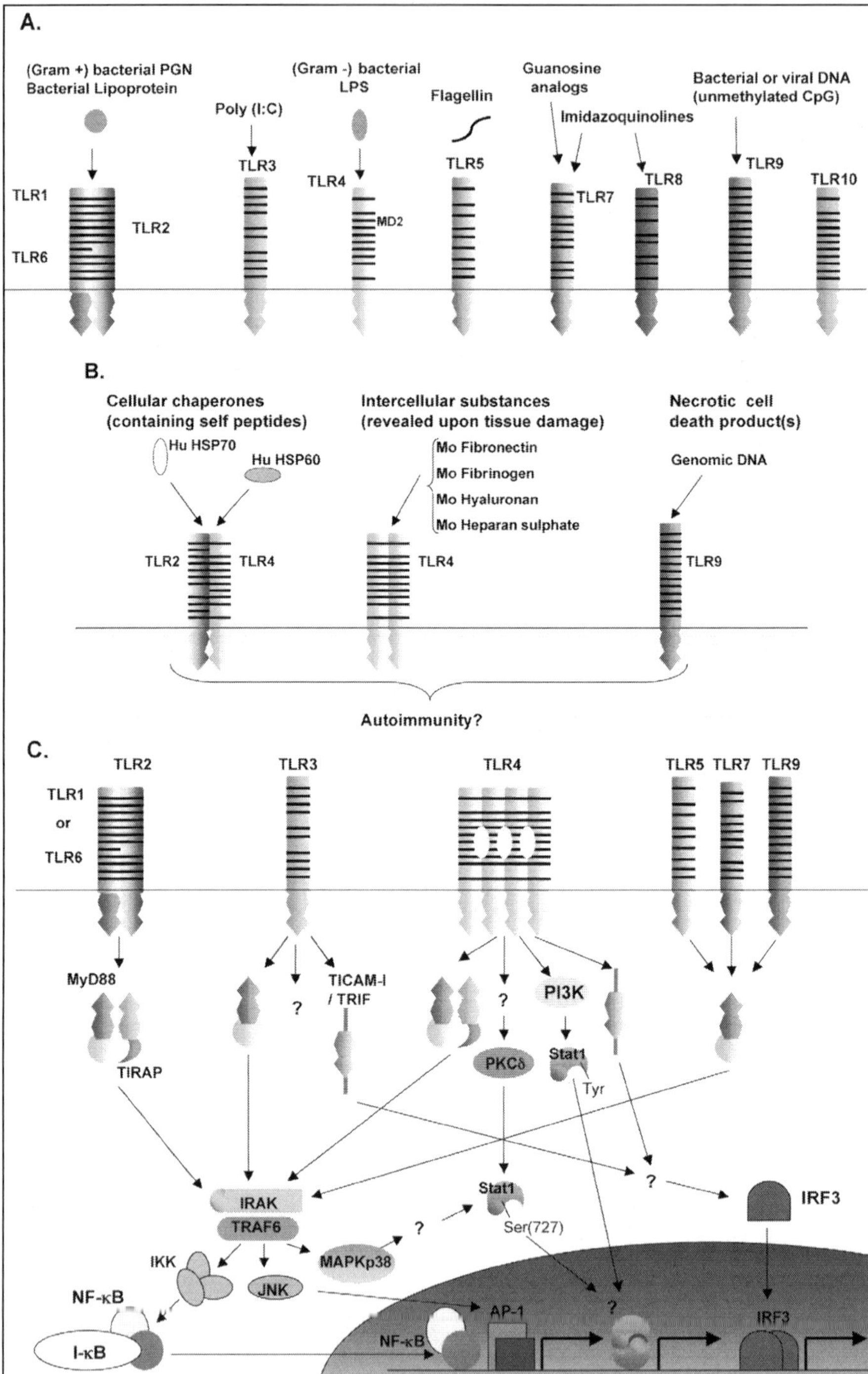

Figure 5. A) Exogenous TLR ligands. B) Endogenous TLR ligands. Mo indicates murine products. Hu indicates human products. C) Intracellular signaling pathways of mammalian TLRs.

Table 1. Divisions of immunity according to various definitions

Division Premise		
Structural	Non-specific	Specific
Developmental	Innate	Adaptive
Functional	Autonomous	Central

p38; and the activation of transcription factors such as NF-κB, Rel, and AP-1 (for a review see refs. 7 and 8). Differences from the IL-1R signaling pathway include a novel signaling molecule TICAM (TRIF), the ERK-1 and ERK-2 kinase pathways and possibly Stat-1 proteins (Fig. 5C). However, a complete picture of TLR signal transduction pathways is yet to emerge. Though the expression of TLRs has been found in many somatic tissues in the human body, the research effort has (as for *Drosophila*) focused on cells of hematopoetic origin and the immune system.

To interpret these findings in relationship to immunity we must review the action of TLR ligands on each cellular effector of the immune system.

Divisions of Immunity

Scientific literature often mentions TLRs as a link between the innate and adaptive wings of the immune system, and discusses their pivotal role in the activation of innate immunity. Perhaps a clearer notion would be that TLRs initiate and modulate central immunity, and play an important part in the activation of autonomous immunity.

The central versus autonomous divisions of immunity will be developed in this chapter to better understand certain aspects of immunity, and especially to explain certain theories of the immune system. In order to avoid confusion by introducing a novel definition to the experienced reader, I will try to alternate the central and autonomous immune designations with the more commonly used terms, innate and adaptive immunity. Although the latter would appear inappropriate for the discussion of some aspects of immunity, it seems that none of the terminologies in present use is perfect. To describe the subtleties of the immune response is difficult, and each division is useful in its own right when addressing particular aspects of immunity (Table 1).

Specific—Nonspecific (Humoral or Cellular) Immunity

A division of the immune system based on the structural features of effector entities was the first to be used. This humoral versus cellular division, each being either nonspecific or specific has helped to explain the clonal selection of lymphocytes (notably without the inclusion of inflammation, which is described as a nonspecific tissue reaction). Each B-cell receptor (BCR) or T-cell receptor (TCR) probably binds many ligands, but we usually consider them to be specific for a single antigen or peptide/MHC combination. Likewise, the cells of nonspecific immunity might be more "specific" than originally thought, as TLRs can bind specifically to certain molecular motifs. Humoral immunity consists of soluble mediators such as complement (non specific) and antibodies (specific) that can neutralize, opsonize or kill pathogens. Cellular or cell mediated immunity is represented by a number of players. First we have the 'nonspecific' phagocytes, which include the macrophages that 'purge' pathogens, then specific cells such as B cells, that become antibody secreting plasma cells and T cells, including the CD4 and CD8 subsets. The CD4 T cells help B and cytotoxic T cells whilst inhibiting other T cells. The CD8 T cell subset kills virus-infected cells. The cellular and humoral divisions cannot satisfactorily describe immunoregulation. This is most evident when we discuss the roles of

CD4 T cells and dendritic cells (DC). Both are effectors and, at the same time, regulators of the immune response by direct (cellular) effects as well as by cytokine production (humoral effects). Therefore, it is hard to separate humoral and cellular effects in DC and CD4 T cells.

Innate—Adaptive Immunity

The next division was suggested on the basis of development: innate or inborn immunity vs. acquired, or adaptive immunity. Here the importance was to understand how we carry, in our germline, a memory of past infection. This 'memory' is embodied by the pattern recognition receptors and their ability to bind to conserved patterns of unusual conformers present only on microbes or distantly related organisms. Such receptors would be oligo specific and could activate the cells that we had previously grouped under the umbrella of nonspecific immunity. The use of the other developmentally defined term—adaptive immunity—quickly became popular, as it cleared the confusion regarding specificity of B and T cells. Thus adaptive immunity came to describe a group of cellular and humoral mediators that each organism could acquire during its somatic development. The main components (B and T cell receptors or BCRs and TCRs) were not encoded in a mature form in the germline, but instead were assembled by rearranging variable and constant gene segments. Because these rearrangements occur randomly, BCR and TCR genes would assemble differently in the B and T cells of individuals of the same species. Therefore, acquired immunity equips every mammalian organism with a different set of randomly assembled TCRs and BCRs.

Problems with this division can be envisaged, if we wish to classify NKT cells, which have germline encoded receptors, but also rearrange their TCRs. Similarly, γδT cells rearrange the variable portions of their TCR genes, yet function at the vanguard of defense. NKT and γδT cells are generally thought to belong to the innate wing of immunity, yet both acquire their receptors in a similar fashion to αβT and B cells.

Autonomous—Central Immunity

Despite a detailed understanding of the cells and mediators of the immune system, the mechanisms that precede the induction of B and T cell effectors have remained elusive. Perhaps the problem lies in the fact that the basic function of the immune system is not only to kill pathogens, but also to tolerate commensals. To better illustrate this we could use a division based on cellular function. Autonomous immunity is the swift firstline defense whereas central immunity can be described as a nodal, regulated, or controlled type of immunity. In my opinion, it is the most suitable division to explain the role(s) of TLRs in the immune system.

Autonomous immunity comprises all but dendritic cells from previous definitions, and would include NK cells, NKT cells and T cells with γδ TCRs. The reasons for this segregation are as follows. Components of autonomous immunity are cells or factors that depend on fast local reaction (Fig. 6A). The activation of cells and components occurs at the site where effector actions are first engaged, which is usually taken to be the locale of tissue damage, inflammation, or antigen challenge (i.e., contact with the micro-organism). Thus, the cellular components require very little communication with cells that are distant from the site of pathogen intrusion.

In contrast, central immunity has afferent and efferent loops of action. As illustrated in Figure 6B, the afferent loop starts with the activation of dendritic cells in damaged tissues by micro-organisms. DCs, having sampled the antigens from their surroundings, migrate to the lymph node, which is the center of the immune response (for a review see ref. 9). There, DCs meet naive CD4 T cells, which they then stimulate. After stimulation, T cells proliferate to become effector (helper) T cells that migrate to B-cell areas of the node. Upon T-B interaction (provided that the B cell has met the same antigen as the T cell), an activatory signal is transferred and B cells go on to proliferate and develop into antibody secreting cells (called plasma

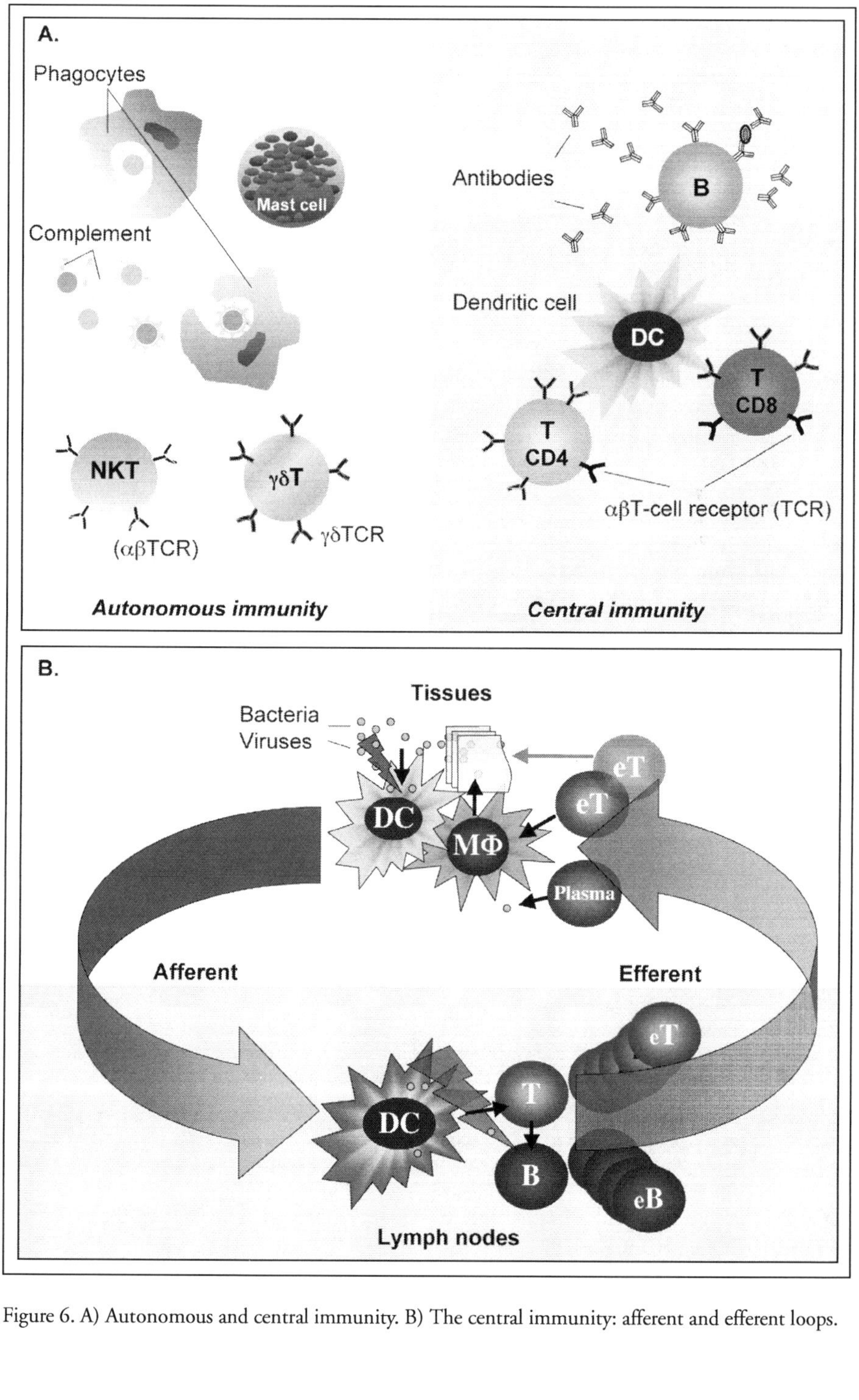

Figure 6. A) Autonomous and central immunity. B) The central immunity: afferent and efferent loops.

cells). Alternatively, CD4 T helper cells could recruit cytotoxic CD8 T cell precursors and license them for action in the periphery.

B cells, in order to receive help from T cells, need to recognize antigen or hapten. There are numerous ways by which the stimulating antigen can be delivered to B cells. Antigen may be derived from the blood stream, usually attached to other proteins, for example an antibody/antigen complex, or may be carried by phagocytic cells, in which case the ingested antigens could be exposed after the death of their carrier. Alternatively, circulating B cells may collect antigens along their recirculation routes to and from the lymph nodes (or spleen). B cell activation by T-dependent antigens leads to the formation of germinal centers in the lymph node cortex. B cells then proliferate and develop into effectors. During peripheral development B cells undergo a selection process by which they can increase the affinity of their BCRs. This process involves hypermutation of BCR genes and the selection of high affinity B cells on follicular dendritic cells (a stromal cell unrelated to DCs). This results in the appearance of effector germinal center (gcB) B cells with a higher affinity for antigen. These effectors can go on to develop into plasma cells.

The efferent loop of action begins with the migration of effector T and B (plasma) cells from the center to periphery, i.e., back to the tissue from whence the DCs first started their trek. Effector cytotoxic CD8 T cells ($\alpha\beta TCR^{+ve}$) will go on to kill virally infected cells, whereas CD4 $\alpha\beta$T cells can activate macrophages via cytokines (interferon γ) and control, or regulate (again with cytokines), other cells of the autonomous system. Plasma cells secrete antibodies and eventually migrate and lodge in a strategically superior locale from where they can ensure a long lasting supply of circulating antibodies (such as the bone marrow or spleen).

Based on these premises, immunity can be subdivided into central and autonomous. Central immunity facilitates our understanding of TLRs as pattern recognition receptors, and emphasizes their additional functions in DCs and memory B cells.

Let us now analyze the distribution of TLRs (Tables 2, 3) and the evidence for their roles in the immune system (Table 3). Later, I shall discuss these (predicted) roles in the context of various immune system models.

Expression and Function of TLRs in Cells of Autonomous Immunity

Two mammalian species are considered in this section; human and mouse. TLR expression can be regulated by various substances or cellular interactions, which differ across animal species, an indication that their sensitizing or desensitizing immune responses might also differ.

Neutrophils (and Granulocytes)

Neutrophils express TLRs 1, 2, 4, 6, and 8. The lifespan of neutrophils is short, but can be prolonged by activation via TLRs 2 and 4. NF-κB, which is implicated in signaling via TLRs 2 and 4, is a known survival factor and was first implicated in the cell death mechanism delivered by TNF and related signal pathways. Only later, with the use of IKK knockout mice, did it become apparent that the survival effects of NF-kB are counter balanced by caspase driven programmed cell death.

Upon binding of TLR2 or TLR4 ligands, neutrophils upregulate the expression of chemokines, downregulate some chemokine receptors, and change their expression of adhesion molecules (integrins and selectins) and respiratory burst mediators.[10,11] All these factors drive the inflammatory response in local tissue.

Mast Cells

LPS and PGN (TLR2 and 4 ligands) differentially activate TNFα and IL-5, IL10, and IL13 in human mast cells. However, the release of TNFα by mast cells requires priming with IL-4 and the presence of serum components such as soluble CD14. Interestingly, regarding the

Table 2. TLR ligand repertoire of human cells

TLR	Ligands: Bacterial G+	G-	Other	Viral	Other Nonself	Self	Cell Types	Refs.
TLR1	+		+				preDC, (preDC2), CD11c+iDC, Mo, PMN, Mϕ Resting B →↑Gc-B, NK	13,15,19,62
TLR2	+		+	+	+	+	preDC, CD11c+iDC, Neutrophils, Mast, Mo, Mϕ,	13,15,19,41
TLR3				+			CD11c+iDC, (Mϕ,↗LPS)	13,15,19,67
TLR4		+	+		+	+	preDC, Neutrophils, Mast, Mo, PMN, Mϕ, (Resting B→↙Gc-B)	13,19
TLR5	+						preDC, (CD11c+iDC), Mo	13,15
TLR6	+		+				(preDC), (preDC2), (CD11c+iDC), Mo, PMN, Mϕ, Resting B→ ≈ Gc-B,	13,19
TLR7					+		preDC2, Mϕ, Resting B→↙Gc-B, (NK)	13,19
TLR8					+		preDC, (CD11c+iDC), Mo, PMN, Resting B→∪ Gc-B,	13,19
TLR9	+	+	+				preDC2, Resting B→↗Gc-B, (NK), (Mϕ↗LPS), (Mo)	13,17,19,25, 67
TLR10							(CD11c+iDC),Resting →↗ –Gc-B,	13,19

Cells that express small amounts of the indicated TLRs are in brackets. TLRs whose expression is regulated via another TLR ligand or signal are underlined. Mo= monocytes; Mast= mast cells; DC= dendritic cells; PMN= polymorphonuclear leukocytes; Mϕ= macrophages; Gc = germinal center; NK= natural killer; G= gram.

possible connection with allergy and atopic diseases, PGN, but not LPS, can induce the release of histamine by mast cells.[12]

Monocytes

Monocytes are precursors of myeloid derived DCs and macrophages. Human monocytes express TLRs 1, 2, 4, 5, 6 and 8[13] and increase their expression of TLR4 when treated by IFNγ,[14] whilst IL-10 has the opposite effect.[15]

Human monocytes can traverse the endothelium and, as for cells of the immune system, become either macrophages or dendritic cells, depending on the cytokine cocktail employed in the culture medium. GM-CSF and IL-4 induce blood monocytes to become precursors of DCs (preDCs) whilst TNFα results in $CD1a^+$ DCs and GM-CSF and IL-15 skew development towards $langerin^+$ DCs (similar to Langerhans cells in skin). On the other hand, the addition of fibroblasts favors the generation of $CD14^+$ macrophages, as the fibroblasts secrete IL-6, which induce the M-CSF receptors that promote macrophage development.

Monocytes can thus be seen as precursors for both the macrophage and DC lineage, or alternatively, as tissue specific precursor cells that appear to be similar but bear different surface markers. Depending on the culture conditions (or that of the tissue), precursors of one cell type might predominate over the other. Thus TLRs may contribute to monocyte differentiation by transducing differentiation specific stimuli.

Table 3. Cellular distribution and function of human TLRs

Cells	TLR Expression	Roles	References
Neutrophils(PMN)	**1**, **2**, **4**, **6**, **8**	Pattern recognition	10,11,19 41
Mast cells	2, 4	Pattern recognition	12
Monocytes	1, 2, 4, 5, 6, **8**	Pattern recognition	13,41,68
Macrophages	1, 2, 3, 4, 6, 7, 9	Pattern recognition (Ag presentation)	18,69
NK cells	1, (7, 9)	Pattern recognition	15,19
Resting B cells	(1, 4), 6, **7**, (8,) 9, 10	Pattern recognition	19,25
Germinal c B cells	1, 2, 6, **7**, 8, **9**, **10**	Pattern recognition	19,25,41
preDC(Monocytes)	**1**, **2**, **4**, **5**, (6,) **8**	Pattern recognition	13
iDC (CD11c$^+$iDC)	1, 2, **3**, (5, 6, 8, 10)	Pattern recognition Ag presentation (Ag crosspresentation)	13,18,31,67,68
plasmacytoid DC (preDC2)	(1, 6,) **7**, **9**	Pattern recognition (Ag presentation) (Ag crosspresentation)	13,18,67

TLRs that are strongly expressed are in bold typeface, whereas weakly expressed TLRs are in brackets. Underlined numbers denote regulation (induction) via an exogenous factor. The brackets in the 'Roles' column denote a requirement for the activation or maturation of the cells indicated. Monocytes appear twice, as some authors report them as preDC1 cells.

Macrophages

Macrophages are generated in the tissues from monocytes, but unlike DCs, do not migrate. Human macrophages are efficient APCs and express a variety of TLRs (see Table 3). The available information about TLRs expressed on mouse macrophages is largely concordant with the human data.

Macrophages can interpret the molecular patterns expressed by intruding microorganisms, and are activated by bacterial products. Their immediate action, at the site of tissue damage, is controlled by various factors that include both the autonomous and the effector arms of central immunity. There is a direct activation of autonomous immunity via TLRs, and an indirect one, via IFNγ. The most potent activator of macrophages is IFNγ, a product of activated NK cells and effector CD4 T helper-1 cells. Activated macrophages perform a series of important functions. These include the production of extracellular oxidative radicals that can damage pathogens (and normal tissue), the production of the vasodilator, nitric oxide (NO), the secretion of proinflammatory cytokines that increase body temperature, mobilize fatty acids, and, in high enough quantities, cause septic shock. Macrophages also act to secrete chemokines that attract effector T, B, plasma and other cells to the site of inflammation, upregulate MHC class I, class II, Fc and complement receptors and kill intracellular bacteria.

Using microchip technology, a unique gene expression profile was found to be induced by Gram negative bacteria (via TLR4) in human macrophages that included the upregulation of IL-12 p70 and type I IFN. This was in stark contrast to the profile induced by Gram positive bacteria acting via TLR2.

Exposing macrophages to LPS (via TLR4) induces a hyporesponsive state to a second challenge with LPS, which is called LPS tolerance. This response is also induced by pre-exposure to TLR2 ligands. LPS signaling involves at least two pathways; a MyD88 dependent cascade that is important for the secretion of proinflammatory cytokines and a MyD88 independent pathway (via TICAM; TRIF) that controls the expression of IFNβ inducible genes.[16]

The direct activation of macrophages via TLRs is exemplified in experiments with TLR gene knockout mice. Macrophages from mice deficient in TLR3 (that binds dsRNA) exhibited specific defects in the secretion of IL-6 and IL-12, activation of NF-κB, and the induction of type I IFN in response to poly(I:C). All of which indicate the role of macrophages in the antiviral response. Experiments with other TLR deficient macrophage cell lines have shown that macrophages also play an important role in antibacterial, antiparasitic and antifungal immunity (for a review see ref. 6). Immunization preparations like bacillus Calmette-Guerin (BCG) can activate murine macrophages and induce the secretion of TNFα. This effect is mediated by both TLRs 2 and 4, as double deficient murine macrophages fail to respond to BCG by secreting TNFα, whereas single TLR2 or TLR4 deletion mutants retain some TNF production.

The expression of certain TLRs can be regulated by receptor ligand interactions that involve other TLR family members. For example, LPS (probably acting via TLR4) can induce mouse macrophages to upregulate TLR9.[17] However, there is a dichotomy in antigen presentation between macrophages and DCs that is initiated by TLR ligand binding. TLR3 and TLR9 ligands are inefficient in inducing crosspresentation in macrophages and macrophages require a thousand fold higher concentration of antigen to generate the same effects as DCs.[18] This would suggest that macrophages are poor stimulators of antigen specific precursor CD8 T lymphocytes (pCTL) when pulsed by TLR3 or TLR9 ligands plus antigen in somatic tissue.

NK Cells

Human NK cells express mRNA of the ubiquitous TLR1 and very low amounts of TLR7 and TLR9.[19] However, TLR9 ligands can activate NK cells[20] and it is possible that pattern recognition by TLRs can induce immediate action in these cells. Activated NK cells can kill somatic cells that lack particular MHC molecules as a result of their viral infection or malign transformation (as detected by killing inhibitory receptors, KIRs). NK cells can also kill cells that express unusual MHC molecules (recognized via killing activating receptors, KARs) and produce IFNγ to activate macrophages.

Expression and Function of TLRs in Cells of Central Immunity

Dendritic Cells (DCs)

Various populations and maturation stages of DCs have been described that depend on their tissue distribution and species of origin. Most DC subtypes share capabilities such as antigen presentation, the costimulation of T cells, migration to and from somatic tissues and cytokine and chemokine secretion. However, some express different sets of TLRs (Tables 2, 3). As DCs can orchestrate immune responses, the function of their TLRs is of paramount interest.

Human DCs

In human blood we can find monocyte like precursor cells (preDC or MoDCs) that develop, after 7 days in culture with GM-CSF and IL-4, into immature DCs. These can develop further, with the addition of TNFα, into mature type 1 DCs. These cells have been used extensively in cancer immunotherapy. Blood also contains immature DCs (iDCs) that are $CD11c^{+}HLA^{+}$ that can be induced to mature into mDCs. Mature DCs display a myeloid phenotype of markers, have upregulated costimulatory markers such as CD80 and CD86 and produce cytokines such as IL-12p70, IL-18 and IL-23 that help Th1 development. This DC pool, called DC1, still contains clearly divisible subpopulations and is probably derived from precursor cell preDCs. Another DC lineage found in blood is the plasmacytoid DC (pDC), with the phenotype $CD11c^{-}IL3R^{+}CD45RA^{+}$ (for review see ref. 21). The plasmacytoid DC lineage resembles that of B plasma cells, without B cell markers but with the CD4 T cell marker. Plasmacytoid DCs secrete IFNα, and can mature into the DC2 subset, which shows a Th2 skewed cytokine secretion profile (IL-4, IL-10).

Human bone marrow cells can also yield various types of dendritic cells in vitro, if incubated with various cytokines. The mature mDC1 and mDC2 subsets probably represent the terminal stages of preDC development and it is thought that these two lineages might have originated from either lymphoid or myeloid precursors. However, it is possible that yet more DC types exist.

DCs express the widest known repertoire of TLRs (Table 3). PreDC, or monocyte like cells (MoDCs) express TLRs 1-8 excepting TLRs 3 and 7, which are expressed on mature DCs. Immature iDCs have a decreased expression of all TLRs, except TLR3. IDCs also seem to lack the expression of TLR4, though TLR4 may be present at a concentration that escapes detection. Interestingly, DCs are equipped with additional pattern recognition receptors that include the mannose receptor and DEC205, a C type lectin, both of which have roles in endocytosis or phagocytosis.

An important albeit transient capacity of immature DCs is endocytosis. It provides the means by which a microbial antigen can be taken up, processed, loaded onto MHC molecules and presented to T cells. This constitutes the first act in T cell activation (and is generally the start of the immune response). The processing and presentation of antigen will be described later. Mature DCs, on the other hand, loose this phagocytic capacity, but instead acquire an increased antigen presentation competence due to their enhanced expression of MHC II molecules. Consequently, DCs can trap their engulfed antigenic loads until they can present them to T cells in the lymph node. This makes them a unique class of cells that can capture a representative antigenic profile of each tissue that they pass.

Mouse DCs

Murine DCs share many markers with their human counterparts (for a review see reference 22). Certain populations have a common marker like CD11c, which is, in association with CD18, an integrin molecule. Murine DCs can also express markers common to the helper and cytotoxic T cells, CD4 and CD8α. The latter is not found on their human counterparts (thought to be pDCs). Based on these markers there could be at least 6 different murine DC populations. DCs can be further subdivided on the basis of their expression of the CD45RA and CD205 markers, a protein phosphatase and C type lectin respectively. Both, the lymphoid and myeloid lineage of DC progenitors have the potential to give rise to all known populations, however at different ratios. In addition, subpopulations differ in their production of cytokines and in their expression of additional cell surface markers, which could relate to their tissue of origin (spleen or bone marrow). The function of these subpopulations is not yet clear.

Precursors of murine homologues of the human preDC1 and preDC2 (pDC) subtypes have also been identified in blood. A $CD11c^{-/+}CD11b^{-}CD45RA^{+}$ mouse population closely resembles human plasmacytoid cells (pDCs or preDC2) on the basis of their morphology and function, as CpG motifs can stimulate these cells to produce IFNα and develop into $CD8^{+}$ DCs. A second population of preDCs has the surface phenotype $CD11c^{+}CD11b^{+}CD45RA^{-}$ and closely resembles the precursors of human preDC1 cells that go on to develop into $CD8^{-}$ DCs after stimulation with TNFα. In the mouse, the precursors of DC2 cells (preDC2) strongly express TLRs 7 and 9. Virus or challenge via TLR9 can trigger these cells to develop into DC2 cells that secrete type I interferon (IFNα/β). These DCs generate IL-4, which has a strong influence on the preDC1 population by enhancing the production of IL-12p70. The preDC1 population has a different TLR profile, and expresses almost all of the TLRs, except 7 and 9.

Spleen or bone marrow derived precursors of the murine preDC1 subset (myeloid precursor) and preDC2 subset (plasmacytoid precursor) can induce the development of both Th1 and Th2 effector cells depending on antigen dose. At high doses Th1 cell development is favored whilst lower antigen doses induce the development of Th2 cells.[23]

DC Maturation

When considering the developmental preDC/DC pathway we should clarify the term maturation when used in the context of DCs. Maturation is a series of changes that can be elicited by LPS, viral or bacterial infection, cytokines, bacterial DNA, extracellular matrix in inflamed tissue or the ligation of molecules on immature DCs (iDCs) such as CD40. Various TLR ligands can induce bone marrow derived DCs to mature. The concentration of ligand that induces maturity (as measured by the induction of costimulatory molecules CD80, CD86) varies greatly. Remarkably, peptidoglycan and flagellin require the highest concentrations, whereas LPS can induce the maturation of DCs at a thousand fold lower concentration than most other ligands. The maturation of iDCs generally leads to downregulation of most TLRs.

Thus, triggering TLRs can induce the maturation of iDCs, in terms of their secretion of cytokines (IL-6, TNFα) and upregulation of costimulatory molecules (Fig. 7A). The intracellular pathways that lead to these events have been extensively studied (Fig. 7A) though are incomplete given that subpopulations of iDCs might differ in their developmental requirements, stimulation, and further maturation. These factors could influence the biological functions of DCs and specific immunocytes. Namely, in addition to costimulatory molecules like CD80 and CD86, there are other B7 family members whose expression on APCs, like DCs, may differentially influence the fate of stimulated T cells. The relationship of these molecules, for example ICOS and the PD-1 receptor and their ligands, ICOS-L and PDL1/PDL2, with TLR signaling is poorly understood.

Maturation of DCs is probably one of the main reasons for successful vaccination against infectious diseases. For example, BCG consists of a purified noninfectious material that includes a cell wall derived peptidoglycan, arabinogalactan, and mycolic acids, all of which can induce the maturation of human preDCs. This is probably mediated via TLR2 and TLR4, as murine macrophages that lack these molecules fail to secrete TNFα in response to BCG.[24] Note that pathogens have also evolved counter stratagems to avoid initiating an immune response. One stratagem adopted by some bacteria is to evade DC pattern recognition receptors by 'hiding' intracellularly.

Another potential function of DC TLRs is to polarize T cells towards either a Th1 or Th2 response. The migration of preDC1 cells from tissues into lymph nodes can be induced by prostaglandin PGE_2 which also inhibits the Th1 polarizing cytokines in DCs. IL-4 can counteract this effect in iDCs. It would be interesting to test whether TLR signaling can influence prostaglandin secretion in somatic tissues.

B Cells

Naive human B ($CD5^-$) cells express significant levels of TLR1 and TLRs 6-10. Activation of B cells in vivo (in the germinal center; Gc-B) results in an induction of TLRs 7, 9 and 10 but not the other TLRs.[19, 25] Thus, central memory B cells can upregulate certain TLRs (Tables 2, 3).

TLR ligands like unmethylated CpG DNA and LPS can also upregulate the CD80 costimulatory protein in B cells. According to some researchers, TLRs are potent, T cell–independent polyclonal activators of murine and human B cells (in terms of proliferation and differentiation). In contrast, others hold that BCR engagement is necessary for TLR mediated B cell activation.[26] Simultaneous crosslinking of the BCR by BCR specific antibody enhances the TLR9 mediated intracellular response of B cells to CpG motifs. Mammalian DNA lacks stimulatory CpG motifs, and many of the remaining motifs are normally methylated. Consequently mammalian DNAs are weak stimulators. Very low affinity DNA binding or IgG binding self reactive B cells usually fall beneath the threshold required to induce anergy, deletion or editing. However, immune complexes in normal individuals (for example, anti DNA antibody bound to self DNA) may be stimulatory to naive B cells because, hypothetically, synergy between the BCR and TLR9 could trigger them. Ordinarily, this does not happen (at least not in naive B

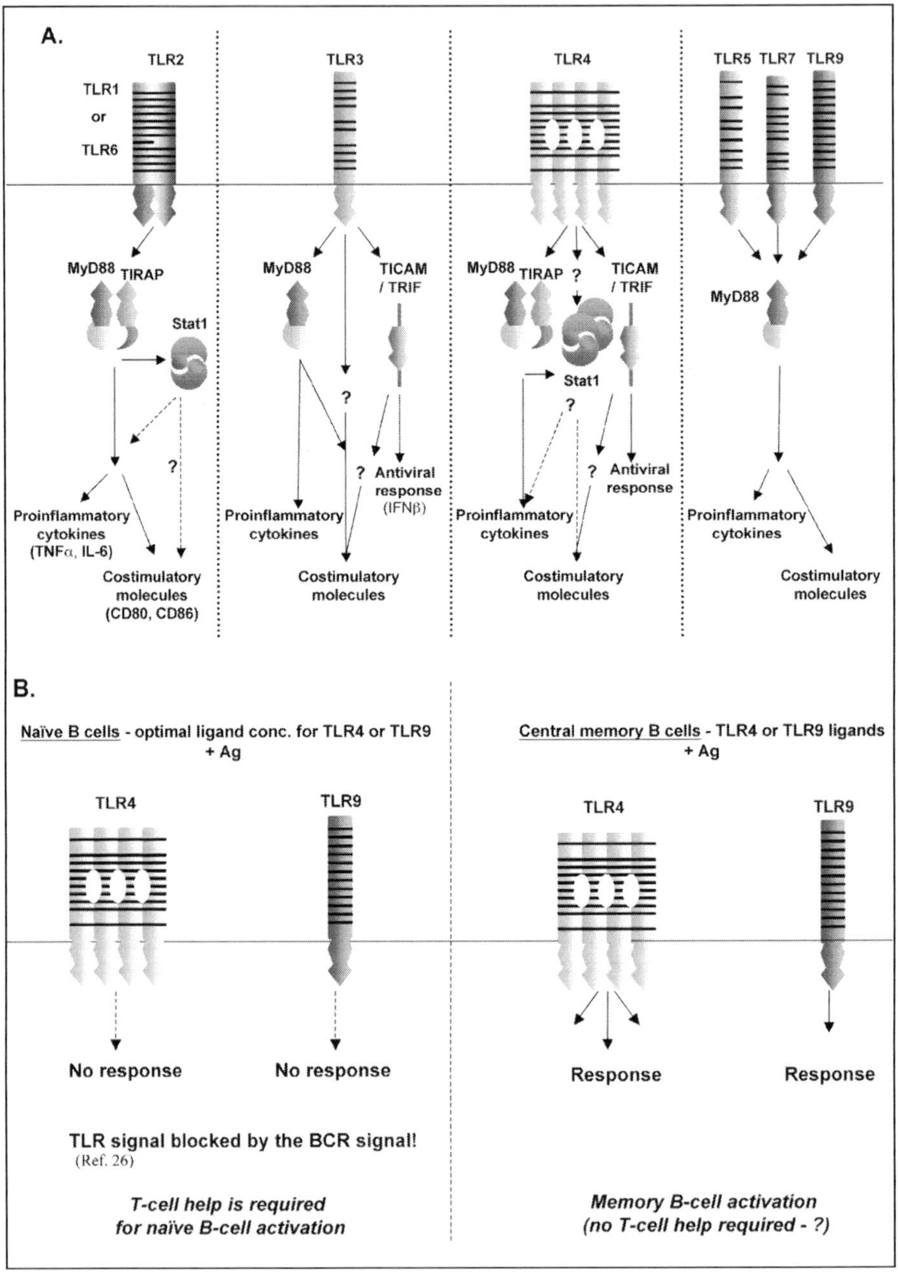

Figure 7. A) Consequences of TLR signal transduction in DCs. B) Consequences of TLR signaling in B cells. Continued on next page.

cells), because BCR signaling mediators inhibit the TLR9 signaling pathway, a situation that also applies to higher affinity DNA specific naive B cells. Analogous to the TLR9 scenario, TLR4 (or LPS) signaling could also cause the stimulation of potentially autoreactive B cells.

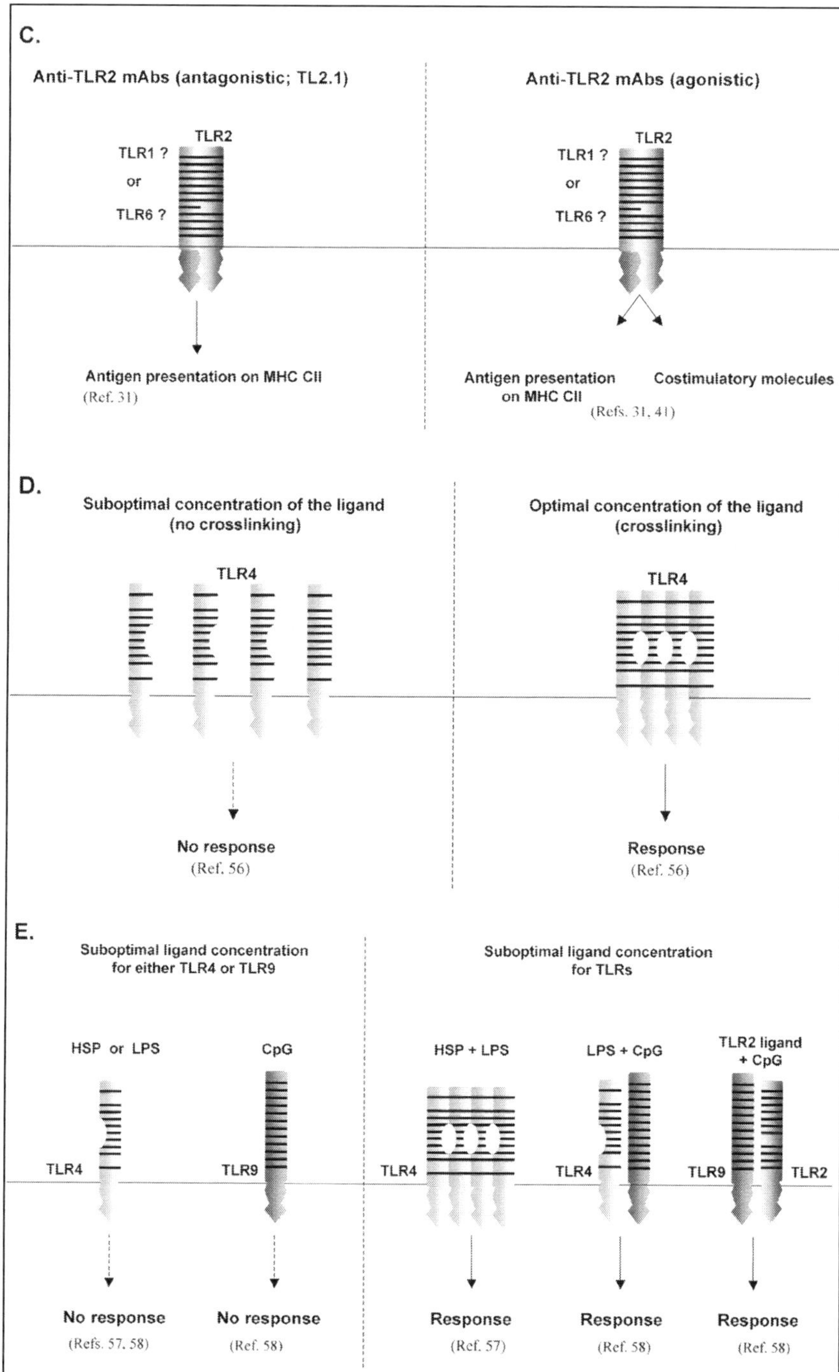

Figure 7. Continued. C) Dissociation of TLR2 signaling in DCs. D) Consequences of TLR4 signal transduction. E) Consequences of TLR4/9 signal transduction.

However, the mechanism that operates in naive B cells to inhibit TLR9 signaling could also inhibit TLR4 signals and suppress autoantibody generation under normal circumstances.[26]

In 1974, Coutinho observed that B cell activation can be triggered by LPS.[27] This prompted the idea that B cells do not necessarily need to signal via the BCR or with T cell help. The hypothesis that a single nonspecific signal could stimulate B cells was subsequently proven wrong as BCR signaling was shown to be necessary during B cell activation.[28] However novel data has shown the increase of, in particular, TLRs 7, 9 and 10 in B memory cells, which leads us to an interesting possibility. Namely, we could envisage that central memory B cells can act similarly to, but not exactly as proposed by Coutinho. Naive B cells do not normally respond to CpG ligands via TLR9, because they uncouple the intracellular pathways that could lead to the development of autoreactive antibodies.[26] However, in memory B cells, a different situation could arise. Renewed pathogen intrusion can cause local tissue damage, which would release double stranded genomic DNA. Memory B cells could bind genomic DNA using TLR9 and pathogenic antigen via their BCRs and become activated, but without T cell help. So, memory B cells would respond swiftly to renewed infection in the presence of recognizable antigen. Thus, TLR9 signaling might increase the speed yet retain the specificity of in vivo B cell responses (Fig. 7B).

Memory B cells are usually formed through a tight T cell controlled response, which prevents the formation of a B cell memory with a high affinity autoreactive antibody repertoire. However, if these controls fail or if B cell tolerance malfunctions, then the normal checks and controls that are exerted over memory could collapse and the potential for autoimmune pathology arise. For example, if a naive B cell clone reactive against self antigen became a memory cell, it could be activated in a T independent fashion via TLRs triggered by the products of local tissue damage. This could lead to the production of autoreactive antibodies.

An experimental autoimmune disease model in which dual engagement involving TLR9 and the BCR could induce naive B cell activation and proliferation has suggested another role for TLR9. B cells were shown to bind antibody-chromatin complexes in the serum of MRL-lpr$^{+/+}$AM14 transgenic mice. Naive AM14 transgenic B cells that are specific for the Fc portion of an IgG2a antibody are normally inactive in AM14 transgenic mice. However, in animals crossed with an autoimmune prone genetic background like MRL-lpr, they become activated, and develop into cells that produce Fc specific antibodies (rheumatoid factors). Their activation was shown to be dependent on the engagement of both TLR9 and transgenic BCR. Firstly, TLR9 binds chromatin, then the transgenic BCR binds the Fc portion of the immune complex formed by chromatin and an antibody specific for chromatin. This dual binding (or crosslinking) triggers AM14 B cell activation and proliferation,[29] leading to rheumatoid factor production and other autoimmune symptoms. This report seems to contradict the finding that naive B cells inhibit TLR9 signaling by contemporaneous BCR signaling.[26] An explanation for this could be that the genetic background of the mice uncouples the "brake" in TLR9 signaling as the AM14 transgenic mice (with non MRL-lpr genetic background) do not normally generate rheumatoid factors. However, there is as yet, no evidence regarding the nature of the molecule(s) involved in these processes.

In mice, TLRs can affect isotype switching in B cells. The TLR9 ligand CpG (but not the TLR4 ligand, LPS) upregulates the intracellular mediator T-bet (Th1 gene regulator) in normal B cells, an effect that was abrogated in mice deficient in TLR9 and MyD88. IL-12 acts synergistically with the TLR9 ligand. In fact, upregulation of T-bet mimics, in part, the Th1 type antibody response, because of the inhibition of IgG1 and IgE switching.[30] IL-4 and CD40 specific antibodies can induce these isotypes in purified B cells from normal mice. A switch to IgE is usually found when a Th2 type response is favored and is epitomized by allergy. Perhaps, TLR9 has immunomodulatory effects in suppressing some allergic responses. In conclusion, the function of TLRs in (CD5 negative) B cells might be to speed up secondary immune response to antigen, and affect isotype switching.

T Cells

Human T cells have either no expression of TLRs or express very low levels of the mRNAs of some TLRs such as TLR6, TLR8, and perhaps TLR2 (Tables 2,3). TLRs have an indirect, though profound influence on T cells which is described in the following section.

The Roles of TLRs in APC—T Cell Interactions

TLR function in antigen presenting cells is implicated in antigen presentation to T cells, the expression of costimulatory molecules and the capability of APCs to polarize T cell responses.

TLRs and Antigen Presentation

Professional APCs like DCs, macrophages and B cells express basal levels of the MHC class I and II proteins at their cell surface. Upon maturation or activation, APCs upregulate the expression of both classes of MHC molecule. The consequence of this is an enhanced presentation of antigens to T cells. There are two basic pathways by which an APC can process and present antigen: endogenous and exogenous.

Endogenous Pathways of Antigen Presentation

The classical endogenous pathway involves MHC class I molecules. These proteins are found on all nucleated cells and usually present peptides derived from intracellular antigens. Proteins synthesized within the cell can be also degraded in the cytoplasm to generate peptide fragments that are transported across the ER membrane and loaded onto newly synthesized chains of the class I molecules in the ER. The MHC class I - peptide complex is then transported to the cell surface via the golgi apparatus and secretory pathway (Fig. 8A). Rarely, endogenously synthesized proteins are also presented on MHC class II molecules.

The Exogenous Pathway of Antigen Presentation

There are two types of exogenous antigen presentation pathway. Peptides from extracellular antigens can be presented on MHC class II molecules via the classical pathway (Fig. 8B). However, peptides derived from extracellular antigens can also be presented by MHC class I molecules in a process called crosspresentation or crosspriming, in reference to the activated T cell readout employed. Crosspresentation represents the nonclassical pathway (Fig. 8C).

The MHC Class II Antigen Presentation Pathway

MHC class II molecules are normally expressed on professional APCs and, under nonphysiologic or pathological conditions, can also be expressed on other tissues. Antigen is internalized by endocytosis into intracellular vesicles, which gradually acidify, degrading their antigen cargos. MHC class II molecules are de novo synthesized in the endoplasmic reticulum (ER) and transported from the ER via the golgi compartments to endosomal vesicles. These fuse with the vesicles in which antigen degradation occurs to form a specialist compartment in which peptide loading can take place. Subsequently, the peptide-MHC class II complexes are transported to the cell surface for presentation to T cells.

Crosspresentation: MHC Class I Antigen Presentation Pathway

Crosspriming was first described as the stimulation of class I restricted precursor cytotoxic T lymphocytes (pCTL; CD8 T cell) by exogenous, cell associated antigens. Since then, soluble forms of antigen have also been shown to induce crosspriming. Therefore, a commonly used definition for crosspriming is the capacity of exogenous antigens to stimulate class I restricted CTL responses (Fig. 8C). Crosspresentation is the presentation of peptides by class I molecules to T cells, whereas crosspriming involves, additionally, the costimulation of naive CD8 T cells. The consequence of these definitions is that peptide-MHC specific effector CTLs can kill cells by crosspresenting

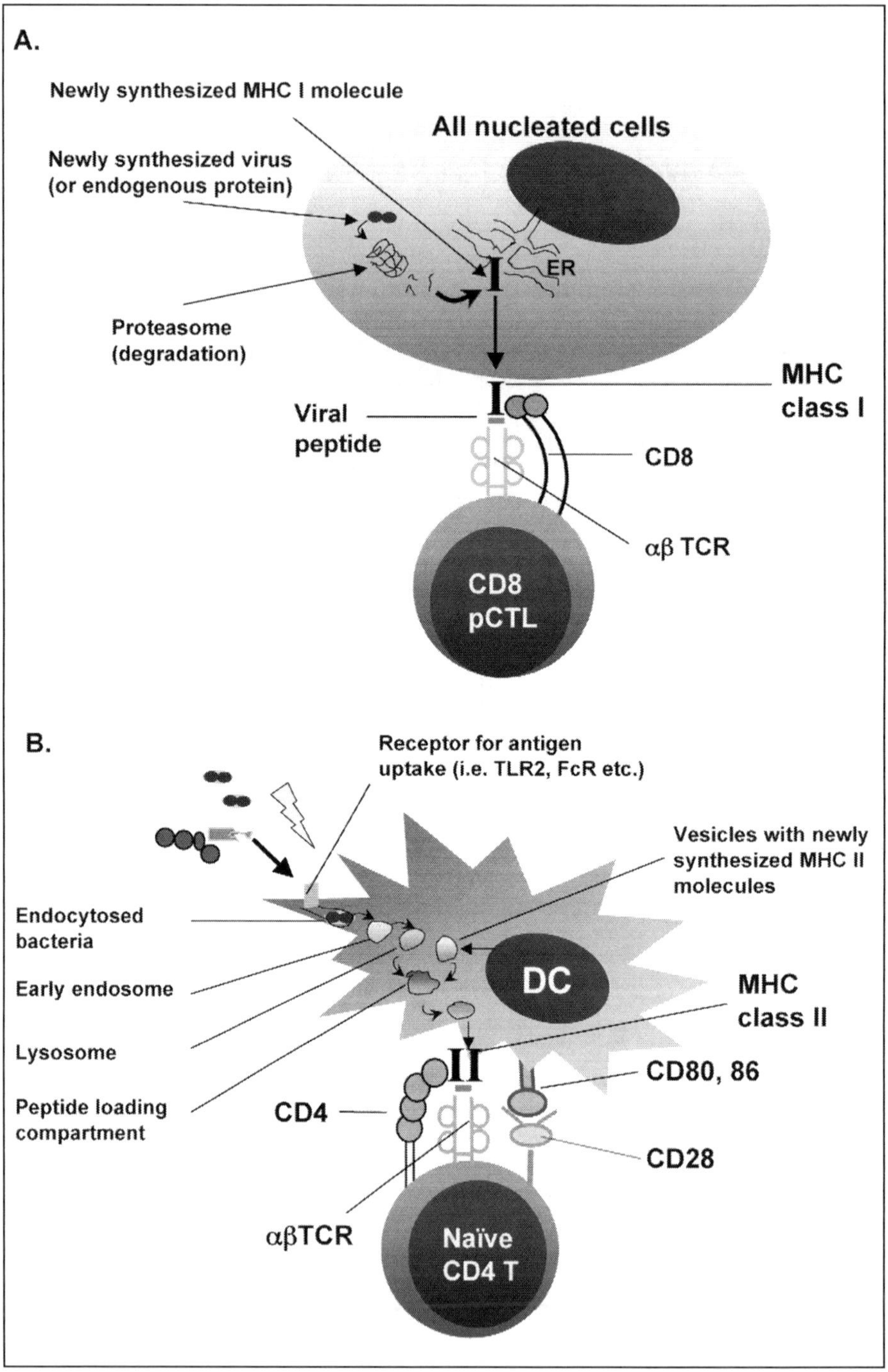

Figure 8. A) MHC Class I (endogenous) pathway of Ag presentation. B) MHC Class II (exogenous) pathway of Ag presentation. Continued on next page.

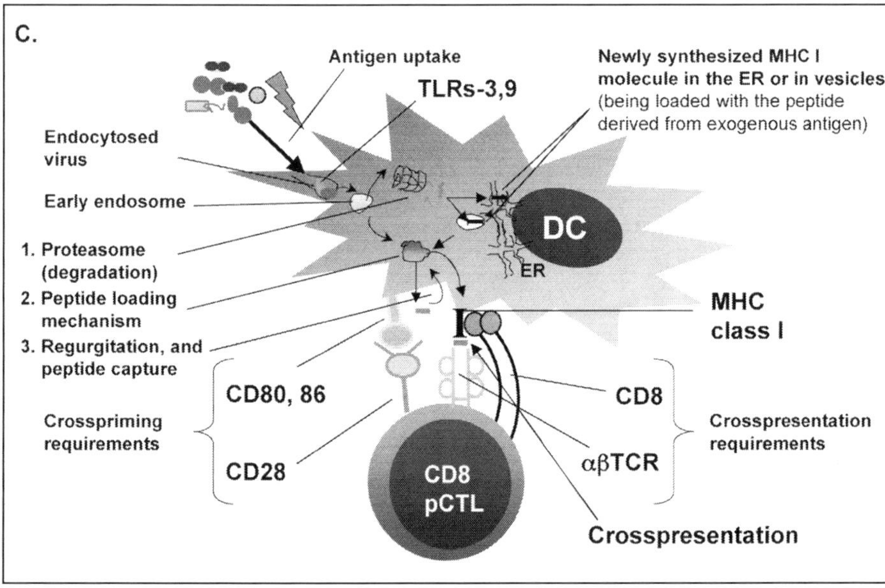

Figure 8. Continued. C) Crosspresentation (crosspriming).

antigenic peptides, even if they lack costimulatory molecules. This is an important distinction: pCTLs require crosspriming for activation and proliferation, whereas effector CTLs require only crosspresentation in order to unleash (activate) their killing machinery.

Protein antigens or their peptide fragments can be taken up by the antigen presenting cell in two ways, endocytosis or direct cell-cell contact (gap junctions). Endocytosis allows APCs to engulf antigens in cellular debris, apoptotic bodies and live or dead micro-organisms. However, peptides derived from neighboring cells in a tissue can, for example, enter a dendritic cell via gap junctions. If the cell died by apoptosis (because of infection or in the course of tumor growth), the apoptotic bodies would contain pathogen specific or tumor derived antigens, and perhaps their peptides. Crosspresentation of these could induce CTL responses against pathogens or tumor cells. Particles taken up by phagocytosis have a tendency to induce crosspriming, but the molecular mechanism is still unclear. Many bacteria live in vacuoles inside phagocytic cells as a part of their life cycle and it has been shown that the antigens derived from these live intracellular bacteria can be crosspresented on MHC class I antigens to induce CTL responses that kill the infected cells. Additionally, antigens in cell debris can, after phagocytosis, be crosspresented on class I molecules. Immature DCs have several receptors for apoptotic bodies that can contribute to crosspresentation, employing the cytoplasmic pathway. These receptors include mannose binding lectin, scavenger receptor and iC3b binding molecules like CD11b, CD11c, CD21 and CD35.

TLR Ligand Linked Antigen Presentation in Immature DCs

Human immature DCs derived from bone marrow, pulsed with antagonistic TLR2 specific mAbs containing κ light chains, could stimulate a Cκ specific $CD4^+$ T cell clone in the absence of maturation effects on iDCs (Fig. 7C). An isotype/light-chain matched control antibody produced a two to three orders of magnitude lower response, indicating enhanced antigen presentation via TLR2. Stimulation was TLR2 specific, as antibodies against other surface molecules such as CD62 and CXCR1 were not stimulatory. Inhibitors of lysosomal degradation, processing and MHC class II presentation like chloroquine, leupeptin or brefeldin A almost completely abolished T cell stimulation. Furthermore, an anti-TLR2 mAb was directly shown to reside in endosomal vesicles in pulsed iDCs.[31] Thus, antigen linked to the TLR2 ligand can be endocytosed after binding TLR2, processed via the classical (exogenous) pathway of antigen presentation, and can enhance the stimulation of T cells. This same route could be exploited to generate more efficacious vaccines.

Unlinked TLR Ligand Antigen Crosspresentation in Mature DCs

TLR9 ligand (oligonucleotides containing CpG motifs; CpG) or TLR3 ligand (poly I:C) were mixed with ovalbumin, incubated with mouse bone marrow derived iDCs, then tested for their ability to crossprime ovalbumin/MHC specific precursors of cytotoxic CD8 cells (pCTL; naive or resting CD8 T cells). The antigen was not linked with the TLR ligand. Ovalbumin was endocytosed, processed, its peptides loaded onto both MHC class I and class II molecules and subsequently presented to T cells. Such pulsed mDCs were able to stimulate syngeneic ovalbumin peptide/MHC class I specific pCTL and this stimulation was MyD88 dependent, showing the importance of the TLR signals in crosspresentation. As a control, these DCs were also able to stimulate autologous CD4 T cells (bearing ovalbumin peptide/MHC class II specific TCR), showing their ability to engage in classical, exogenous antigen presentation using MHC class II molecules. A $CD8^+$ subset of mouse DCs was shown to be able to crosspresent and present ovalbumin peptides.[18]

Interestingly, if DCs were first pulsed with TLR9 ligand, washed, then incubated with ovalbumin an unexpected result occurred. TLR9 binding could sensitize DCs to take up ovalbumin for crosspresentation for several hours after they had been pulsed with CpG. In contrast, TLR9 ligand pulsed, then ovalbumin pulsed DCs, were unable to present ovalbumin to autologous CD4 T cells. It seems, therefore, that TLR9 triggering engages a crosspresentation machinery (Fig. 8C), whilst at the same time uncoupling the classical, exogenous MHC class II pathway of antigen presentation in mDCs (Fig. 8B). Macrophages, on the other hand, do not possess the ability to crosspresent ovalbumin after being pulsed with CpG.[18]

In these experiments, antigen was not endocytosed by TLRs. DCs have plenty of molecules that facilitate antigen capture, for example, Fcγ receptors. There is however, a dichotomy between the crosspresentation pathway of FcγR and those of TLRs 3 and 9. All three use cytoplasmic pathways, but TLR mediated crosspresentation does not require acidification in an endosomal compartment. It is possible therefore that TLRs trigger an unknown mechanism that operates early in endosomes, shuttling bound antigen into the cytoplasm of DCs for further processing and crosspresentation on MHC class I molecules.

Presentation of Self Antigens that Are TLR Ligands in DCs

In mice, TLR2 and TLR4 binds exposed ends of fibronectin,[32] hyaluronan,[33] and heparan sulfate,[34] which are present in the extracellular matrix. TLR4 has been implicated in binding mouse extravascular fibrin(ogen),[35] which is found in tissues only if vascular permeability is increased, as in inflammation. Interestingly, human DCs can mature in the presence of chondroitin sulphate and hyaluronic acid (hyaluronan) when cultured together with GM-CSF,[36] suggesting that human TLRs bind inflamed tissue extracellular matrix. Furthermore, human HSP60 and 70 can also act as ligands for TLRs 2 and 4.[37-39] Human genomic DNA may also be modified in inflamed tissues, such that CpG islands can bind TLR9 (Fig. 5B). Even without modifications, mammalian genomic DNA is weakly stimulatory.[40] All these TLRs were shown (using different antigen response models) to be able to present antigens to T cells. Thus it is possible that self-antigens can be captured and processed by DCs expressing TLRs, and used to stimulate autoreactive T cells.

TLRs and T Cell Costimulation

The antigenic peptide/MHC complex binding to TCRs of naive or memory T cells is a prerequisite for the initiation of the immune response. There is ample evidence that TLRs can indirectly provide costimulatory signals to T cells via the activation or maturation of APCs.

TLR signals in immature DCs can, under certain conditions, provide only an antigen presentation facility to T cells. In other words, antigen presentation can be dissociated from costimulation (upregulation of CD80 and CD86) in immature DCs. Evidence for this comes from experiments in which authors used TLR2 as a point of antigen entry and showed that an antagonistic human TLR2 specific mAb (TL2.1) containing κ–light chains could be taken up by iDCs, processed into peptides, and subsequently loaded onto MHC class II molecules to stimulate a Cκ specific $CD4^+$ T cell clone.[31] The TL2.1 mAb was antagonistic, because it had no stimulatory effect on macrophages in comparison to agonist mAb (Fig. 7C).[41] The authors showed that TL2.1 could bind to TLR2 molecules on iDCs, but neither upregulation of CD80 and CD86 nor secretion of TNF was observed. There was nothing wrong with the maturation capability of TL2.1 pulsed iDCs, because iDCs could be induced to mature and to express costimulatory molecules by the addition of TLR ligands like Pam_3Cys (TLR2) or LPS (TLR4). Then, they showed that TL2.1 was endocytosed, processed into peptides, loaded onto class II molecules, and finally expressed on the cell surface for presentation, as determined by proliferation and interferon γ release of a responder T cell clone. The T cell clone did not require costimulatory molecules for these actions, because it was an effector cell. The result indicated that iDCs could dissociate antigen presentation from the costimulation of T cells. Thus, antigen internalization and processing can be initiated by binding to TLR2 for presentation on MHC class II molecules. Further, it does not necessarily hold that every time a ligand binds to TLR2, an iDC will be activated to provide both a stimulus and costimulus to naive T cells (Fig. 7C). Though the experiment with naive T cells is needed to confirm this suggestion, the result by Schjetne et al is consistent with this hypothesis. It would be interesting to know whether the actions of other human TLRs on iDCs can be similarly dissociated. If so, perhaps the missing signal (needed for full activation of iDCs) could be called co-initiation or co-activation. Thus, TLRs can provide two kinds of signals, one that fully activates DCs, propelling them towards a more mature phenotype, and a second that only partially activate DCs, rendering them devoid of costimulatory molecules, but capable of presenting antigen.

Is there other evidence that TLRs have two modes of action? The answer is yes and comes from experiments using crosslinked TLR4 to test signaling under suboptimal ligand concentration. Crosslinking (not just binding) of TLR4 by LPS is necessary for downstream signaling from macrophages.[42] Ligands such as HSPs act on TLR4 by enhancing the stimulatory effects of otherwise substimulatory concentrations of LPS (Fig. 7D).[43] This could, perhaps, be understood as a dissociation of TLR4 signaling, similar to that seen with TLR2. Furthermore, evidence that intracellular TLR signaling cascades can be synergistically triggered by ligands for TLRs 2 and 9, or 4 and 9, under suboptimal conditions (Fig. 7E),[44] supports the idea for two-level TLR signaling.

TLRs and Polarization of the T Cell Response

In human DCs, a TLR4 agonist specifically promoted the production of IL-12p70, which is associated with the Th1 responses. In contrast, TLR2 stimulation resulted in the secretion of the IL-12 inhibitory $p40_2$ homodimer, producing an environment that would favor Th2 development.[45] CpG DNA, as a TLR9 ligand, also has immunomodulatory effects, such as induction of the Th2 type cytokines (IL-4 and IL-10) in DCs.

Induction of costimulatory molecules on APCs is probably not the only mechanism that can control naive T cell activation. T cell responses can also be regulated by $CD4^+CD25^+$ regulatory T cells (Tr cells) (for a review see ref. 46), and TLRs might play a role in their generation. These cells are thought to be important for the maintenance of peripheral T cell tolerance, as their depletion leads to organ specific autoimmune diseases. Though the molecular mechanism of Tr cell mediated suppression is unknown, it seems to be cell contact dependent. The inhibition of IL-2 transcription in responder T cells is one of the results of such suppression.

The culture medium of DCs stimulated by TLR4 and TLR9 ligands (LPS and CpG) has been shown to inhibit the generation of regulatory T cells. This has been shown to be dependent on IL-6.[47] It seems likely that the generation of regulatory T cells depends on the lack of certain TLR signals, however the precise mechanism remains unclear.

Various kinds of TLR signaling pathways may influence the expression of cytokines in APCs, which may, in turn, regulate the formation of various types of T helper (Th1, Th2 and Th3) or regulatory T cells.

TLRs and Theories about the Function of the Immune System

A critical question in immunology concerns the initiation of the immune response. How do the current theories of immune system function tackle the role of TLRs? Further, how useful are the current immunological theorems and how can we put them to the test?

Ideally, theories about immunity should find fundamental and, hopefully simple rules, to explain and clarify the biological, cellular and molecular functions of immunity. The theory then becomes a "map" upon which we can plan our research. It is obvious that such a "map" will only be complete when we have obtained an all-encompassing knowledge of all the structures and molecules involved, and their interactions. Obviously, this will take some time! In the meantime we can take a short cut by making some educated guesses. At the same time, it stands to reason that an oversimplistic theory will require multiple additional explanations or rules in order to accommodate new or controversial results. This has the effect of generating a rather complicated and unwieldy theorem of limited use in explaining, or predicting, the workings of the immune system.

Self-Nonself Discrimination and Associated Antigen Recognition

The "Self-nonself" discrimination (S-NS) model has evolved considerably since it was first conceived by Bretscher and Cohn in 1970.[48] Its basic tenet is that cells of the immune system can recognize exogenous molecules by clonally distributed receptors on immunocytes. These exogenous or "nonself" molecules are mostly derived from pathogenic microorganisms. That TLR ligands are polyclonal activators of the immune system neither fits the original theory, nor its latest incarnation as the Associated Antigen Recognition (AAR) model (Langman and Cohn).[49] This model describes the activation of the specific immune system solely by nonself antigen. This occurs by the generation of two major signals that can activate naive B or T cells. The first signal (signal-1) is derived from the antigen receptor (BCR or TCR), and causes programmed cell death (clonal deletion), which results in tolerance to a particular antigen. The second signal (signal-2) rescues the cell from death. For B cells, this signal is T cell help. For T cells, this is also T cell help, but from effector T cells. How the latter are formed is not known, but the authors predict that each naive T cell would, by default, differentiate after some period of time into an effector. Thus, in short, only novel antigen, never before encountered by the immune system can initiate specific immune responses. The model suggests that tolerance to self-antigens occurs during embryonal development or very early in the life of an individual and predicts that any novel antigen detected by the adult immune system would be seen as nonself.

The induction of antigens in the lactating breast is an example for which this model lacks a simple explanation. Why do these proteins not cause autoimmunity? The concept of regulatory or suppressor T cells is also incompatible with this model as the AAR model presupposes that simultaneous antigen driven inhibition would counteract the response.

Cytokine Burst

This model suggests that the control of the immune system rests with cytokines.[50] An increase in the local concentration of certain cytokines would stimulate T cells to initiate the immune response. This hypothesis could be adapted to suggest that pattern recognition leads to the secretion of cytokine "soups" that upregulate the antigen presenting capacity of APCs, and costimulatory molecules, both of which are required (in addition to the cytokines) to stimulate T cells.

Antigen Localization (Ignorance)

This model proposes that the regulation of the immune response lies outside of the specific immune system, and is principally controlled by antigen localization.[51] This model can be adapted to include TLRs as mediators of signals that mobilize adaptive immunity and implies that migrating (non effector) T cells are ignorant of antigens in somatic tissues. In other words, such T cells would not be stimulated to proliferate as they cannot see antigenic peptide/MHC complexes, at least not in the appropriate costimulatory context. The model also predicts that somatic tissues lack antigen crosspriming (for pCTLs) and the expression of costimulatory molecules in tissue residing APCs (for naive CD4 T cells). Naive and resting memory T cells would only respond to antigenic peptide/MHC ligands by proliferating in lymph nodes. In support of this model, immature DCs (mainly found in tissues) cannot crosspresent antigens, whereas mature DCs (mostly lymph node resident) can.[18] The failure of macrophages to crosspresent after stimulation via TLRs 3 or 9[18] is also consistent with this hypothesis. Inconsistent is evidence showing that macrophages can express costimulatory molecules upon activation and are therefore capable of activating naive CD4 T cells. However, naive CD4 T cells mainly home to lymph nodes, and are not often seen in somatic tissues. Another prediction of the antigen localization model is that there would be no clonal deletion of pCTLs in somatic tissues as a mechanism of peripheral T cell tolerance (in contrast to the Danger model).

Pattern Recognition

Janeway originally proposed that infectious "nonself" substances are initiators of the immune response.[52] The TLRs, according to this model are called pattern recognition receptors (PRRs). Consequently, their ligands are pathogen associated molecular patterns (PAMPs). The model explains the function of TLRs as sensors of pathogenic "nonself" and TLRs would raise the alertness of innate immunity, and prime adaptive immunity to discriminate self from nonself (Fig. 9A). The model implies that TLRs are receptors for nonself molecular patterns that are evolutionarily foreign to a species and, as such, pattern recognition receptors serve as a memory of past infection. Basically, the model suggests that antigen alone, or in conjunction with the PRR (TLRs) signal, can activate adaptive immunity. This hypothesis can be seen as an extension of the self-nonself model. Recently, it was suggested that a lack of TLR signaling in DCs might induce regulatory T cells (Fig. 9B), whereas TLR induced maturation of DCs would cause T cell activation and thence an immune response.[47] This model neglects the possible influence of somatic tissues on the activity of the adaptive immune system, lacks simple explanations for the existence of endogenous (autologous) TLR ligands, and does not explain the dissociation between antigen presentation and the expression of costimulatory molecules on iDCs brought about by antagonistic TLR2 mAb.

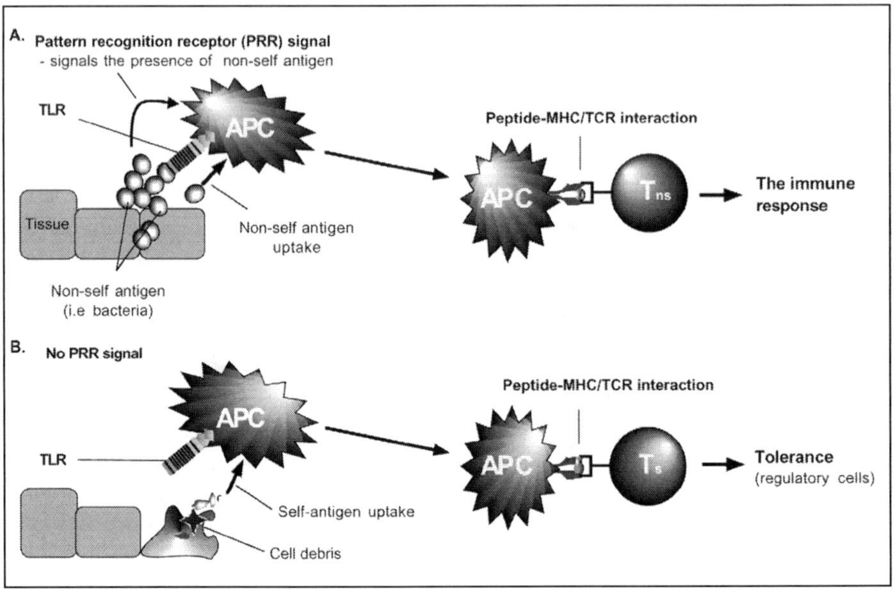

Figure 9. The Pattern recognition model.

Danger

Matzinger's 'Danger' model takes a fresh look at signal-1 and signal-2. It proposes that 'danger' signals act as an alarm to kick start the specific immune response (signal-0), and that these signals include pathogen specific molecules.[53, 54] TLR signaling is but one of the prototype danger signals (exogenous danger) (Fig. 10A). Other danger signals include tissue distress, disruption and necrotic death (endogenous danger), all of which should activate the specific (adaptive) immune response (Fig. 10B). Although the difference between "infectious nonself" and danger seems semantic, this was the first model that allowed us to shift our understanding about the control of specific immunity away from the strict self-nonself paradigm. According to Matzinger, specific B or T cells could still react to nonself, but would only do so if danger were signaled. Her model predicted that danger signals would upregulate costimulatory molecules on APCs, and hence activate naive T cells. Without danger, even nonself antigens would be tolerated (by clonal deletion of reactive T or B cells) (Fig. 10C). Therefore, a nonspecific, 'alarm' type danger signal emanating from tissues that surround APCs and lymphocytes would have ultimate control of the immune response, which is quite different from Janeway's model. However, the molecular definition of danger remains elusive. Furthermore, danger represents a conserved alarm signal, which could, in itself, be "dangerous", as pathogens could evolve to avoid tripping the alarm. Consequently evolution could deselect nonspecific warning systems in higher vertebrates. A solution for this problem might lie in the variability of the proposed danger signal, which may be sufficient to overcome its eradication in evolutionary terms. Perhaps this is the reason for the diversified group of pattern receptors that we see in higher vertebrates. Another problem with the Danger model is the nature of "endogenous" danger signals. For example, if the binding of bacterial DNA to TLR9 represents "exogenous danger", why does the binding of autologous genomic DNA to the same receptor not constitute "endogenous danger"? Wouldn't this predispose us to autoimmune disease? Similar problems occur in explaining the dissociation of stimulation and costimulation of T cells via TLR2 binding in iDCs (see section "TLRs and T Cell Costimulation"). Lastly, the danger hypothesis lacks an explanation for regulatory T cells.

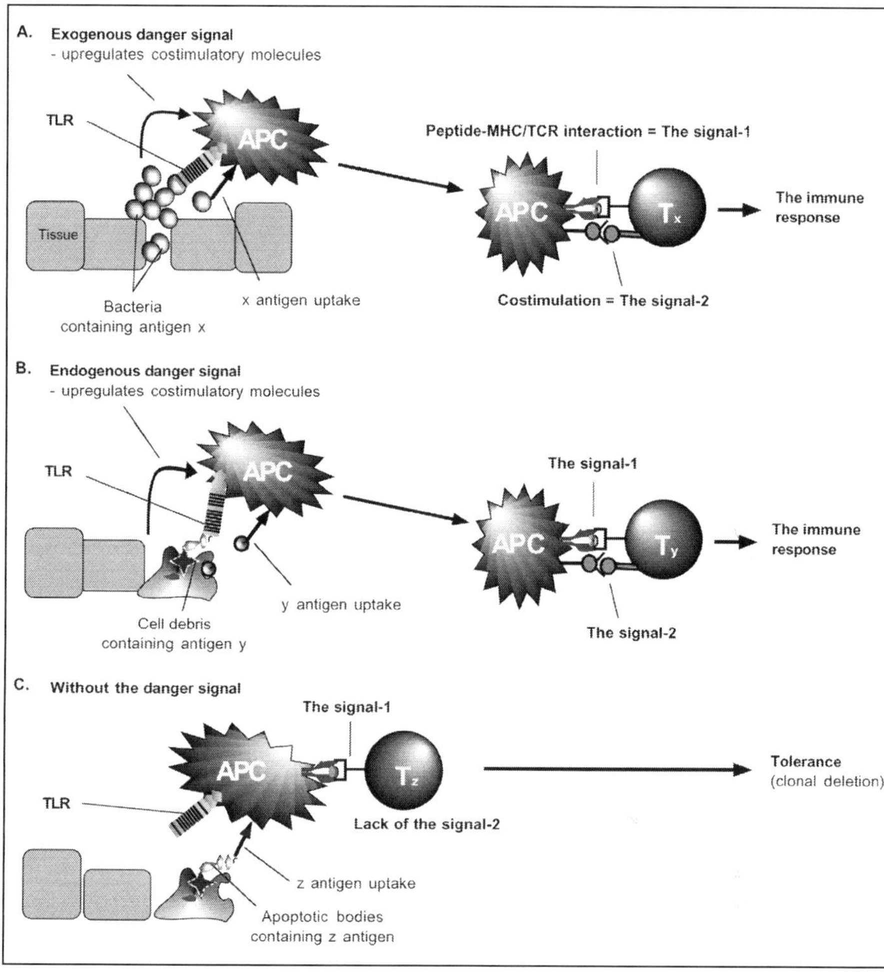

Figure 10. The Danger model. A) Exogenous danger signal. B) Endogenous danger signal. C) Induction of tolerance.

Integrity

The Integrity model suggests that three signals control the cells of central immunity (DC, T and B cells).[55-57] In general, it is well known that any cell can divide, grow, differentiate, die, or lie dormant, but we are far from knowing all of the signals that regulate these outcomes. Each tissue may have its own set of regulatory signals, perhaps overlapping with one another. Many specialized cells, like those of central immunity, have evolved another property, namely, a specific effector function that requires activation. Activation, though an ill-defined term, is used in many different contexts in cell biology. In my view, activation should be regarded as a process by which a cell acquires the capacity to unleash an effector function. Therefore, to become an effector, a cell must make a number of decisions after its initial activation. These decisions would depend on the cellular microenvironment and state of differentiation. Related intracellular signals that can perform these functions could be grouped into several clusters such that the transmission of a main activating signal would be modulated by the influence of two (in its simplest form) auxiliary ones (Fig. 11). Hence, we can distinguish three groups of

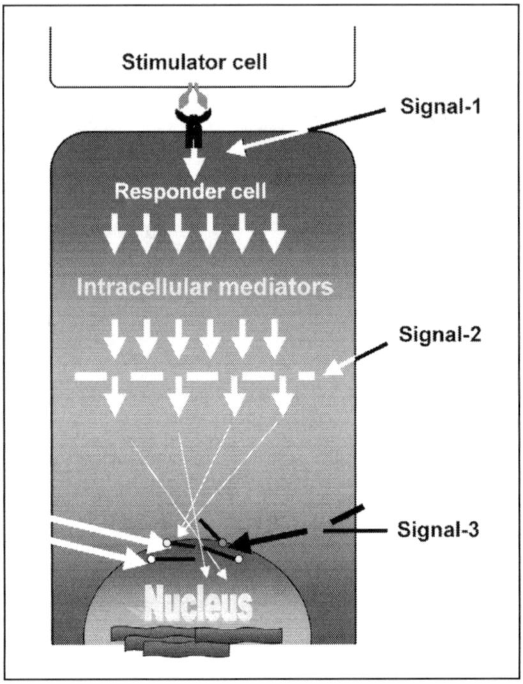

Figure 11. The cellular signaling model

stimuli. Those that descend from the main effector-function related receptor and engage various cytoplasmic intracellular mediators until the nuclear factor level. A second that assists receptors to modulate signal-1 either within the cytoplasm or nucleus, by engaging a different set of intracellular mediators and/or DNA-binding factors. These signals could compete with each other. A third set of signals could act to regulate the availability of certain nuclear factors and/or DNA accessibility (in terms of DNA binding, chromatin organization and cell-differentiation stage)(Fig. 11). These signals represent a necessarily simplified view of complex intracellular signal transduction pathways. In its most rudimentary sense, the cells of central immunity are thought to operate by receiving, modulating and transmitting incoming signals to the nucleus. These result in the expression of a set of specific genes that causes the cell to engage in a series of specific functions. The three-signal concept allows for the creation of cellular messages that could be transmitted to and thereby influence neighboring or distant cells.

In cells of central immunity, the main signal (signal-1) is provoked by antigen, peptide/MHC or TLR ligand in B, T and DCs, respectively. In naive and resting B and T cells the receipt of signal-1 alone would cause clonal deletion. Signal-2, arising from auxiliary receptors would modulate signal-1. Signal-3 would derive from the acceptor signal from supplementary inputs and facilitate, modify or prevent the acceptance of signal-1 (e.g., nuclear translocation - DNA binding effects). Examples of signal-2 can be envisaged. For B cells, signal-2 would be T cell help. For T cells it would be costimulation (if originating from CD28), or inhibition (if generated via CTLA4 or some other inhibitory B7 family receptor). Possibilities for signal-3 could be predicted and their modifying action may transform stimulatory signals into homeostatic ones for cells of central immunity. Consequently, these signals would play an important role in the regulation of the class of the immune response and the generation of regulatory T cells. Thus, for B and T cells, signal-1 and signal-2 are, in part, similar to those described in the previous models (S-NS, Danger), whereas signal-3 is a novel "integrity disruption" input that may derive, in part, from TLR signaling.

The signals for DCs would be different from those for B and T cells, in that DCs would use TLRs for pattern recognition (signal-1), whereas signal-2 would give an estimate of integrity. Thus the receipt of signal-1 and signal-2 in DCs would provide a strong stimulus for activation (maturation) together with signal-3 (Fig. 12A). Signal-3 for DCs could be provided by soluble factors such as prostaglandins and / or by a break in the cell-cell or cell-intercellular matrix interaction and homeostatic signaling.

The Integrity model proposes that TLRs take part in sensing tissue integrity as well as in pattern recognition. DCs perform these functions more efficiently than any other antigen presenting cell, and would sense aberrant molecular patterns (signal-1) together with the extent of damage (signal-2). Evidence suggests that TLR4 can engage endogenous ligands that appear after tissue damage, for example, fragments of fibronectin, hyaluronan, heparan sulphate in intercellular matrix, and extravascular fibrin(ogen).[32-35] Based on these findings, some authors have proposed that TLRs are surveillance receptors for tissue damage.[58] Furthermore, additional endogenous ligands have been reported for the TLR2/4 heterodimer and TLR9 such as the stress induced heat shock proteins HSP60 and 70, and mammalian genomic DNA, respectively.[37-40] In programmed cell death, apoptotic bodies would shield the release of such products and prevent their binding to TLRs. Apoptotic bodies could also inhibit DC maturation via the C3d fragments with which they are coated. During necrotic cell death these same endogenous TLR ligands can stimulate DC maturation. However, the Integrity model predicts that this would happen only when endogenous ligands were present at optimal concentrations, as the level of tissue damage would be "measured" by the strength of TLR signaling and the synergy between signals. When signal-1 is limiting (i.e., monomeric TLR ligands, insufficient crosslinking of TLRs), a second (coactivating) signal for iDCs would be required for the expression of costimulatory molecules. Maturation would require, additionally, signal-3. On the other hand, bacterial DNA would provide a sufficiently strong signal to iDCs for maturation (via TLR9, thus linking signal-1 with signal-2, provided that the putative signal-3 is also present). Supporting evidence includes data showing that crosslinking (not just binding) of TLR4 by LPS is necessary for macrophages to signal downstream.[42] Further, ligands such as HSPs can act on TLR4 by enhancing the effects of otherwise substimulatory concentrations of LPS (Fig. 7D).[43] The intracellular TLR signaling cascade can also be synergistically triggered by the ligands of TLRs 2 and 9, or TLRs 4 and 9, under suboptimal conditions (Fig. 7E).[44] In all of these experiments, the in vitro handling of cells would provide signal-3. Differences between this and the Danger model are best illustrated in the explanation of the experiment described in the '*TLRs and T Cell Costimulation*' section.[31] This experiment showed that iDCs, pulsed with a TLR2 specific mAb (TL2.1), could present a TL2.1 derived antigenic peptide but failed to upregulate their costimulatory molecules (Fig. 7C). The Danger model cannot, in simple terms, explain why iDCs did not upregulate costimulatory molecules when they received a "danger signal" via TLR2. The Integrity model explains it by proposing that DCs need more than one signal for maturation. Hence, TL2.1 mAb binding to TLR2 would transmit signal-1 to iDCs, which would be insufficient to upregulate costimulatory molecules. The prediction is that the missing "coactivating" signal (signal-2) could be provided by other TLR ligands under suboptimal concentrations.

In the afferent loop of central immunity, a weak TLR signal like danger without tissue damage (or vice versa) would cause DCs to migrate to the lymph node, where they would tolerize rather than activate the immune response (Fig. 12B). Such tolerogenic DCs have been found to exist (reviewed in ref. 59). Using this mechanism, TLRs (on DCs) could detect commensals, but regulatory T cells would alleviate any deleterious responses to them. This would allow the individual to benefit from, for example, vitamin K producing bacteria in the intestine.

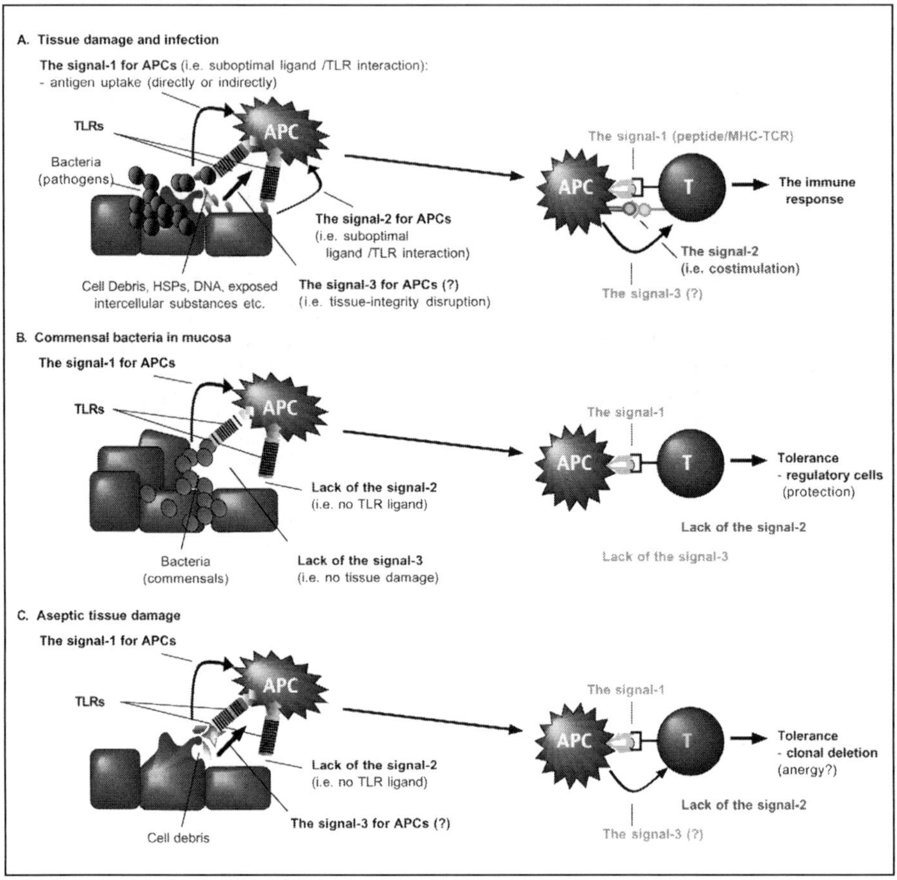

Figure 12. The Integrity model. A) Tissue damage and infection with pathogens. B) Induction of tolerance regarding commensals. C) Induction of tolerance (clonal deletion in the periphery).

The DCs that would initiate such a reaction would be stimulated by TLRs under suboptimal conditions and, as such, would not elicit T cell activation, but would instead generate regulatory T cells. The latter would be able to protect commensals more assiduously than the rather passive intervention offered by clonal deletion. Clonal deletion would depend on an alternative signal-3, or the lack of it, (Fig. 12C). Evolutionarily selected to preserve commensal microorganisms, regulatory T cells might also have been selected for their protection of vital organs, as a last barrier against autoimmunity. The mechanism by which TLRs might regulate the generation of regulatory T cells would, perhaps, depend on clonal competition between various TLR signals during the initiation of the immune response. DCs would generate two kinds of responses, one that activates T cells specific for a particular antigen, and another that would generate regulatory T cells (with different antigen specificity). The immune response might thus depend on the predominant TLR signal (i.e., how much tissue damage there is).

In summary, the Integrity model conceives the immune system as a delicate web of cells that senses the integrity of tissues and reacts to these as well as to external threats. The immune system actively responds to information about the internal milieu and distinguishes between three possible actions: destruction of the harmful, protection of the useful and neglect of the rest (non dangerous microorganisms). Destruction would involve the activation of autonomous and central immunity effector cells and soluble mediators. Protection possibly includes the selection of regulatory immunocytes in central immunity and neglect could operate by the deletion of potentially autoreactive clones in the thymus, bone marrow or peripheral lymphoid organs.

TLRs in Health and Disease

Targeting TLRs could be an efficient way to prepare vaccines against infectious diseases and for cancer immunotherapy. The use of agonistic ligands coupled (linked) to antigen would greatly enhance the potency of vaccine preparations. In addition, the vaccine might be designed to target various subpopulations of DCs in order to produce the desired effect; presentation or crosspresentation that would in turn yield regulatory cells or a strong cytotoxic T cell response. In the case of the former, using monomeric TLR ligands or antagonistic mAbs like anti-TLR2 (see "TLRs and T Cell Costimulation" section) linked to autologous antigens might help in the treatment of autoimmune diseases. For the latter, antigens mixed with ligands for TLR3 or TLR9 could cause crosspriming in pDCs and be useful in cancer immunotherapy. The predicted therapeutic advantage could be enormous.

Likewise, the animal models of infectious diseases using TLR deficient mouse strains will undoubtedly be useful in studying the role of TLRs in infection and immunity.[6] The detection of polymorphisms in patients with various diseases (including cancer) might also yield valuable information concerning risk profiles for the susceptibility to bacterial infections, autoimmunity and perhaps other diseases not necessarily linked to immunity.

In humans, many studies have attempted to identify allelic variants of TLRs. Thus far, several allelic variants of TLRs 2, 4 and 9 have been identified, and used in case control studies to identify any predisposition to infectious diseases, atherosclerosis or autoimmunity.[60] The detection of allelic variants of other TLRs in human populations is also rapidly expanding. It is interesting that hypomorphic (underactive) variants of TLR4 genes are found in increased frequencies in individuals prone to develop toxic septicemia in the course of infectious disease with meningococcus.[61]

Two polymorphisms of the TLR-2 gene have been described: Arg753Gln, which correlates with sepsis in a Caucasian population and Arg677Trp, which correlates with lepromatous leprosy in an Asian population. Leprosy is caused by *M. leprae,* an intracellular bacterium that lives in the Schwann cells around the axons of peripheral nerves. There are two types of leprosy. A localized tuberculoid form of the disease that causes disfigurement. Comparatively few bacteria populate the lesions and a pronounced cell mediated host response is characteristic. The second type is a lepromatous form, which can be distinguished from the former by disseminated (non tuberculoid) lesions with a large bacillary load and a relatively weak host immune response. TLRs 1 and 2 are strongly expressed in lesions from the localized tuberculoid form compared with the lepromatous type of disease.[62] TLR1/2 heterodimers can bind *M. leprae* products. However, the TLR2 Arg677Trp mutation abolishes the recognition of mycobacterial products when transfected into HEK293 cells.[63]

Several allelic variants have been described for TLR4 in humans, however two seem to have functional significance, TLR4 Asp299Gly and TLR4 Thr399Ile. Both are less efficient in signaling than the wild type allele. Tested in vitro, the effects produced by Asp299Gly and Thr399Ile were as low as 5-10% and 20-30% that of the wildtype. These variants lead to a shortened immunologic response to inhaled LPS and to lower levels of proinflammatory cytokines, acute phase reactants, and soluble adhesion molecules. Interestingly, they are also associated with a reduced extent and progression of carotid atherosclerosis.[64] Atherosclerosis is associated with chronic infection or inflammation, and may have an infectious origin. The TLR4 Asp299Gly polymorphism is associated with a risk of coronary artery disease. Strikingly, patients that carry this allele (in particular those with elevated CRP levels) benefit more from prevastatin treatment (to prevent cardiovascular events), than patients that express the wild type variant. On the other hand, the TLR4 Asp299Gly genotype is not associated with disease progression.[65] One explanation for these clinical findings might be that the oxidized low density lipoprotein (LDL), which is elevated in such patients, is a potent upregulator of TLR4. TLR4 proinflammatory activity might therefore be reduced in patients with the Asp299Gly allele, which diminishes the extent of vascular tissue damage. The TLR4 Asp299Gly allele has also been suggested to have a beneficial role in the systemic inflammatory response syndrome.

Premature birth can be a consequence of urogenital infection, which is often caused by Gram negative bacteria. The Asp299Gly allele was associated with an increased risk of premature birth. The same allelic variant may also predispose to septic shock with Gram negative bacteria though does not influence the susceptibility to, or severity of, meningococcal disease. Data showing that several otherwise healthy children, with an inherited deficiency in the IRAK4 molecule, developed infections with pyogenic bacteria show that deficiencies in the downstream mediators of TLR signaling can also predispose to infectious diseases.

TLRs have also been implicated in the pathogenesis and severity of some autoimmune diseases. For example, the overexpression of TLR2 has been reported in the synovial fluid cells of patients with rheumatoid arthritis.[66] DNA specific IgG can trigger low affinity autoreactive antibodies like rheumatoid factor when complexed with autologous CpG DNA in the circulation of autoimmune prone mice.[29] Moreover, one of the defects that is associated with susceptibility to systemic autoimmune diseases is the defective clearance of dying cells and self DNA. This would exacerbate the potential for an autoimmune response.

In conclusion, hypomorphic TLR allelic variants may predispose to many infectious diseases and perhaps autoimmunity. The genetic analysis of allelic frequencies in patients with various diseases is a research field in its infancy. Future studies will undoubtedly show the extent of TLR functions in health and disease. The knowledge that will be accumulated will allow us to improve health care by the genetic tailoring of therapies and the development of novel pharmacological and biological agents.

Acknowledgements

I thank Bjarne Bogen and Karoline Schjetne for sharing information about their work and comments, Hanne Quarsten for the inspiration for Figure 6A, and Trond S. Halstensen for comments on the text.

References

1. Belvin MP, Anderson KV. A conserved signaling pathway: the Drosophila toll-dorsal pathway. Annu Rev Cell Dev Biol 1996; 12:393-416.
2. Hultmark D. Drosophila immunity: paths and patterns. Curr Opin Immunol 2003; 15:12-9.
3. Hoffmann JA, Reichhart JM. Drosophila innate immunity: an evolutionary perspective. Nat Immunol 2002; 3:121-126.
4. Remillieux-Leschelle N, Santamaria P, Randsholt NB. Regulation of Larval Hematopoiesis in Drosophila melanogaster. A role for the multi sex combs gene. Genetics 2002; 162:1259-1274.
5. Pujol N, Link EM, Liu LX et al. A reverse genetic analysis of components of the Toll signaling pathway in Caenorhabditis elegans. Curr Biol 2001; 11:809-821.
6. Qureshi ST, Medzhitov R. Toll-like receptors and their role in experimental models of microbial infection. Genes Immun 2003; 4:87-94.
7. Akira S, Hemmi H. Recognition of pathogen-associated molecular patterns by TLR family. Immunol Lett 2003; 85:85-95.
8. Barton GM, Medzhitov R. Toll-like receptor signaling pathways. Science 2003; 300:1524-1525.
9. Lanzavecchia A, Sallusto F. Progressive differentiation and selection of the fittest in the immune response. Nat Rev Immunol 2002; 2:982-987.

10. Kurt-Jones EA, Mandell L, Whitney C et al. Role of toll-like receptor 2 (TLR2) in neutrophil activation: GM-CSF enhances TLR2 expression and TLR2-mediated interleukin 8 responses in neutrophils. Blood 2002; 100:1860-1868.
11. Sabroe I, Prince LR, Jones EC et al. Selective roles for Toll-like receptor (TLR)2 and TLR4 in the regulation of neutrophil activation and life span. J Immunol 2003; 170:5268-5275.
12. Varadaradjalou S, Feger F, Thieblemont N et al. Toll-like receptor 2 (TLR2) and TLR4 differentially activate human mast cells. Eur J Immunol 2003; 33:899-906.
13. Kadowaki N, Ho S, Antonenko S et al. Subsets of human dendritic cell precursors express different toll-like receptors and respond to different microbial antigens. J Exp Med 2001; 194:863-869.
14. Bosisio D, Polentarutti N, Sironi M et al. Stimulation of toll-like receptor 4 expression in human mononuclear phagocytes by interferon-gamma: a molecular basis for priming and synergism with bacterial lipopolysaccharide. Blood 2002; 99:3427-3431.
15. Muzio M, Bosisio D, Polentarutti N et al. Differential expression and regulation of toll-like receptors (TLR) in human leukocytes: selective expression of TLR3 in dendritic cells. J Immunol 2000; 164:5998-6004.
16. Sato S, Takeuchi O, Fujita T et al. A variety of microbial components induce tolerance to lipopolysaccharide by differentially affecting MyD88-dependent and -independent pathways. Int Immunol 2002; 14:783-791.
17. An H, Xu H, Yu Y et al. Up-regulation of TLR9 gene expression by LPS in mouse macrophages via activation of NF-kappaB, ERK and p38 MAPK signal pathways. Immunol Lett 2002; 81:165-169.
18. Datta SK, Redecke V, Prilliman KR et al. A subset of Toll-like receptor ligands induces cross-presentation by bone marrow-derived dendritic cells. J Immunol 2003; 170:4102-4110.
19. Bourke E, Bosisio D, Golay J et al. The Toll-like receptor repertoire of human B lymphocytes: inducible and selective expression of TLR9 and TLR10 in normal and transformed cells. Blood 2003; 10:10
20. Stacey KJ, Sweet MJ, Hume DA. Macrophages ingest and are activated by bacterial DNA. J Immunol 1996; 157:2116-2122.
21. Banchereau J, Paczesny S, Blanco P et al. Dendritic cells: controllers of the immune system and a new promise for immunotherapy. Ann NY Acad Sci. 2003; 987:180-7.
22. Shortman K, Liu YJ. Mouse and human dendritic cell subtypes. Nat Rev Immunol 2002; 2:151-161.
23. Boonstra A, Asselin-Paturel C, Gilliet M et al. Flexibility of mouse classical and plasmacytoid-derived dendritic cells in directing T helper type 1 and 2 cell development: dependency on antigen dose and differential toll-like receptor ligation. J Exp Med 2003; 197:101-109.
24. Tsuji S, Matsumoto M, Takeuchi O et al. Maturation of human dendritic cells by cell wall skeleton of Mycobacterium bovis bacillus Calmette-Guerin: involvement of toll-like receptors. Infect Immun 2000; 68:6883-6890.
25. Bernasconi NL, Onai N, Lanzavecchia A. A role for Toll-like receptors in acquired immunity: upregulation of TLR9 by BCR triggering in naive B cells and constitutive expression in memory B cells. Blood 2003; 30:30
26. Rui L, Vinuesa CG, Blasioli J et al. Resistance to CpG DNA-induced autoimmunity through tolerogenic B cell antigen receptor ERK signaling. Nat Immunol 2003; 4:594-600.
27. Coutinho A, Moller G. Editorial: Immune activation of B cells: Evidence for 'one nonspecific triggering signal' not delivered by the Ig receptors. Scand J Immunol 1974; 3:133-146.
28. Mamchak AA, Hodgkin PD. Absence of lipopolysaccharide high-dose paralysis in B-cell responses: implications for the one-signal theory. Immunol Cell Biol 2000; 78:133-141.
29. Leadbetter EA, Rifkin IR, Hohlbaum AM et al. Chromatin-IgG complexes activate B cells by dual engagement of IgM and Toll-like receptors. Nature 2002; 416:603-607.
30. Liu N, Ohnishi N, Ni L et al. CpG directly induces T-bet expression and inhibits Ig(gamma1) and Ig(epsilon) switching in B cells. Nat Immunol 2003; 25:25
31. Schjetne KW, Thompson KM, Nilsen N et al. Link between innate and adaptive immunity: Toll-like receptor 2 internalizes antigen for presentation to CD4+ T cells and could be an efficient vaccine target. J Immunol 2003; 171:32–36.
32. Okamura Y, Watari M, Jerud ES et al. The extra domain A of fibronectin activates Toll-like receptor 4. J Biol Chem. 2001; 276:10229-10233.

33. Termeer C, Benedix F, Sleeman J et al. Oligosaccharides of Hyaluronan activate dendritic cells via toll-like receptor 4. J Exp Med 2002; 195:99-111.
34. Johnson GB, Brunn GJ, Kodaira Y et al. Receptor-mediated monitoring of tissue well-being via detection of soluble heparan sulfate by Toll-like receptor 4. J Immunol 2002; 168:5233-5239.
35. Smiley ST, King JA, Hancock WW. Fibrinogen stimulates macrophage chemokine secretion through toll-like receptor 4. J Immunol 2001; 167:2887-2894.
36. Yang R, Yan Z, Chen F et al. Hyaluronic acid and chondroitin sulphate A rapidly promote differentiation of immature DC with upregulation of costimulatory and antigen-presenting molecules, and enhancement of NF-kappaB and protein kinase activity. Scand J Immunol 2002; 55:2-13.
37. Vabulas RM, Ahmad-Nejad P, da Costa C et al. Endocytosed HSP60s use toll-like receptor 2 (TLR2) and TLR4 to activate the toll/interleukin-1 receptor signaling pathway in innate immune cells. J Biol Chem 2001; 276:31332-31339.
38. Asea A, Rehli M, Kabingu E et al. Novel signal transduction pathway utilized by extracellular HSP70: role of toll-like receptor (TLR) 2 and TLR4. J Biol Chem 2002; 277:15028-15034.
39. Vabulas RM, Ahmad-Nejad P, Ghose S et al. HSP70 as endogenous stimulus of the Toll/interleukin-1 receptor signal pathway. J Biol Chem 2002; 277:15107-15112.
40. Ishii KJ, Suzuki K, Coban C et al. Genomic DNA released by dying cells induces the maturation of APCs. J Immunol 2001; 167:2602-2607.
41. Flo TH, Halaas O, Torp S et al. Differential expression of Toll-like receptor 2 in human cells. J Leukoc Biol 2001; 69:474-481.
42. Latz E, Visintin A, Lien E et al. Lipopolysaccharide rapidly traffics to and from the Golgi apparatus with the toll-like receptor 4-MD-2-CD14 complex in a process that is distinct from the initiation of signal transduction. J Biol Chem 2002; 277:47834-47843.
43. Wallin RP, Lundqvist A, More SH et al. Heat-shock proteins as activators of the innate immune system. Trends Immunol 2002; 23:130-135.
44. Equils O, Schito ML, Karahashi H et al. Toll-like receptor 2 (TLR2) and TLR9 signaling results in HIV-long terminal repeat trans-activation and HIV replication in HIV-1 transgenic mouse spleen cells: implications of simultaneous activation of TLRs on HIV replication. J Immunol 2003; 170:5159-5164.
45. Re F, Strominger JL. Toll-like receptor 2 (TLR2) and TLR4 differentially activate human dendritic cells. J Biol Chem 2001; 276:37692-37699.
46. McGuirk P, Mills KH. Pathogen-specific regulatory T cells provoke a shift in the Th1/Th2 paradigm in immunity to infectious diseases. Trends Immunol 2002; 23:450-455.
47. Pasare C, Medzhitov R. Toll pathway-dependent blockade of CD4+CD25+ T cell-mediated suppression by dendritic cells. Science 2003; 299:1033-1036.
48. Bretscher P, Cohn M. A theory of self-nonself discrimination: paralysis and induction involve the recognition of one and two determinants on an antigen, respectively. Science 1970; 169:1042-1049
49. Langman RE, Cohn M. Self-nonself discrimination revisited. Semin Immunol 2000; 12:159-162.
50. Weigle WO. Immunologic tolerance: development and disruption. Hosp Pract (Off Ed) 1995; 30:81-4, 89-92
51. Zinkernagel RM. Immunology taught by viruses. Science 1996; 271:173-8
52. Janeway CA Jr. Approaching the asymptote? Evolution and revolution in immunology. Cold Spring Harb Symp Quant Biol 1989; 54:1-13
53. Matzinger P. Tolerance, danger, and the extended family. Annu Rev Immunol 1994; 12:991-1045
54. Matzinger P. An innate sense of danger. Semin Immunol. 1998; 10:399-415.
55. Dembic Z. Do we need integrity? Scand J Immunol 1996; 44:549-550
56. Dembic Z. Immune system protects integrity of tissues. Mol Immunol 2000; 37:563-569.
57. Dembic Z. About theories and the integrative function of the immune system. The Immunologist 2000; 8:141-146
58. Johnson GB, Brunn GJ, Tang AH et al. Evolutionary clues to the functions of the Toll-like family as surveillance receptors. Trends Immunol 2003; 24:19-24.
59. Steinman RM, Hawiger D, Liu K et al. Dendritic cell function in vivo during the steady state: a role in peripheral tolerance. Ann NY Acad Sci 2003; 987:15-25.
60. Lorenz E, Mira JP, Cornish KL et al. A novel polymorphism in the toll-like receptor 2 gene and its potential association with staphylococcal infection. Infect Immun 2000; 68:6398-6401.

61. Smirnova I, Mann N, Dols A et al. Assay of locus-specific genetic load implicates rare Toll-like receptor 4 mutations in meningococcal susceptibility. Proc Natl Acad Sci USA 2003; 100:6075-6080.
62. Krutzik SR, Ochoa MT, Sieling PA et al. Activation and regulation of Toll-like receptors 2 and 1 in human leprosy. Nat Med 2003; 9:525-532.
63. Bochud PY, Hawn TR, Aderem A. Cutting edge: a toll-like receptor 2 polymorphism that is associated with lepromatous leprosy is unable to mediate mycobacterial signaling. J Immunol 2003; 170:3451-3454.
64. Kiechl S, Lorenz E, Reindl M et al. Toll-like receptor 4 polymorphisms and atherogenesis. N Engl J Med 2002; 347:185-192.
65. Boekholdt SM, Agema WR, Peters RJ et al. Variants of toll-like receptor 4 modify the efficacy of statin therapy and the risk of cardiovascular events. Circulation 2003; 107:2416-2421.
66. Seibl R, Birchler T, Loeliger S et al. Expression and regulation of toll-like receptor 2 in rheumatoid arthritis synovium. Am J Pathol 2003; 162:1221-1227.
67. Kadowaki N, Antonenko S, Liu YJ. Distinct CpG DNA and polyinosinic-polycytidylic acid double-stranded RNA, respectively, stimulate CD11c- type 2 dendritic cell precursors and CD11c+ dendritic cells to produce type I IFN. J Immunol 2001; 166:2291-2295.
68. Visintin A, Mazzoni A, Spitzer JH et al. Regulation of Toll-like receptors in human monocytes and dendritic cells. J Immunol 2001; 166:249-255.
69. Takeuchi O, Sato S, Horiuchi T et al. Cutting edge: role of Toll-like receptor 1 in mediating immune response to microbial lipoproteins. J Immunol 2002; 169:10-14.

Chapter 2

Structures and Motifs Involved in Toll Signaling

Monique Gangloff, Phumzile L. Ludidi and Nicholas J. Gay

Introduction

In this chapter we aim to provide a detailed overview of Toll-like receptor (TLR) structure. Initially we focus on the different ways in which the related vertebrate and invertebrate TLRs recognize pathogen patterns, with particular reference to the LPS sensor (hTLR4/CD14/MD2) and *Drosophila* Toll/Spätzle. In the second part of our chapter we discuss the evidence that signal transduction involves receptor driven recruitment of multitypic protein complexes and the activation and autoregulation of the IRAK family of Serine/Threonine protein kinases.

Components of the Extracellular Pathway: The Extracellular Leucine-Rich Repeat Domain of TLRs

The extracellular region of TLRs is composed of a tandem array of leucine-rich repeats (LRRs), versatile building blocks for the recognition of endogenous proteins and/or pathogen-associated molecular patterns (PAMPs). The N-terminal extremity contains a signal peptide which directs TLRs towards the cell surface or to endosomal vesicles in the case of TLR9 (for a review see ref. 1). The size of the mature extracellular region ranges from 552 residues in TLR6 (63kDa), to 813 residues in TLR7 (93kDa), but post-translational modifications such as glycosylation increases the molecular weight significantly. These N-linked glycosylations are crucial not only for the proper folding and cellular localisation of TLRs but also for their signaling, as has been suggested for TLR4.[2] The extracellular domain of TLRs also contains a number of cysteine residues engaged in disulfide bridges whose connectivity will be analysed in the light of recent structural data.

Leucine-Rich Repeats

Up to 19% of the extracellular domain of TLRs are leucine residues (L) that are arranged periodically in a 24 residue-long sequence (Fig. 1). Tandem arrays of this motif span the entire ectodomain, interrupted only by internal cysteine-rich regions (CRR). Considerable variation in the sequence and length of LRRs has however resulted in a certain degree of difficulty in detecting them in TLRs.

The three-dimensional structures of several LRR-proteins are now available (Table 1).[3-9] Two recently solved structures are of human LRR-proteins, the platelet-receptor glycoprotein Ibα (GpIbα) and Nogo receptor (NgR).[3,8] Both share structural similarities. Interestingly,

Toll and Toll-Like Receptors: An Immunologic Perspective, edited by Tina Rich.

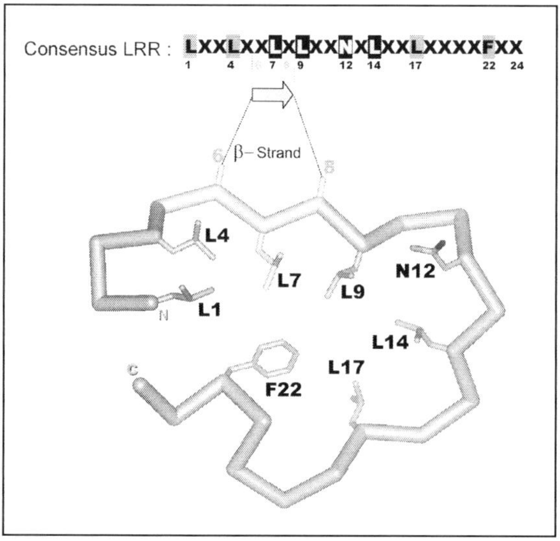

Figure 1. Primary and secondary structure adopted by an LRR motif. L denotes leucine; N= asparagine; F= phenylalanine and X= any amino acid. The motif spans 24 residues. Numbering is shown for conserved residues only. The degree of conservation is highlighted by black and gray boxes, which denote strong and weak conservation respectively. In the backbone trace of a consensus repeat only the side chains of conserved residues are shown. Atomic coordinates were taken from NgR LRR5 (region L153-D176).

TLR repeats are generally the same length and also express the same conserved features as GpIbα and NgR.

Each repeat is a structural unit, rolled up as a spiral in which the conserved residues are stacked. The hydrophobic consensus residues form the core of the protein and make intra- and inter-repeat interactions. The asparagine residue at position 12 has a central role in the LRR structure and is therefore highly conserved. Its side chain forms a hydrogen bond with the peptide backbone both from the previous repeat and within the same repeat. The resulting hydrogen bond network or "asparagine ladder" is crucial in stabilizing the turn.[10] The molecular fold generated is called an extended right-hand superhelix.

Secondary structure predictions for TLRs suggest that each LRR begins with a loop followed by a β-strand formed by at least three residues XLX at positions 6-8, where L stands for leucine, valine, or isoleucine, and X for any amino acid. The β-strands combine to form a tight parallel β-sheet. The remaining sequence does not adopt secondary structure in most GPIbα and NgR repeats, whereas α-helices have been observed in the ribonuclease inhibitor[7] and 3_{10} helices in, for instance, the listerial internalin protein (Fig. 2).[4] It is not yet clear if secondary structures such as helices are found in TLRs though predictions using the PSIPRED algorithm suggests that some regions may be helical.[11] The presence of either loops or helices on one side introduces an overall curvature, the extent of which depends on the secondary structures present. Bulkier α-helices induce a much tighter curvature than 3_{10} helices or loops (compare Figs. 2A, B and C). The parallel β-sheet lies on the inner concave side of the solenoid, which is bent and distorted to different extents depending on the protein. The overall architecture of LRR-proteins is reminiscent of a "banana" in NgR, a "sickle" in internalin, or a "horseshoe" in the ribonuclease inhibitor, according to the number of repeats and their relative curvature. The extracellular region of TLRs contains up to 27 repeats (Fig. 3). To date there is no known structure for a protein with such a high number of repeats. The largest LRR-repeat containing proteins for which we have structural data are internalin[9] and cytotoxin YopM from *Yersinia pestis*[4] which

Table 1. Structures of LRR-proteins

LRR Proteins	PDB
A leucine-rich repeat variant with a novel repetitive protein structural motif	1LRV
Crystal structure of Nogo-66 receptor and structure of the nogo receptor ectodomain	**1P8T, 1OZN**
Crystal structure of Ran-GDP-RanBP1-RanGAP complex in the presence of aluminum fluoride	***1K5G***
Crystal structure of the complex of glycoprotein Ibα and the von Willebrand factor A1 domain	**1M10**
Crystal structure of the RNA-binding domain of the mRNA export factor tap	1FO1, 1FT8, 1KOH, 1KOO
Crystal structure of the von Willebrand factor binding domain of glycoprotein Ibα	**1GWB, 1M0Z**
Crystal structure of tropomodulin C-terminal half	1100
Insights into Scf ubiquitin ligases from the structure of the Skp1-Skp2 complex	***1FQV, 1FS2***
Internalin B leucine rich repeat domain	1D0B
Internalin B: crystal structure of fused N-terminal domains	1H6T
Internalin H: crystal structure of fused N-terminal domains	1H6U
Porcine ribonuclease inhibitor	2BNH
Ribonuclease inhibitor complexed with ribonuclease A	***1DFJ***
Ribonuclease inhibitor-angiogenin complex	***1A4Y***
Solution structure of Chlamydomonas outer arm dynein light chain	1 1DS9
The crystal structure of Rna1p (RanGAP)	1YRG
U2 B''/U2 A'/RNA ternary complex	***1A9N***
Internalin (*Listeria monocytogenes*)/ E-cadherin (human) Recognition complex	***1O6S***
Internalin (Inla,*Listeria monocytogenes*) - functional domain, uncomplexed	1O6T, 1O6V
Crystal structure of Rab geranylgeranyltransferase from rat brain	1DCE
Crystal structure of polygalacturonase-inhibiting protein (PGIP), a LRR-protein involved in plant defense	<u>1OGQ</u>

The PDB code of extracellular human proteins, GPIba and Nogo receptors, are bolded. The recent extracellular plant LRR-protein structure is underlined. Complexes are bolded and italicized, with the exception of the complex of GPIbα and the von Willebrand Factor A1 domain.

contains fifteen and a half turns. Interestingly the repeats in these proteins are not planar as in the ribonuclease inhibitor, but instead the screw-like rotation of each repeat dictates a slightly twisted structure. It is possible that TLRs with more than 15 consecutive repeats adopt a twisted LRR fold, reminiscent of a corkscrew.

Cysteine-Rich Flanking Regions

LRR blocks are often flanked by cysteine-rich domains called LRRNT and LRRCT when located at the amino- and carboxy- terminal ends of LRR regions respectively.[12-14] These crucial capping structures bury the otherwise exposed hydrophobic residues at the ends of LRR superhelices. Until recently no structural information was available for these regions, as

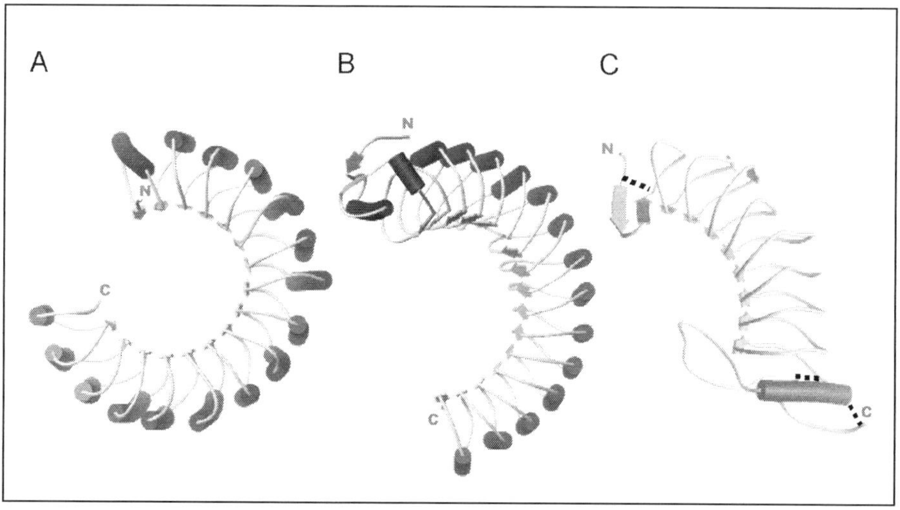

Figure 2. Overall architecture of LRR-proteins. A) The ribonuclease inhibitor (PDB code 1A4Y) adopts a horseshoe shape with α-helices on its outer face. B) The listerial protein internalin Inla LRR domain (PDB code 1O6S) adopts a sickle shape, in which the repeats are not planar as in A, but display a helical twist. The convex side is built up with 3_{10} helices. C) Glycoprotein Ibα (PDB code 1MOZ) does not have secondary structure elements on its convex side. Disulfide bridges are shown as black dotted lines, β-strands as arrows and helices as cylinders.

intracellular LRR-proteins do not display these features. However, we now know that the cysteines in these capping structures are involved in disulfide bridges whose connectivity has been elucidated.

LRRNT

There are at least two types of LRRNT (Fig. 4). Type I structures, exemplified by GPIbα, form one disulfide bond, whereas type II motifs, as in NgR, are linked by two disulfide bonds. Both types of LRRNTs are located in very similar environments. In GPIbα, the amino-terminal sequence forms a 14-residue β-hairpin delimited by a disulfide bond between the cysteines at positions 4 and 17. The two antiparallel β-strands are stabilized by this disulfide bridge at their base. The second β-strand joins the parallel β-sheet formed by the LRR on the concave side. Interestingly, this region, which is disordered in one of the two molecules in the asymmetric unit, participates in the binding interface between GPIbα and Von Willebrand Factor.[8] This suggests a certain degree of flexibility in a type I LRRNT and a potential role in ligand binding. However, it is not clear if the LRRNTs of TLRs can bind ligands. For example, mutagenesis experiments showed that the LRRNT of TLR2 was dispensable for peptidoglycan (PGN) binding.[15]

In NgR,[3] two disulfide bonds stabilize the amino-terminus, which adopts a compact structure, similar to but not as extended as the β-finger in GPIbα. The first disulfide bond between cysteines 27(CI) and 33(CIII) forms "a small knot" and holds the amino-terminal loop in a closed conformation on top of the tandem segments of LRRs. The second connection occurs between cysteines 31 (CII) and 43 (CIV). Interestingly, the spacing between these residues (11 residues instead of 12 in GPIbα) and their environment (a β-hairpin protruding to the side of the LRR concave face, instead of the β-finger in GPIbα) are similar in both LRRNT structures.

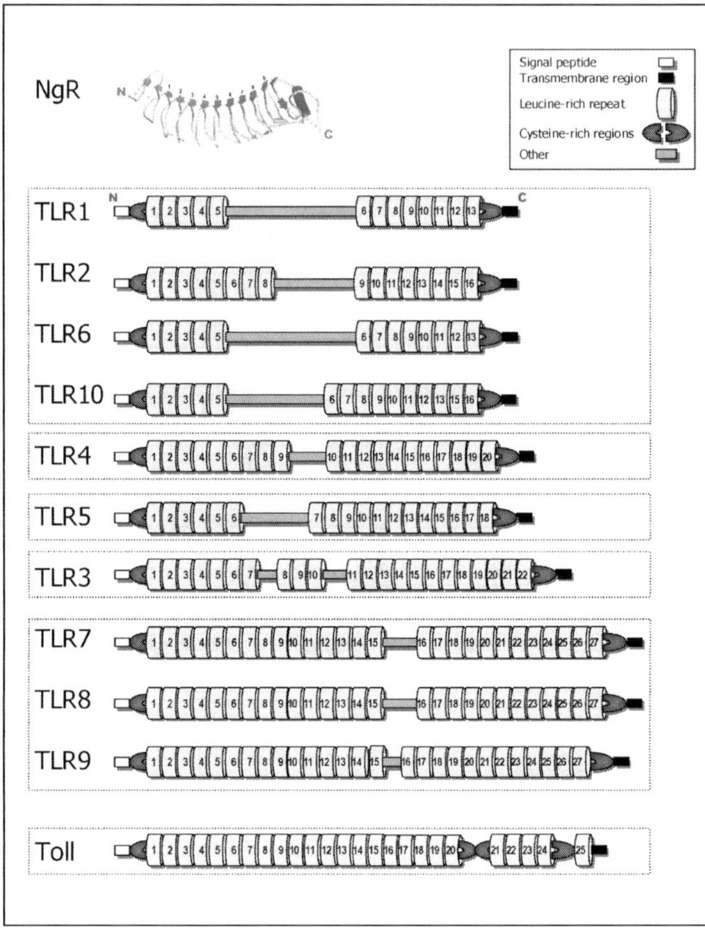

Figure 3. Overall architecture of TLR ectodomains. NgR at the top of the schematic is given as a reference structure. TLRs are divided into subgroups, according to phylogenetic analysis. The repeats are shown as stacked cylinders, cysteine-rich regions (CRR) are dark gray and the signal peptide and transmembrane regions are indicated by white and black rectangles respectively. The regions shown as gray bars in between LRRs do not display consensus motifs, although they are likely to contain either degenerate repeats and/or CRRs.

The LRRNT of TLRs can be modeled on either the GPIbα or NgR structures more or less reliably, given the number and the spacing of cysteine residues. Secondary structure predictions corroborate a potential small β-structure at the amino-terminal end of TLRs. *Drosophila* Tolls are the only members of the family to display a type II LRRNT. Although human TLRs have type I LRRNTs they can be quite different from GPIbα. Indeed, the spacing ranges from a maximum of 14 residues in TLR7 to 5 residues in TLR2, which is too narrow to accommodate a β-finger structure, but resembles instead the connection between CI and CIII in the NgR structure. All the members of the TLR2 subgroup show a high degree of heterogeneity in their amino-terminal capping structures. TLR2 possesses a type II LRRNT that lacks the second disulfide bridge. In contrast, TLRs 1 and 6 do not have any cysteine residues and TLR10 has only one. It is worth noting that this subgroup of receptors is known to cooperate with each other in signaling. For example, TLR2 acts as a common binding partner for both TLR1 and

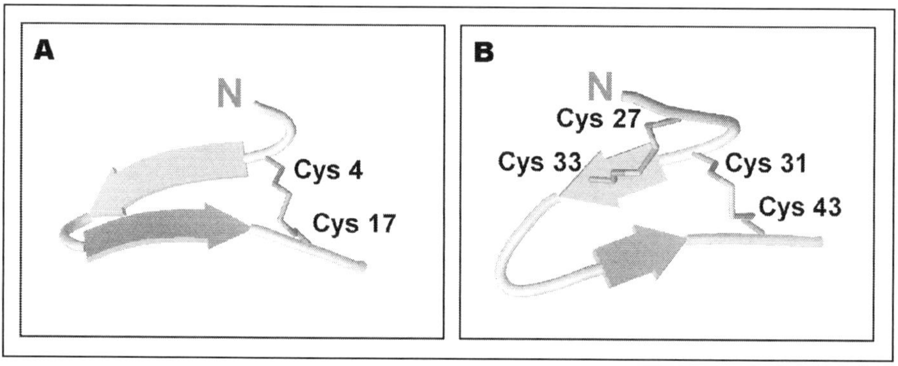

Figure 4. Two types of LRRNTs, as revealed by the crystal structures of GPIbα (A) and NgR (B).

TLR6.[16] This leads us to ask whether there can be a link between their mode of dimerization and the features of the amino-terminal capping structures? Further experiments are required in order to answer this question.

LRRCT

The LRRCT domain is located at the carboxy-terminal end of LRR proteins. Its signature sequence contains four cysteine residues that are conserved in various extracellular LRR proteins. The disulphide connectivity of CI-CIII and CII-CIV is the same as for the type II LRRNT but the spacing of the cysteines is much greater (between 22 and 52 residues). In contrast to the β only LRRNT, LRRCT is an α/β structure, with an α–helix, several short 3_{10}-helices and β-strands that cap the hydrophobic core, forming hydrogen bonds with the last repeat. The first two cysteines of the LRRCT motif occur in the middle of the last repeat at positions 15 and 17, respectively. The NgR LRRCT is slightly larger than its GPIbα counterpart but lacks the so-called β-switch, a protruding loop that is involved in Von Willebrand Factor binding.[8] Upon ligand binding, GPIbα undergoes a conformational change in which the protruding loop adopts a β-hairpin structure. The antiparallel β-strands participate in a continuous antiparallel β-sheet, shared between the two molecules.

Drosophila Toll contains an internal (CF1) and a carboxy-terminal (CF2) LRRCT juxtaposed to the transmembrane region. The structural integrity of CF2 is crucial for receptor function as revealed by mutagenesis studies. Three separate cysteine to tyrosine mutations have been shown to generate a constitutively active receptor.[17] Constitutive receptor activity may occur as a result of intermolecular disulfide bond formation via the now unpaired cysteine residues; a hypothesis first proposed for the fibroblast growth factor receptor.[18] However, any mutation that destroys the disulfide bonding network of CF2 could also affect protein stability and increase its sensitivity to proteolytic cleavage. This latter hypothesis is supported by the isolation of cleavage products and the absence of disulfide-linked mutant dimers in mutant embryos (our unpublished data).

Structural Diversity of TLR Ligands

A great deal of interest in the TLR family centers on their role in sampling and signaling in response to pathogens and/or endogenous ligand molecules. In contrast to *Drosophila* Toll, which is activated by an endogenous protein ligand Spätzle, mammalian TLRs are thought to be true pathogen-recognition receptors able to recognize a wide range of pathogen-associated molecular patterns (PAMPs) (Table 2). These will be described at the structural level in the following section.

Table 2. Ligands recognized by TLRs

TLRs	Ligands	Origin of Ligands and References
TLR1/2	Tri-acyl lipopeptides	bacteria, mycobacteria[115]
	Soluble factors	*Neisseria meningitides*[116]
	OspA	*Borrelia burgdorferi*[117]
TLR2	Lipoprotein/lipopeptides	a variety of pathogens[118-121]
	Peptidoglycan	Gram-positive bacteria (not accessible in Gram-negative)[122-124]
	Lipoteichoic acid	Gram-positive bacteria[122,125]
	Lipoarabinomannan	mycobacteria[126,127]
	A phenol-soluble modulin	*Staphylococcus epidermidis*[128]
	Glycoinositolphospholipids	*Trypanosoma Cruzi*[129]
	Glycolipids	*Treponema maltophilum*[130]
	Porins	*Neisseria meningitidis*[131]
	Zymosan	fungi[127]
	Atypical LPS	*Leptospira interrogans*[132]
	Atypical LPS	*Porphyromonas gingivalis*[133]
	Hsp70	host[134,135]
TLR3	Poly (I-C) double-stranded RNA	Virus[136]
TLR4	LPS	Gram-negative bacteria[2,87,98,137,138]
	Flavolipin	*Flavobacterium meningosepticum*[139]
	ER-112022, E5564, E5531	Synthetic compounds[24]
	Taxol	Plants[45]
	Fusion protein	Respiratory Syncytial Virus [140,141]
	Envelope proteins	Mouse Mammary Tumor Virus [142]
	Hsp60	*Chlamydia pneumoniae*[143,144]
	Hsp60	host[145]
	Hsp70	host[134,135,146]
	Type III repeat extra domain A of fibronectin	host[147]
	Oligosaccharides of hyaluronic acid	host[148]
	Polysaccharide fragments of heparan sulfate	host[149]
	Fibrinogen	host[150]
TLR5	Flagellin	bacteria[26,42,151]
TLR6/2	Di-acyl lipopeptides	mycoplasma[37]
TLR7	Imidazoquinolines (imiquimod, R-848)	synthetic compounds[152]
	Bropirimine	synthetic compounds[1]
	guanosine analogs	synthetic compounds[44]
TLR8	R-848	synthetic compounds[44]
TLR9	Unmethylated CpG DNA	Bacteria, virus, yeast, insects[23,152]
	Chromatin-IgG complexes	host[153]

Cystine-Knot Structure of Spätzle

Spätzle is composed of an amino-terminal prodomain and a carboxy-terminal domain C106, the active fragment. Numerous alternatively spliced isoforms have been identified[19,20] which vary in the length of their pro-sequences. It is not clear how the prodomain prevents the active fragment from binding to its receptor, as this domain is thought to be "natively unstructured".[21] The inactive precursor of Spätzle undergoes proteolytic processing by the serine protease Easter to generate the active form.

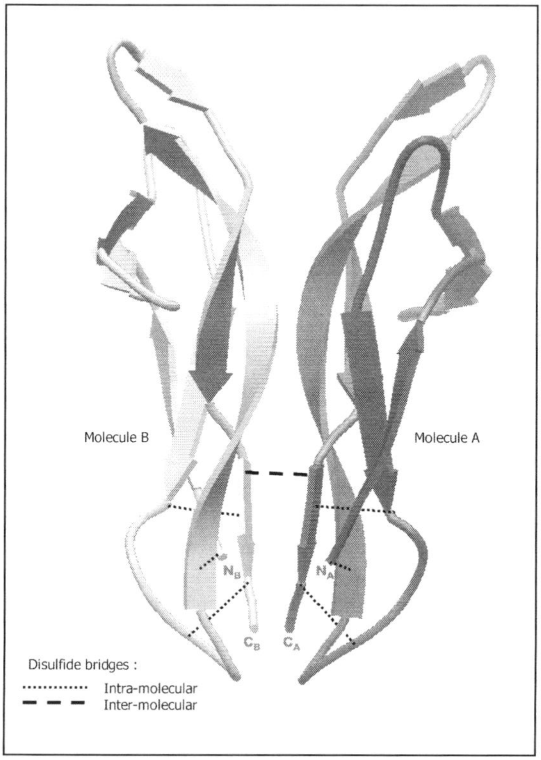

Figure 5. Three-dimensional model of the active Spätzle C-106 dimer.[22]

The active form of Spätzle possesses a signature sequence with 7 cysteines, whilst the remainder of its sequence shows no similarity to any proteins of known three-dimensional structure. The spacing of these cysteines is similar to that found in several growth factors including nerve growth factor (NGF), transforming growth factor β, platelet-derived growth factor and human chorionic gonadotropin. These growth factors/hormones adopt a unique fold called the cystine knot, which is characterized by an elongated β-strand structure and three disulphide bridges with unusual connectivity. Models of Spätzle based on NGF and coagulogen (Fig. 5),[22] confirmed that Spätzle formed a cystine-knot fold and could also dimerize, with individual protomers bound by an intermolecular disulfide bridge. This suggested that *Drosophila* Toll, like other Type 1 cytokine receptors, is activated by receptor dimerization.

A Variety of Structurally Unrelated Ligands for Mammalian TLRs

In contrast to *Drosophila* Toll, members of the vertebrate TLR family recognize a wide variety of PAMPs (Table 2). PAMPs are structurally unrelated molecules, ranging from small molecules to proteins, lipids, carbohydrates and nucleic acids. The paradigm is that each TLR receptor displays a pathogen-specific activity. Gram-negative bacteria are recognized by TLR4, Gram-positive bacteria by TLR2, and flagella by TLR5. Immunostimulatory bacterial DNA binds to TLR9 in endosomal vesicles whereas viral RNA binds to TLR3.

An increasing number of reports suggest direct ligand binding by TLR ectodomains.[23-26] There are currently 37 potential ligands and these are listed in Table 2. One wonders how specific the recognition mediated by receptors such as TLR2 and TLR4 is. On the one hand, they seem to bind multiple PAMPs. On the other, both receptors are also thought to bind to the same ligands, namely lipoteichoic acid (LTA) and lipopolysaccharide (LPS). One possible explanation could be the relative impurity of some ligand preparations used. Alternatively, molecular recognition may involve a number of non-TLR binding partners, as will be discussed later.

The Structure of LPS

Lipopolysaccharide (LPS) is the main constituent of the outer leaflet of the outer membrane in Gram-negative bacteria.[27] For bacteria, the outer membrane represents the first line of defense against environmental stress or chemical attack, such as antibiotics or bile salts and its integrity is critical for survival, which would imply that LPS molecules must be highly conserved. On the other hand, LPS molecules flag the bacteria as foreign when they engage the PAMP receptors of the host innate immune system.

LPS is an amphipathic molecule composed of lipid A and a core oligosaccharide region (Fig. 6), which in many bacteria carries a polysaccharide chain called O-antigen. The conserved lipid A domain of LPS comprises a large portion of the outer membrane bilayer of Gram-negative bacteria, and is a mono- or di-phosphorylated $\beta 1 \rightarrow 6$-linked glucosamine disaccharide to which up to seven fatty acids are attached by amide or ester bonds. The inner core oligosaccharide region of LPS contains two or three KDO molecules (see below) that are linked to two heptose residues that may contain phosphate or other substituents. KDO is an 8-carbon saccharide 2-keto-3-deoxyoctonate that is exclusive to LPS. LPS displays negative charges derived from the diglucosamine phosphates and from the KDO carboxylate anions. In terms of its structure, LPS is highly variable at its O-antigen portion,[28] and to a lesser extent, at the LPS core.[29] However the lipid A portion is well conserved across species.[30] LPS has been available in a pure biologically active form since 1952[31] and its chemical synthesis in 1985 was used to prove that lipid A is the active ingredient.[32]

Lipid A is the molecular pattern recognized by PRRs such as TLR4 and its synthetic analogs display either agonist (in the case of ER-112022) or antagonist (for E5564 and E5531) activities,[24] though the structural basis for these differences is unclear.

Lipoteichoic Acid

Lipoteichoic acid (LTA) (Fig. 7) is a surface-associated adhesion molecule in Gram-positive bacteria. Like LPS, it is an amphiphilic, negatively charged glycolipid. Released principally from bacterial cells after bacteriolysis, it binds to target cells either non-specifically (via membrane phospholipids) or specifically to CD14 and Toll-like receptors 2 and/or 4. This binding can be blocked by phospholipids and antibodies specific for CD14 and TLR4. LTA triggers a number of immune processes such as the activation of the complement cascade and the stimulation of neutrophils and macrophages. Consequently, LTA shares many of the pathogenic properties of LPS. In vitro, LTA release can be inhibited by non-bacteriolytic antibiotics and by poly-sulphates such as heparin, which probably interfere with the activation of autolysis.

LTA from *S. aureus* has been purified and characterized.[33] It is a diacylated amphiphilic molecule with a D-alanine substitution on its polyglycerophosphate backbone. Deacylation by alkaline hydrolysis results in its failure to induce any cytokine release in human whole blood, which indicates the importance of the lipid moiety in immune responsiveness. The role that this moiety (also called "Lipid B") plays in receptor binding is demonstrated by the failure of deacylated LTA to display any antagonist activity on LTA stimulation.

Lipopeptides

Bacterial lipoproteins (BLP) trigger immune responses via TLR2 and their immunostimulatory properties can be attributed to their lipid portion. Most BLPs are triacylated at their N-terminal cysteine residue, though mycoplasma macrophage-activating lipopeptide-2 kD (MALP-2) (Fig. 8) is only diacylated.[34] Synthetic lipoprotein analogs, such as tripalmitoyl cysteinyl lipopeptide

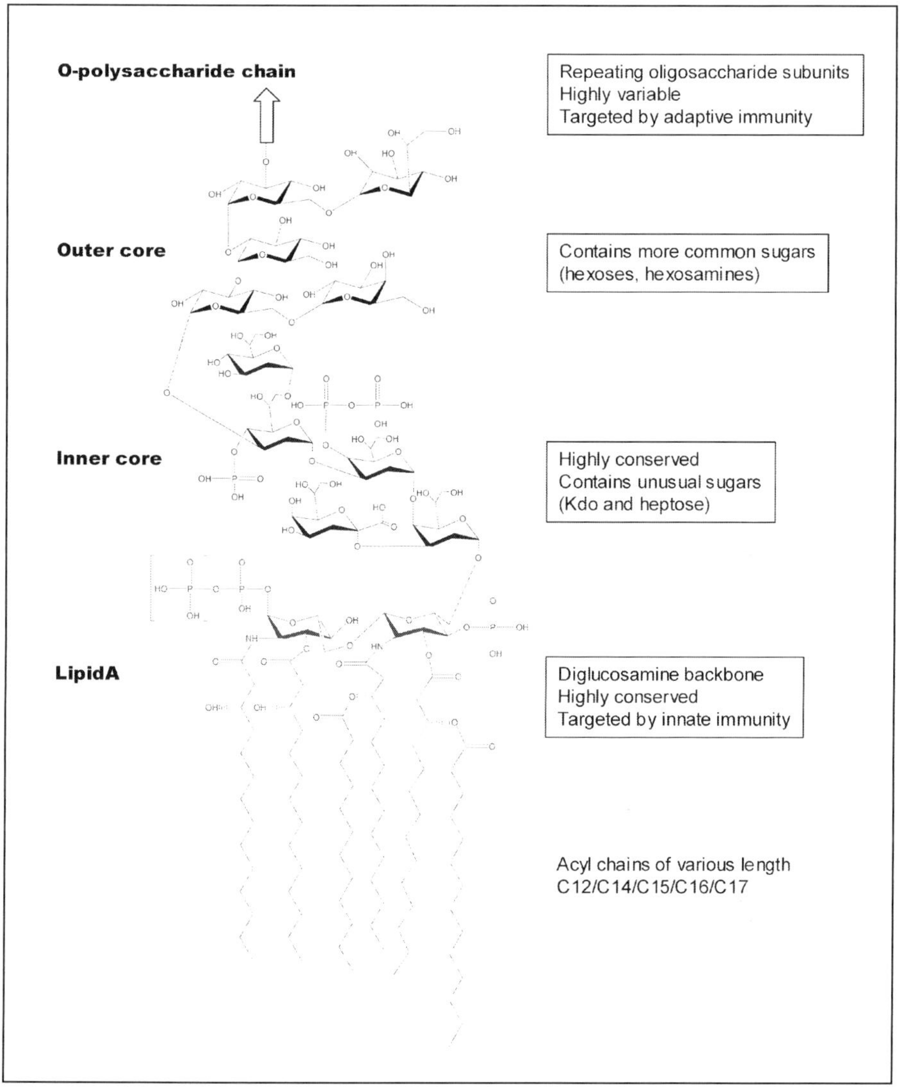

Figure 6. Structure of LPS. Reprinted from Ferguson AD, Welte W, Hofmann E et al. A conserved structural motif for lipopolysaccharide recognition by procaryotic and eucaryotic proteins. Structure Fold Des 2000; 8(6):585-592. ©2000, with permission from Elsevier.

Pam_3CSK_4 and dipalmitoyl MALP-2, have been shown to mimic the proinflammatory properties of BLP.[35-37]

Flagellin

Flagellin is the primary protein component of the bacterial flagellum, a highly complex filamentous structure that extends from the outer membrane of Gram-negative bacteria. Flagella serve as propellers, driven by a rotary motor at their base, to move the bacterium through its aqueous environment. They also aid in the attachment of bacteria to host cells, assisting in bacterial invasion and thereby contributing to the virulence of pathogenic bacteria.

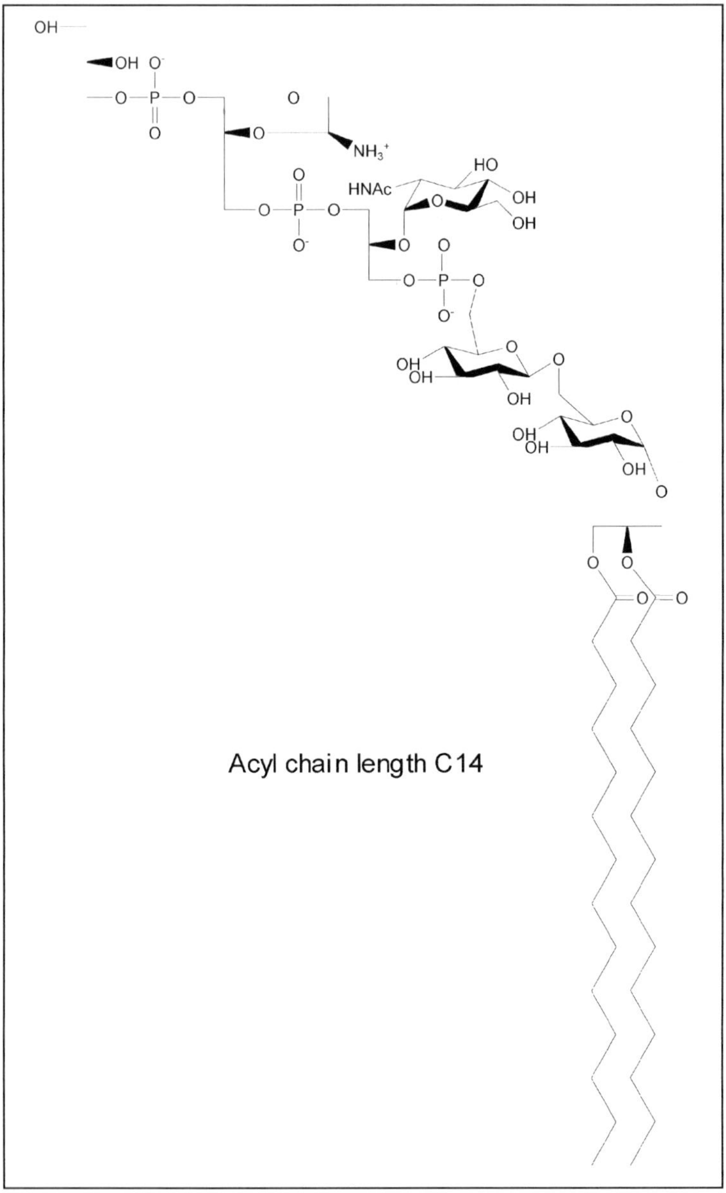

Figure 7. Structure of LTA from *Staphylococcus aureus*.[33]

The flagellin protein has been crystallized and its structure solved at 2.0 Å resolution (Fig. 9).[38] It is a multidomain protein displaying α/β structures. The first domain is an α-helical bundle followed by two β folds. The protein adopts an L-shape, in which residues from the amino and carboxyl-termini are juxtaposed to participate in a hairpin structure. Eaves-Pyles and collaborators[39] showed that the pathogen-associated molecular pattern of flagellin is localized to this region, which constitutes an elongated innate-immunoreactive surface. The other branch of the "L" is in fact the middle hypervariable domain.[40-42] Monomeric flagellin is

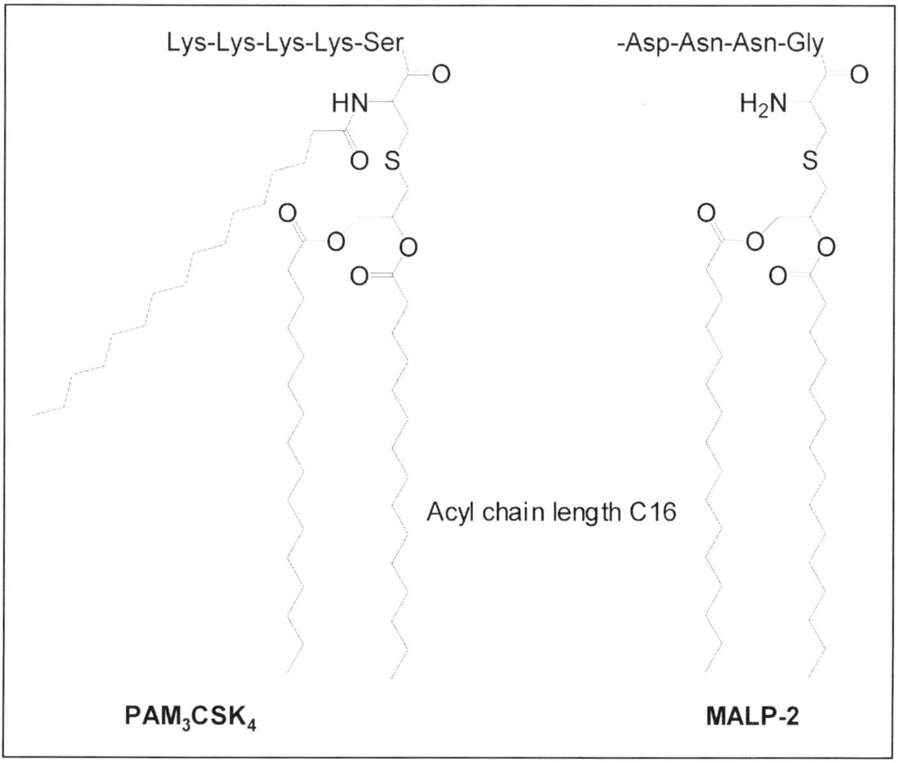

Figure 8. Structure of the bacterial and mycoplasma lipoproteins, PAM_3CSK_4 and MALP-2 respectively.

polymerized into protofilament structures that assemble with different helical symmetries (left- or right-handed) to form an 11-stranded filament. Interestingly, the conserved terminal regions of flagellin are located in the core of the filament, where axially aligned α-helical bundles are densely packed. The variable middle domain is solvent-exposed and targeted by the adaptive immune system.

Small Immunostimulatory Molecules

A number of synthetic molecules have been found to modulate the activity of different members of the TLR family (Fig. 10). TLR7, and to some extent TLR8, are activated by anti-viral and anti-tumor compounds such as the imidazoquinolines which are referred to as immune response modifiers (IRMs).[43] Though the natural ligands for TLRs 7 and 8 have not yet been identified, the use of synthetic ligands that display agonist or antagonist activity should help in their identification. Guanosine analogs have been studied over the years for their ability to stimulate the innate immune system in mouse and human models and the structural requirements for their immunostimulatory activity have been characterized. In this respect the pattern of purine ring substitution at the 7th and 8th positions appears to be absolutely critical. It is worth noting that imidazoquinolines and guanosine analogs are nucleic acid-like structures,[44] prompting us to ask whether there is a link between ligand discrimination and TLR classification that groups TLRs 7, 8 and 9 together?

It is not clear if TLRs bind small immunomodulatory molucules directly. Genetic evidence can be confusing and does not adequately compensate for direct ligand binding experiments. This is illustrated by the example of the mode of action of taxol. Taxol is a plant dipertene

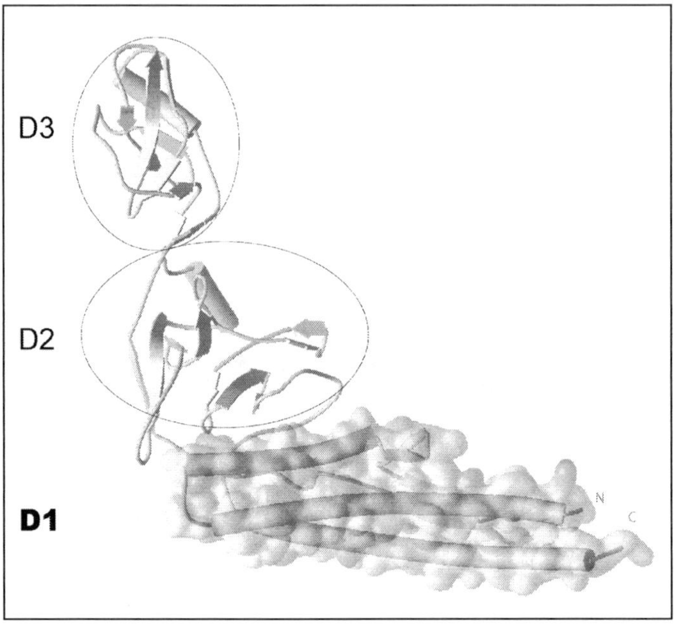

Figure 9. Structure of the multidomain Flagellin protein.

with anti-cancer properties that modulates murine TLR4 activity though is not a TLR ligand per se. First, it was shown that taxol mimics LPS-like activity, though solely in mice.[45,46] Subsequently, it was reported that taxol activity required the presence of mouse MD-2 (mMD-2), but not human MD-2, whichever species of TLR4 was expressed.[45] Therefore the likelihood is that mMD-2 is the true binding site for taxol, although signaling requires TLR4.

Nucleic Acids

DNA from bacteria, viruses, yeast and insects stimulate the mammalian immune system. In contrast to cell-wall components, DNA is probably invisible to the immune system until liberated during processes that affect pathogen integrity. Phagocytosis of pathogens by macrophages is known to be one of these triggers. It comes as no surprise then that internalization, endosomal maturation and acidification are required for bacterial DNA to acquire its immunostimulatory properties.[47,48]

It has also been noticed that the stimulatory effects of bacterial DNA depends on the presence of unmethylated 2′-deoxyribo(cytidine-phosphate-guanosine) CpG dinucleotides.[49-51] In contrast, methylated DNA or inversion of the central CG dinucleotide abolishes immune responses. A correlation has been made between the frequency of CpG dinucleotides, its methylation state, the presence of palindromic sequences and the stimulatory activity of oligonucleotides. Mammalian DNA, with its low frequency of CpG sequences, most of which are methylated, is simply not competent to trigger host immune responses.

Several lines of evidence suggest that unmethylated CpG oligonucleotides are true TLR9 ligands. Firstly, TLR9 localizes to endosomal vesicles in contrast to other members of the TLR family that are found on the cell surface. Secondly, TLR9 displays species specificity. Mouse cells respond maximally to 'GACGTT', the "mouse motif"[52] whilst for humans the optimal sequences are 'GTCGTT' and 'TTCGTT', (human motifs).[23] Complementation experiments with CpG-unresponsive 293 cells transfected with either human or mouse TLR9 revealed that

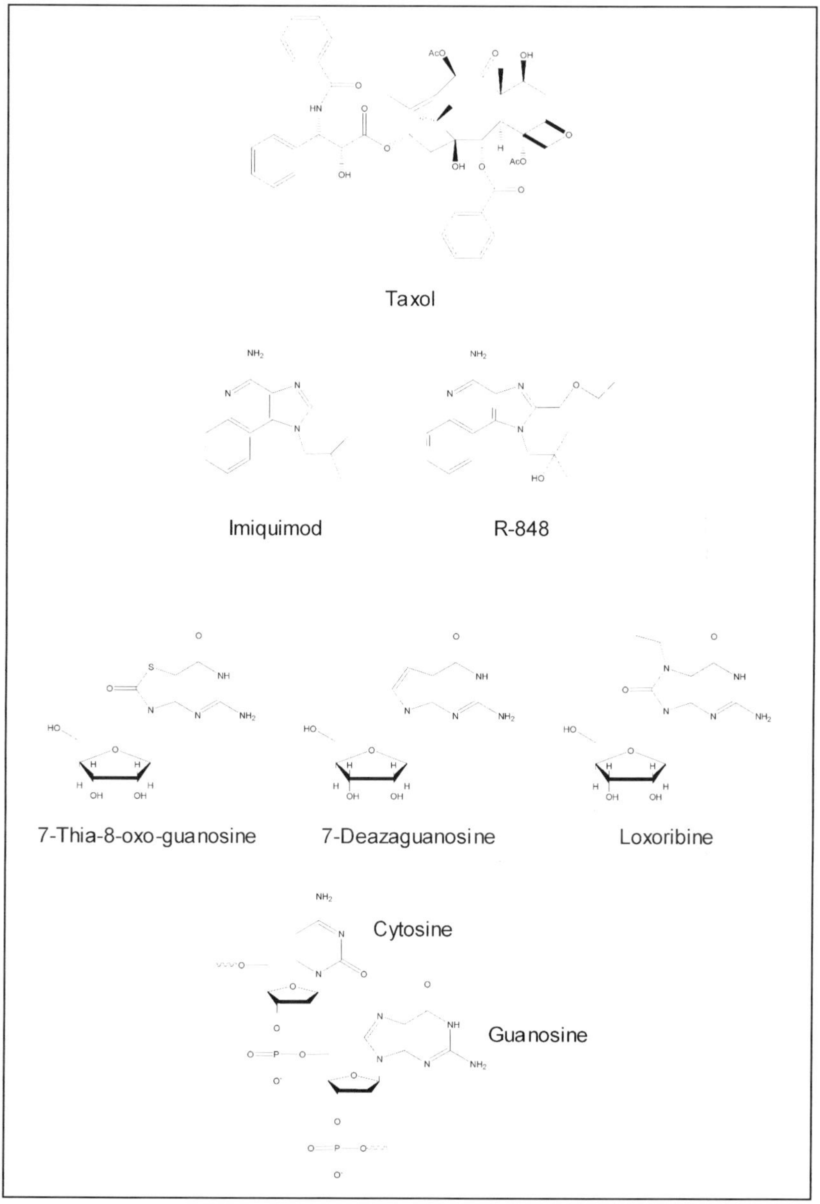

Figure 10. Small immunostimulatory molecules. The natural plant molecule taxol is a diterpene. Synthetic compounds such as imidazoquinolines and guanosine derivatives (but not guanosine), and unmethylated deoxyribo-CpG oligonucleotide, are ligands for the TLR7/8/9 subgroup.

human TLR9 preferentially recognized human CpG motifs whereas murine TLR9 preferred the mouse CpG motif. This suggests a direct interaction between TLR9 receptors and CpG DNAs. A palindromic 'AACGTT' motif induces immune responses in both mouse and human systems.[53] Although TLR9 recognizes CpG motifs in various flanking sequences and secondary structures, including double stranded bacterial DNA, palindromic DNA[50] and single-stranded synthetic DNAs, the downstream effects may differ. It is also worth noting that TLR9 recognizes both natural phosphodiester and synthetic phosphorothioate DNAs.[52] Recently Kandimalla and collaborators[54] showed that divergent synthetic nucleotide motifs displayed potent immunomodulatory activities with distinct cytokine induction profiles. This could be of great therapeutic value in the treatment of a number of diseases, such as cancer, asthma, rheumatoid arthritis and systemic lupus erythematosus (SLE).

Mechanism of Ligand Binding and Signal Transduction

LRRs are found in a wide range of proteins displaying different functions and cellular locations. All LRR proteins appear to be involved in diverse molecular recognition processes such as signal transduction, cell adhesion, cell development, DNA repair and RNA processing. They are often involved in protein-protein interactions which is the case for *Drosophila* Toll which forms hetero-tetrameric complexes with the cytokine-like ligand, Spätzle.[21] Like TLRs *Drosophila* Toll is activated by exposure to PAMPs but recognition is mediated by intermediary binding proteins.[55] In contrast to *Drosophila* Toll the interactions of many PAMPs with mammalian TLRs appears to be direct and occasionally involves co-operativity between TLRs and other receptors. This is a new field of research and the mechanism of ligand binding and signal transduction remains to be elucidated. Given the medical importance of TLRs, the issue of ligand binding and signal transduction is of primary importance if one is to consider TLRs as drug targets.

Spätzle Binding is Necessary and Sufficient for Toll Activation

Genetic evidence for the direct binding to *Drosophila* Toll by Spätzle C-106 has recently been confirmed by our laboratory using purified proteins.[21] Spätzle C106 binds specifically to the extracellular domain of Toll, but not to mammalian TLRs such as human TLR2, with a binding affinity (in solution) in the nanomolar range and a dissociation constant of about 90nM. The search for the binding site has been narrowed by using Toll5B, a dominant negative form of Toll[17] truncated in the second LRR block. Consequently, in *Drosophila* Toll, which contains two LRR blocks in contrast to mammalian TLRs, only the amino-terminal block contributes to the binding interface.

Pathogen Recognition by TLRs

The ectodomains of TLRs are divergent, displaying very low sequence identity. This suggests that they have evolved to bind the wide range of chemically unrelated molecules described in the previous section. LRRs seem particularly suited to the rapid evolution of diverse specificities and mutagenic fingerprinting of the receptor-ligand interface has already revealed that different regions in the TLR ectodomains are involved. Unfortunately this also infers that accurate modelling of these regions will be difficult.

TLR2-PGN

There is evidence for the direct binding of zymosan[56] and of peptidoglycan[15] to TLR2. Using a mutagenesis strategy in which portions of the extracellular region of TLR2 were deleted, Mitsuzawa and collaborators identified a region in TLR2 that was critical for peptidoglycan recognition. This region, mapping to the first LRR, encompasses residues Ser 40 to Ile 64 (S40-I64). The cysteine-rich amino-terminal flanking region is not required for ligand binding and deletion of this region did not affect TLR2 function. A three-dimensional model of the S40-I64 region was generated using the crystal structure of GPIbα (Fig. 11)[57] though the significance of the solvent exposed residues Glu 52, Lys 55, Asp 58 and Arg63 needs further analysis. TLR2 potentially recognizes the polar part of PGN whereas TLRs 6 or 1 are required to distinguish the number of acyl chains on its ligands.

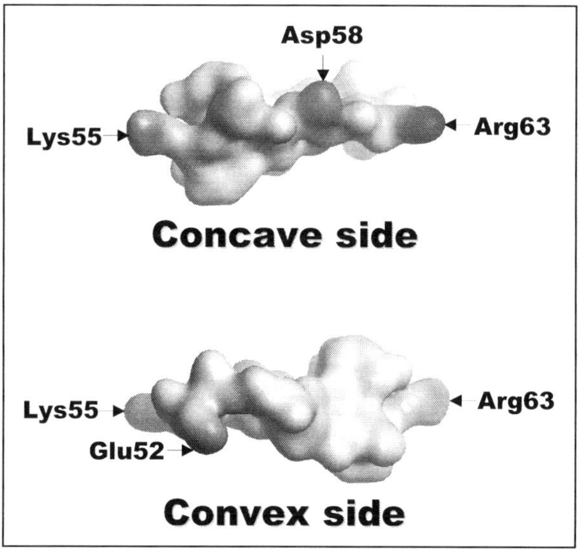

Figure 11. TLR2 N-terminal region modelled on GpIbα: the molecular surface represents the region crucial for peptidoglycan recognition.[15] The figure was generated with MSViewer.

TLR5-Flagellin

Mizel and collaborators have recently identified a region crucial for bacterial flagellin binding using an in-vitro binding assay.[26] Truncated forms of the TLR5 extracellular domain were generated and their ability to bind flagellin analyzed using pull-down and sedimentation velocity centrifugation experiments. Two overlapping fragments encompassing residues 1 to 407 and 386 to 636 retained flagellin binding. In contrast, binding was not observed with an amino-terminal fragment stretching from residue 1 to 386. It was concluded that flagellin binding is conferred by the region between residues 386 and 407, which is part of a consensus LRR that spans residues 383 to 407 (referred to as LRR9 in the NCBI/Entrez database, with the accession number O60602 for TLR5). As for TLR2's peptidoglycan recognition site, it is not possible to map the binding site to either the flank, the concave or convex side of the repeat. Further mutagenesis studies will be required to accurately map the binding site.

A Receptor Cluster for LPS

Although genetic studies point towards TLR4 as the major LPS receptor, there is accumulating evidence to suggest that LPS recognition and signaling is mediated by a supramolecular protein cluster. The composition of this cluster is cell type specific which confers specificity to LPS stimulation. Briefly, lipopolysaccharide-binding protein (LBP) binds and transfers LPS to membrane-bound CD14 and this transient complex recruits signaling molecules. LPS is then released from CD14 and binds to a complex of receptors that includes the chemokine receptor 4 (CXCR4), heat shock proteins (HSPs) 70 and 90, growth differentiation factor 5 (GDF5) and possibly CD55.[58] Signal transduction requires Toll-like receptor 4 (TLR4) associated with MD-2. In B cells, TLR4 also associates with RP105 (also called CD180) complexed with MD-1. RP105 is closely related to TLR4, displaying about 30% sequence identity, but lacks the intracellular TIR domain. Integrins, such as CD11 or CD18 have also been found to colocalize with TLR4 in clusters at the surface of monocytes. Recent studies have identified a new family of LPS responsive proteins. The mammalian NOD proteins (nucleotide-binding oligomerization domain) are a family of intracellular proteins homologous to plant R gene

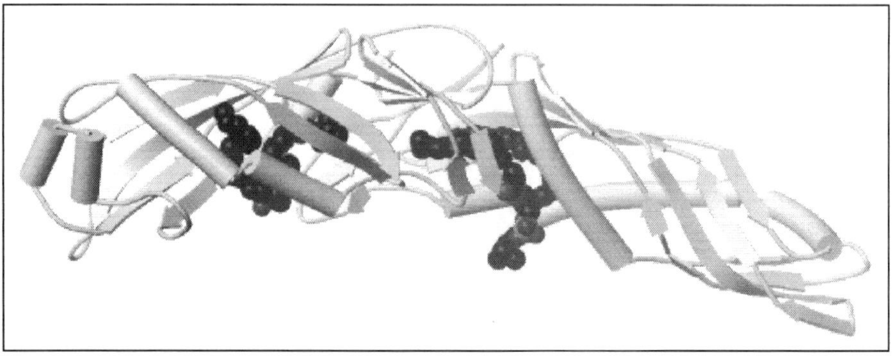

Figure 12. Ribbon diagram of the BPI structure (PDB code 1EWF). The two bound phosphatidylcholine molecules seen in the BPI crystal structure are shown in CPK representation. BPI and LBP share high sequence identity, which suggests that LBP adopts a similar overall architecture and ligand-binding mode.

products.[59] These proteins include the cytosolic LRR-proteins Nod1 and Nod2, and confer responsiveness to LPS in the absence of TLR4. The following section discusses only LBP, CD14, MD2 and the TLR4 ectodomain.

LPS-Binding Protein (LBP)

LBP is essential for the rapid induction of an inflammatory response to LPS or Gram-negative bacteria as LBP deficient mice lack such responses.[60] LBP is a 60-kDa serum glycoprotein that interacts with LPS micelles and delivers LPS monomers to other host proteins, such as CD14. Extracellular LBP and intracellular bactericidal/permeability-increasing protein (BPI) display a high sequence identity (about 45%), suggesting a similar fold. The crystal structure of BPI is known (Fig. 12)[61,62] and reveals that the protein adopts a "boomerang" shape formed by two similar domains. Two apolar binding pockets occupied by lipid molecules were found in the structure. LPS binding is thought to occur in a similar way in LBP, with the LPS acyl chains buried in the core of the protein and the polysaccharide part of the molecule solvent exposed. Biochemical evidence supports a stoichiometry of one or two LPS molecules for each LBP, bound with nanomolar affinity.[63]

CD14

Years before the discovery of TLR4,[64,65] CD14 was known to be a LPS receptor.[66] Direct binding data indicates that each CD14 molecule binds one or two molecules of LPS with a nanomolar affinity (as for LBP) and that the binding of LPS to CD14 is facilitated by LBP.[67] CD14 is an extracellular LRR-protein unrelated to the TLR superfamily. Composed of 375 amino acids, CD14 possesses a signal peptide, several heterogeneously occupied glycosylation sites, and a carboxyl-terminal signal sequence for glycosyl phosphatidyl inositol (GPI) anchorage. Both membrane-bound (mCD14) and soluble (sCD14) forms are found, but the function of sCD14 is not fully understood. A soluble form could compensate for the lack for mCD14 in CD14-negative cells which is the case for endothelial and epithelial cells.[68-71] On the other hand, sCD14 will compete with mCD14, as shown by reports in which LPS stimulation of mCD14 is inhibited by sCD14.[71,72]

Mutagenesis data suggests that the LPS binding site is located at the amino-terminal region of CD14. The first 152 amino acids contains the LPS transducing capacity for a variety of different cell types.[73] Limited proteolysis experiments using sCD14 alone and complexed with LPS, revealed a region protected from proteolysis by LPS binding.[74] This region includes amino acids 57 to 64 (with the following amino acid sequence : DADPRQYA) which precedes the consensus LRR region in CD14. Interestingly, LPS binding was not accompanied by any major structural rearrangement as measured by circular dichroism (CD) spectroscopy. However it is not clear how CD14 interacts with the acyl chains of its ligands, as the binding site is located in a region that cannot be modeled on any protein with a known three-dimensional structure.

MD-2

MD-2 is a secreted glycoprotein of 160 amino acids, harboring a signal peptide and two N-linked glycosylation sites. Shimazu and his collaborators showed that MD-2 is necessary for LPS signaling through TLR4.[75] Indeed a point mutation in a conserved region of MD-2, changing a cysteine residue at position 95 into a tyrosine, abolishes LPS-induced signaling.[76] MD-2 not only interacts with TLR4, but also directly binds LPS.[2,77] The binding is in the nanomolar range and does not require the presence of either LBP or CD14.

The connectivity of the 7 cysteine residues found in MD-2 is not clear, though the odd number of cysteines raises the possibility of at least one inter-molecular and up to three intra-molecular disulfide bridges. At present the biochemical data are puzzling and suggest an array of potential intra- and intermolecular disulfide bridges.[78-80] These disulfide bridges may explain the ability of MD-2 to polymerize. Polymerization through the formation of intermolecular disulfide bridges shows that the cysteine residues are solvent exposed and reactive. Further structural insights for MD-2 may be gained by sequence analysis. It is also unclear as to which molecular form of MD-2 is active. Re and Strominger[78] showed that monomeric MD-2 bound TLR4 and mediated LPS signaling more efficiently than MD-2 multimers. Conversely both Visintin and Mullen reported that MD-2 formed multimers based on stable dimeric subunits that could enhance the responsiveness of TLR4 reporter cells to LPS, regardless of the presence of monomers.[79,80] MD-2 is the founding member of a new family called ML, which stands for MD-2-related lipid recognition.[81] Members of this family are characterized by a single domain architecture, belonging to the all-beta class, and display an immunoglobulin-like β-sandwich fold. 3D structures of three members of the ML family, Der f2, Der p2 and GM2-AP (PBD codes 1AHM, 1KTJ and 1G13, respectively) have been solved. They all form a β-sandwich of at least 7 strands grouped in 2 β-sheets and display a greek key motif (Fig. 13).

A mechanism for ligand binding can be extrapolated from the information provided by the 3D structures of ML proteins. Indeed a putative lipid-binding cavity was found in the crystal structure of mite allergen protein Der p2, the closest structural homolog of MD-2. This cavity, located in the core of Der p2, surrounded by β-strands, is filled with two distinct elongated fragments of high electron density, each able to accommodate an aliphatic chain of 14-16 carbon atoms (Fig. 13). It could thus be possible that MD-2 is involved in binding LPS by burying the acyl chains in its core. Further experiments are required to confirm this hypothesis. Using lipid IVa, a lipidA analog, which is an agonist for mouse TLR4 and an antagonist of human TLR4, Akashi and coworkers showed that MD-2 influenced the specificity of TLR4.[82] Another member of the ML family is the human ganglioside activator protein GM2-AP, which displays a cavity suitable for binding 18-carbon lipid acyl chains.[83] Together these results suggest that MD-2 is necessary for ligand binding.

TLR4

TLR4 is concerned with the recognition of evolving pathogens. It is thus not surprising that its ectodomain has evolved to optimally fulfill this function. The divergence is obvious in TLR4 paralogs (i.e., the different members of the TLR family). More importantly, variability is also observed in TLR4 orthologs (i.e., TLR4 that evolved by speciation). Sequence identity is about 40% between human, dog and chicken TLR4 ectodomains, which is surprisingly low compared to the 62% identity between human and mouse. A tempting explanation is that

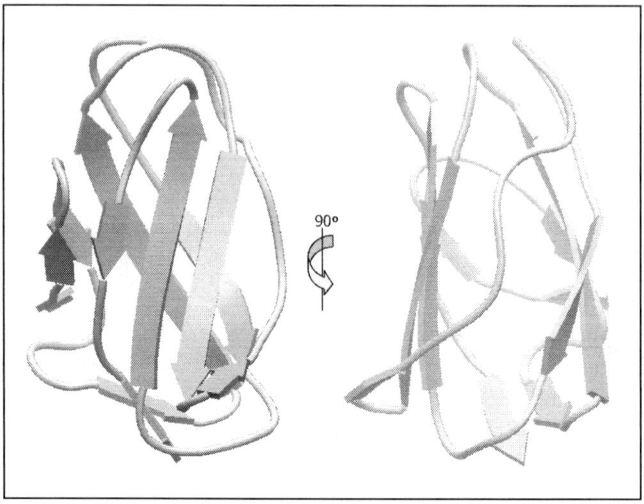

Figure 13. A central cavity was found in the crystal structure of Der p2 (PDB code 1KTJ), the closest structural homolog of MD-2.

different species come across different pathogens, necessitating the generation of divergent TLR structures. A hypervariable region has been identified and this appears to correlate with ligand-binding and signaling functions.[84]

A naturally occurring human mutation has been identified that is associated with endotoxin hyporesponsiveness.[85] This mutation changes an aspartic acid to a glycine at position 299, which is located in LRR9 (as described in the NCBI database, with the accession number O00206). Whether this mutation disrupts a region crucial for protein folding or function, i.e., ligand binding, dimerization or signal transduction, is not yet clear. Biophysical considerations would favor the former hypothesis, as replacement of the aspartic acid side chain with glycine would disrupt an α-helical structure. Position 299 is located at the convex side of the LRR, which is highly variable in secondary structure content and could accommodate a helix. Arguing against the disruption of ligand binding in the mutant are data showing that ligand binding sites are generally found on the concave side of LRR-containing proteins. At present we cannot accuratedly discribe the structural basis of the LPS hypresponsive phenotype.

Mechanism of Signal Transduction

There are several potential ways for TLRs to transduce signals: ligand binding might trigger a conformational change or alter the oligomeric state of the receptor. In this model TLRs are sufficient for signaling. Alternately, signal transduction might depend on the assembly of a multi-receptor complex in which several components besides TLRs contribute, as for the LPS complex.

TLRs are thought to form homo- or heterodimers upon activation. This notion is supported by experimental evidence using chimeric receptors, in which the extracellular region of TLRs are substitued by CD4 (CD4-TLR constructs) which promotes homodimerization of the molecule.[86] In addition chimeras of the extracellular domain of *Drosophila* Toll and the TIR of human TLR4 can activate NFκB in response to Spätzle C-106.[21] Chimeras composed of the extracellular domain of CD4 fused with the transmembrane and intracellular region of TLR4 are constitutively active.[65,87] Ozinsky and collaborators[16] found that constitutively active

CD4-TLR2 required the co-expression of either CD4-TLR1 or CD4-TLR6 chimeras. Heterodimerization as a functional mode for TLR2 was further supported by coimmunoprecipitation experiments. The interaction between TLRs 2 and 6 was shown to occur between extracellular domains, without any need for the cytoplasmic sequence. Moreover, the association between TLR2 and TLR6 was suggested to be ligand-independent but was augmented by the addition of ligand.

On a structural level, there are, as yet, no data to shed light on how TLR dimerization is achieved. A head-to-tail and side-by side configuration has been observed in a number of LRR protein/protein interactions (Fig. 14) though such a configuration is unlikely to occur in type I transmembrane TLRs as the cytoplasmic domains would be too far apart. Thus, a head-to-head orientation seems more plausible, which leaves several possibilities for the interface arrangement. The interface could occur to the side, at either the convex or concave face. However, as TLRs form dimers, the head-to-head and side-by-side orientation should not be considered as such an arrangement would lead to protein oligomerization. Clearly then, further studies will be required to distinguish between the convex and concave interfaces.

TLRs have generally been modeled like cytokine receptors in which ligand-induced dimerization is sufficient to trigger intracellular signaling. This comparison has been made because TLR signal transduction is almost identical to IL-1R signaling. The IL-1R and the IL-1R accessory protein dimerize to form a high-affinity binding site for IL-1, an event that triggers the association with adaptor molecules that will be discussed in the following section.

Components of the Intracellular Pathway

Introduction

The ability of Toll-like receptors to specifically recognize a broad range of antigens via their diverse extra-cellular domains is crucial for mounting a specific immune response. To ensure that this response is appropriate, the signal that is propagated intracellularly must be regulated such that it drives expression of the appropriate inflammatory cytokines and immune co-stimulatory molecules.

There is now evidence to suggest that extra-cellular receptor diversity is mirrored intra-cellularly by a myriad of effector molecules that can act singularly or in combination to generate highly specific signals. For example, there are now at least five adaptor molecules that have been reported in Toll-like receptor pathways, Myd88, Mal/TIRAP, TRAM, SARM and TICAM-1/TIRF,[88-92] with some displaying striking receptor selectivity.

The intracellular signaling system contains some of the most conserved features of Toll pathways, with components shared between the different Toll/IL-1 signaling pathways across a broad range of species. The mechanism of signal transduction from the extra-cellular domains of Toll-like receptors to post-receptor complexes is poorly understood. What can be stated is that protein-protein interactions, mediated by a subset of conserved adaptors and coupled to phospho-transfer events, are key to the activation and regulation of this pathway. The Toll/Interleukin-1 receptor domain, the death domain and the catalytic kinase domain are some of the most important structural facets of these classes of mediators.

Toll/Interleukin-1 Receptor (TIR) Domains

TIR domains form the intracellular trans-activation domains for all members of the Toll/IL-1 receptor superfamily. The 200 amino acid domain was first identified in the human IL-1 and *Drosophila* Toll receptors [93] and has since been reported in a broad range of plant, *Xenopus*, insect, and mammalian proteins where it is involved in mediating immune responses and developmental pathways (Fig. 15).

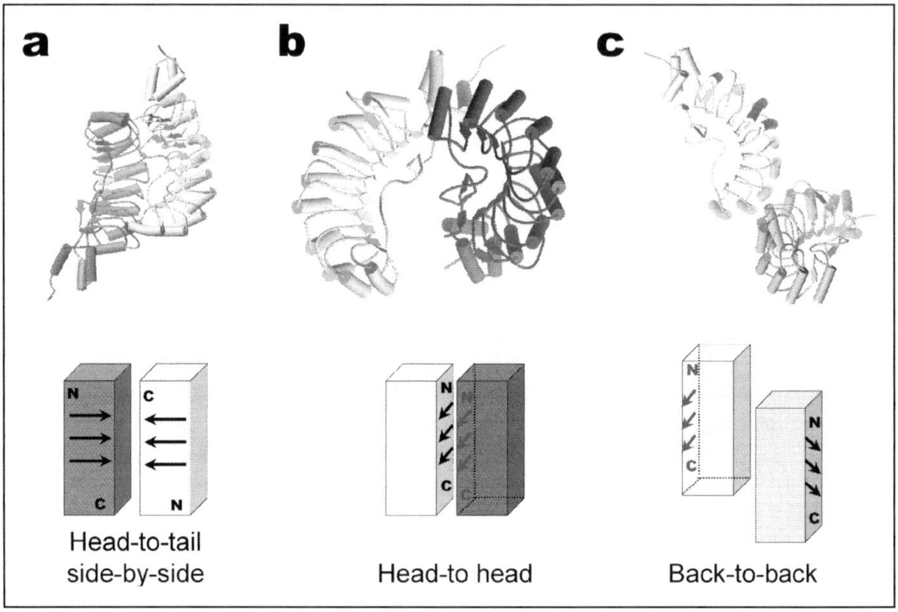

Figure 14. Different LRR-protein interfaces. In the Skip2 crystal packing (PDB code 1FQV), different interfaces are involved suggesting different modes of dimerization for LRR-proteins. In a) a head-to-head arrangement is observed and the interface is formed by the flanking loops on either sides of the LRR motif. In contrast, in b and c, both configurations are head-to head arrangements. In b) the concave face is involved in protein-protein contacts, with the C-terminal tails of Skip2 occupying the central space usually defined as the ligand-binding site in LRR proteins. In c) the interface is on the convex « back» side of the horseshoe structure.

Interestingly, TIR domains also appear to be present in bacteria and viruses where they may act as decoys to allow evasion of the host defense.[94,95] This discovery could have some important evolutionary implications since it suggests that these domains may have arisen before the divergence of eukaryotes and prokaryotes. It is conceivable that TIR domains could have existed as a general protein-protein interaction domain that subsequently acquired specialized functions in development and later, in immunity.

TIR Domain Classification

The TIR domain superfamily can be classified into three distinct sub-groups according to their associated functional domains. The first group comprises TIR domains that are linked to three extra-cellular immunoglobulin-like domains. Members of this group are mainly cytokine-activated members of the IL-1 receptor superfamily. The activation mechanism involves the formation of a high affinity complex between the receptor and an accessory molecule, the IL-1 receptor accessory protein, that allows the recruitment of post-receptor molecules through the TIR domains.[96]

The second subgroup is composed of TIR domains linked to extra and intracellular leucine rich repeats. The Toll receptor family and plant defense R proteins belong to this group. The receptor activation mechanism in Toll receptors is believed to involve the formation of homo and heterodimeric receptor complexes that bring TIR domains into close proximity, followed by the recruitment of post-receptor complexes by an unknown mechanism.

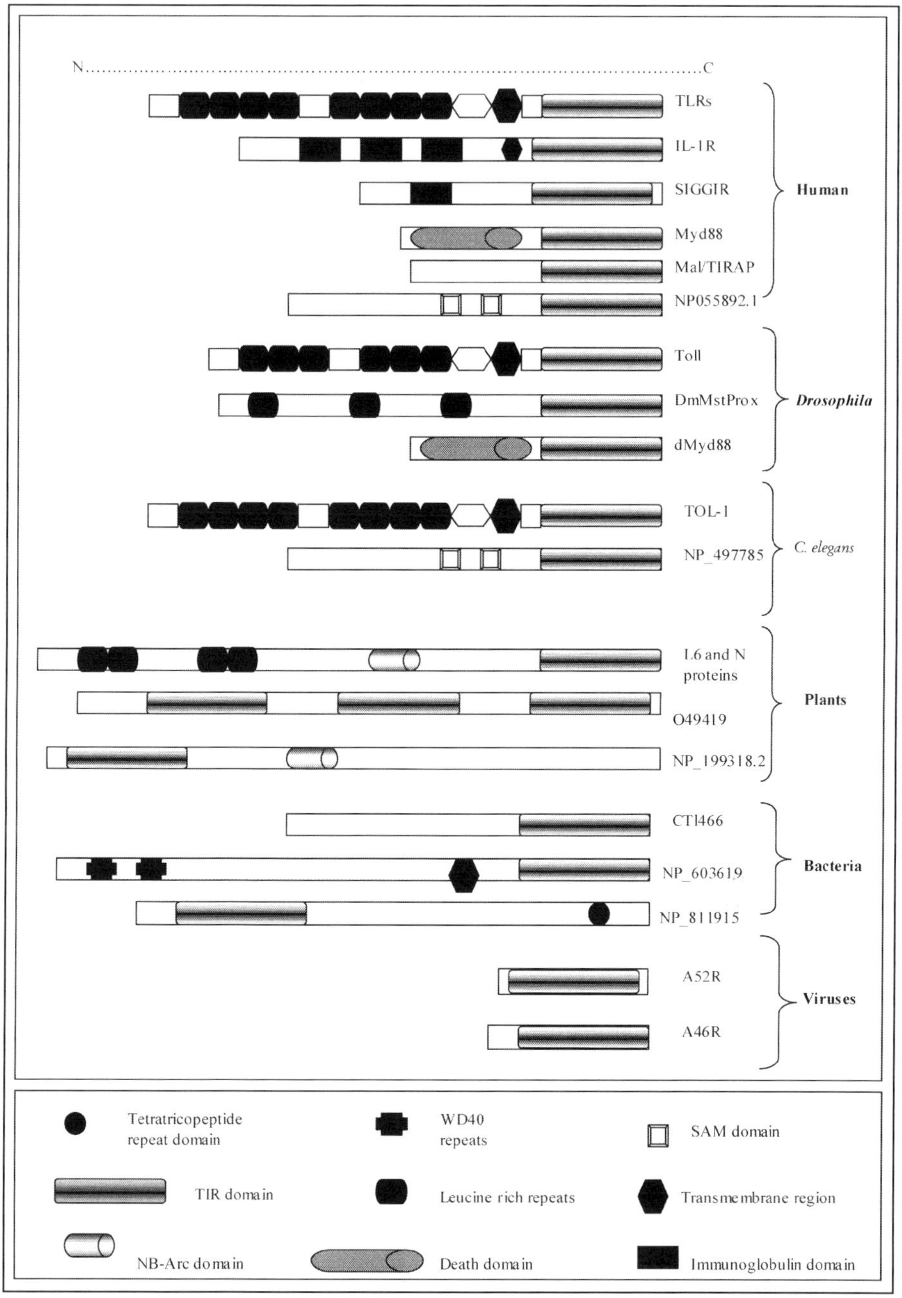

Figure 15. Domain organization in representative TIR domain-containing proteins in humans, *Drosophila*, *C. elegans* and plants.

The third sub-group is comprised principally of cytosolic adaptor molecules with solo TIRs or TIRs linked to other functional domains. Their function appears to involve mediating protein-protein interactions with mounting evidence to suggest a role in signal regulation through phosphorylation. This subgroup is the largest and is especially well represented in *Arabidopsis.*

TIR Primary Structure

TIR domains are characterized by three conserved sequence boxes, designated Box 1, 2 and 3, spread across the length of the module (Fig. 16). Outside these boxes TIR domains are divergent with sequence similarities in the range of 20-30 percent amongst superfamily members.

Box 1, the most conserved of the three, has the signature sequence F/YDAF-Y, mutation of which leads to disruption but not total abrogation of signal transduction.

Box 2 has the consensus sequence G-LC-RD-PG. Mutational studies have mapped several residues that are crucial for receptor signaling to this region.[97] Substitution of the conserved proline to histidine in murine hTLR4 renders mice insensitive to LPS.[98] A similar mutation in TLR2 abrogates signal transduction in response to yeast and gram-positive bacteria. Swapping the conserved arginine with an alanine in the human IL-1 receptor also abolishes Toll/IL-1 signaling.[99]

Box 3 is characterized by the conserved FW motif. This region is poorly conserved in some of the downstream adaptors suggesting that it could play an important role in specific signal routing. Mutations in box 3 have, thus far, only been shown to affect the subcellular localization of the IL-1 receptor.[100]

TIR Tertiary Structure

Our current understanding of the TIR domain fold has been accrued from studies of the crystal structures of the TLR1 and TLR2 proteins (Fig. 17).[101] These structures suggest that the region spanned by box 1 to 3 forms a conserved core of 135 to 160 amino acids. This core adopts an overall β/α fold composed of a central 5-stranded parallel β-sheet surrounded by 5 α-helices.

A similar fold is seen in CheY, a bacterial chemotaxis protein. CheY protein consists of a single regulatory domain that is activated by a conformational change induced by a phospho-transfer reaction from a histidine kinase to a conserved aspartyl residue. Structural superposition of CheY, TLR1 and TLR2 shows remarkable topological and tertiary structure conservation, especially over the central beta sheet region. Similar signal response regulators with a β/α fold have been identified across species suggesting that this could be an evolutionarily conserved signal regulatory conformation.

Whether this structural similarity has any functional significance for Toll/IL-1 receptor signaling is unclear. There is however, preliminary evidence, to suggest that some TIR domain containing adaptors could be phosphorylated in vivo. Phosphorylation of CheY regulates its association with other complexes in the pathway, e.g., the unphosphorylated form has up to 6-fold higher affinity than the phosphorylated form. It is possible therefore that a similar phosphorylation event in TIR signaling could trigger a conformational change that regulates the interaction of the TIR domain with the Toll-like receptor signalosome.

Residues of the conserved TIR box mostly form the core of the molecule though some are solvent exposed on one face of the protomer. Box 2 forms part of a protruding loop, termed the BB loop. This protruding surface is highly conserved between the two known structures and has been proposed to be the interaction surface for receptor dimerization.[102]

Analysis of the electrostatic surfaces of the BB loop of TLR1 and TLR2 (Fig. 18) suggests some charge conservation but also striking variation that could be crucial for determining specificity with different adaptor molecules.

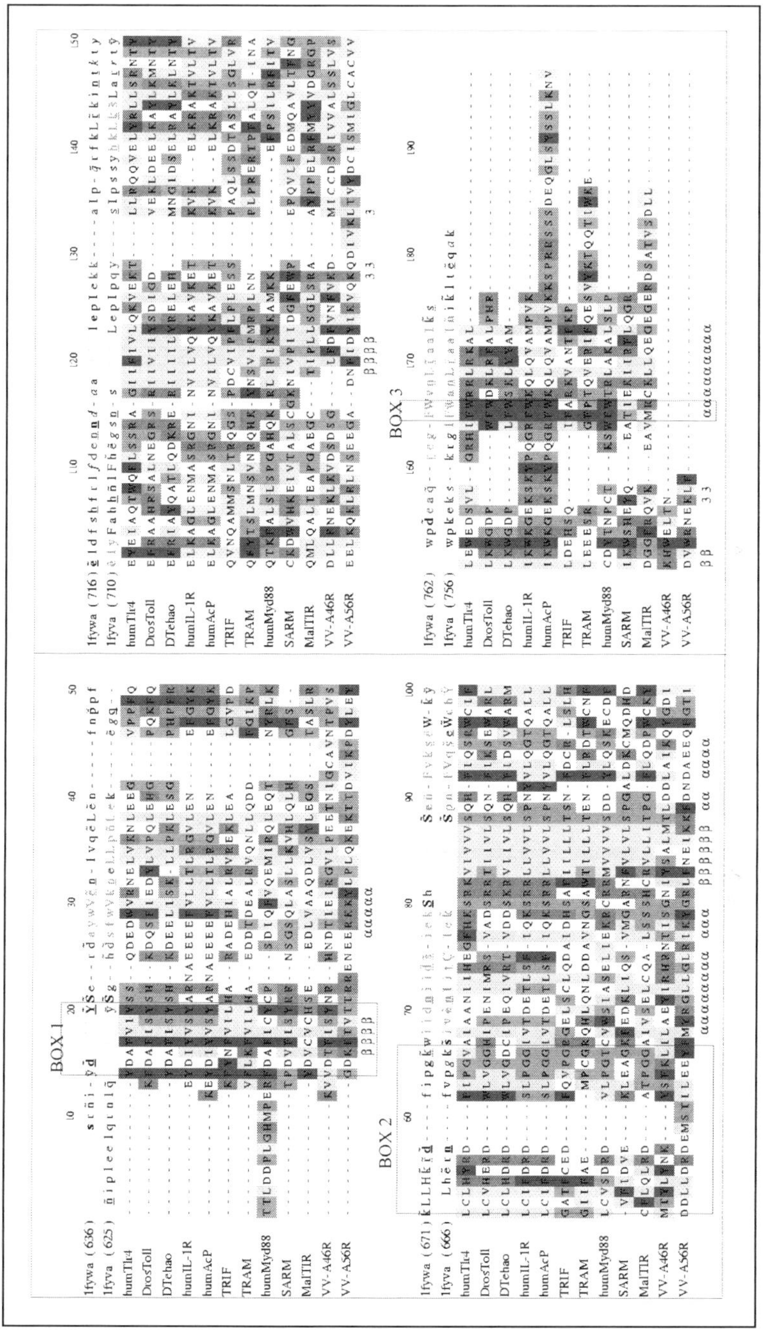

Figure 16. Structure based sequence alignment of representative TIR domain sequences, to the crystal structures of TLR1 and TLR2. The figure was generated with JOY[154]. The key to Joy annotation: an alpha helix is shown in medium gray, a beta strand in dark gray, a 3_{10} helix in very dark gray, solvent accessible residues in lower case, solvent inaccessible in upper case, in bold for hydrogen bond to main-chain amide, underlined for hydrogen bond to mainchain carbonyl, disulfide bond suggested with a cedilla, and a residue with positive phi torsion angle is written in italics.

A recent structure of a TLR2 disulphide linked dimer suggests that the BB loop might adopt different conformations in the bound or unbound form, to allow optimal association.[102] In the monomeric form and in one form found in the dimeric structure, but not involved in the dimerization interface, the BB loop protrudes away from the rest of the domain (conformation

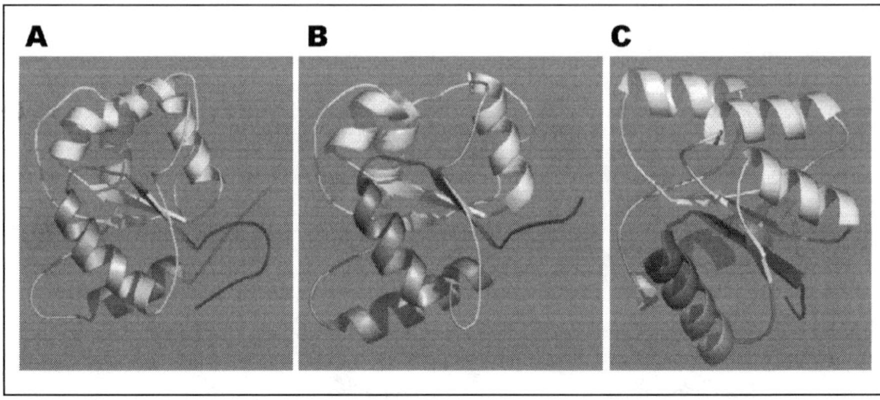

Figure 17. TLR1 (A) and TLR2 (B) TIR domain structures. These adopt an α/β fold similar to bacterial CheY (C) regulatory domain. TIR domains are shown with the BB loop facing the viewer.

B). However, in the BB loop associated dimer, the loop is located closer to the TIR domain (conformation A).

The TLR1 and TLR2 structures suggest that box 2 residues may make important contacts with the conserved signature residues in box 1. Analysis of the proline to histidine TLR2 mutant shows no major structural changes that could explain its insensitivity to LPS stimulation. This would suggest that the observed effect could be due to the disruption of key interactions between the receptor and downstream molecules rather than conformational rearrangements over the BB loop region.

TIR domain interactions are clearly important for regulating the Toll/IL-1 signal transduction pathway. How the specificity of adaptor TIR domains for certain receptor domains is determined seems to lie in the structural variability of the TIR fold. Although the two known TIR domain structures have a sequence similarity of 50 percent they display some striking structural variability. Since the sequence similarity between members of the TIR domain superfamily is usually in the order of 20 to 30 percent, it is conceivable that we will find even more significant structural differences between different TIR domains.

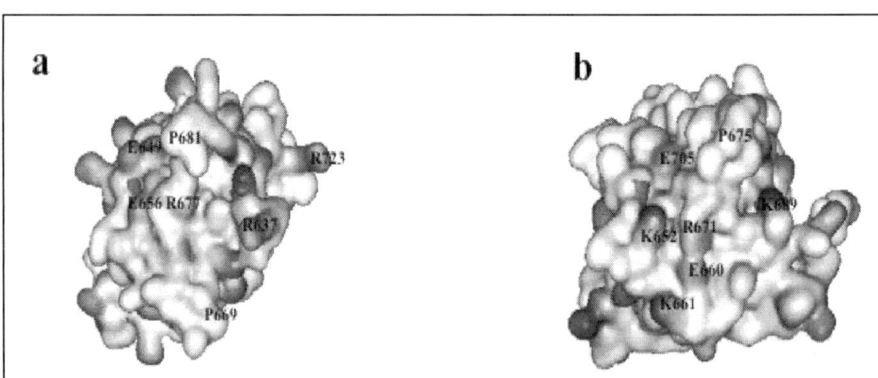

Figure 18. Electrostatic surface representation of the TIR domain S face, a) TLR1, b) TLR2. Used with permission from Dunne A, Ejdebäck M, Ludidi P et al. Structural complementarity of Toll/Interleukin-1 receptor domains in toll-like receptors and the adaptors Mal and MyD88. J Biol Chem 2003; 278(42):41443-41451.

TIR Domains form Multitypic Complexes

As noted above, TIR domain proteins fold into a conserved structure but the surface characteristics suggest that they may be able to assemble into multitypic complexes. Recent work in our laboratory[155] has shown that the Mal and MyD88 adaptors interact with different regions of the TIR domains of TLRs 2 and 4, and that these interactions do not involve the BB loop (Fig. 19). In addition the two adaptor TIRs were shown to form heterodimers and this binding is predicted to involve areas of the molecules that do not overlap with the receptor binding sites. These studies suggest an activation mechanism that involves dimerization dependent assembly of a post-receptor complex.

Death Domains

The multitypic receptor complexes nucleated by the TIRs are coupled to downstream effectors by a second protein-protein interaction motif, the death domain; so called because it is also found in signaling pathways that lead to apoptosis.[103] The adaptors MyD88 and Tube contain death domains (see Fig. 15) as do the IRAK and Pelle kinases (Fig. 20).[104] The Tube and Pelle death domains associate to form stable heterodimers that are requisite for signaling.[105,106] Recruitment of the protein kinase into the post-receptor complex is accompanied by autophosphorylation.[107] The death domains themselves consist of about 120 amino acids and are characterized by a conserved six α-helical bundle fold.[103,108] They typically associate via homotypic interactions. Crystal structures of the Tube/Pelle death domain heterodimer (Fig. 21)[109] and, more recently, a solution structure of the isolated Pelle death domain, have been solved (Moncrieffe et al, submitted). These studies reveal that the binding interface between Tube and Pelle is extensive, and involves at least 27 of the residues in the fold. Helix-4 of Pelle interacts with a groove formed by helices 1, 2 and 6 of Tube and additionally, the C-terminal tail of Tube interacts with a groove on Pelle created by helices 2,3,4 and 5. In addition to this stable interface, solution studies suggest that a second, unstable association occurs. These additional weak interactions may contribute to the stability of the post-receptor activation complex. Equivalent structural studies are not available for the vertebrate MyD88/IRAK death domains although biochemical evidence suggests that signaling is mechanistically similar.[110]

Kinase Domains in Toll-Like Receptor Signaling

Protein kinases play a crucial role in the activation and regulation of Toll-like receptor signaling systems. Serine/threonine are often involved in biochemical processes like cell growth and development. In general serine/threonine kinases are involved with cytosolic signaling whilst tyrosine kinases tend to be associated with cell surface receptor molecules.

Figure 20 shows some of the protein kinases implicated in Toll receptor pathways. The IL-1 receptor associated kinase (IRAK) family of serine/threonine kinases are recruited to the receptor complex following activation. On activation by association with adaptor molecules downstream of TLRs they induce a signaling cascade that culminates in the activation of IκB kinase and NFκB mediated transcription.

As noted above, IRAKs are multi-domain proteins composed of a conserved N-terminal death domain linked to a typical serine/threonine kinase domain. Five have been described so far in mammals and insects, IRAK-1, IRAK-2, IRAK-M, IRAK-4 and Pelle.[111] Overall they adopt a similar domain organization except for a unique C-terminal region that is absent in IRAK-4 and Pelle. Current evidence suggests that this region could be important in mediating TRAF6 interactions through the conserved QxPxE motif.

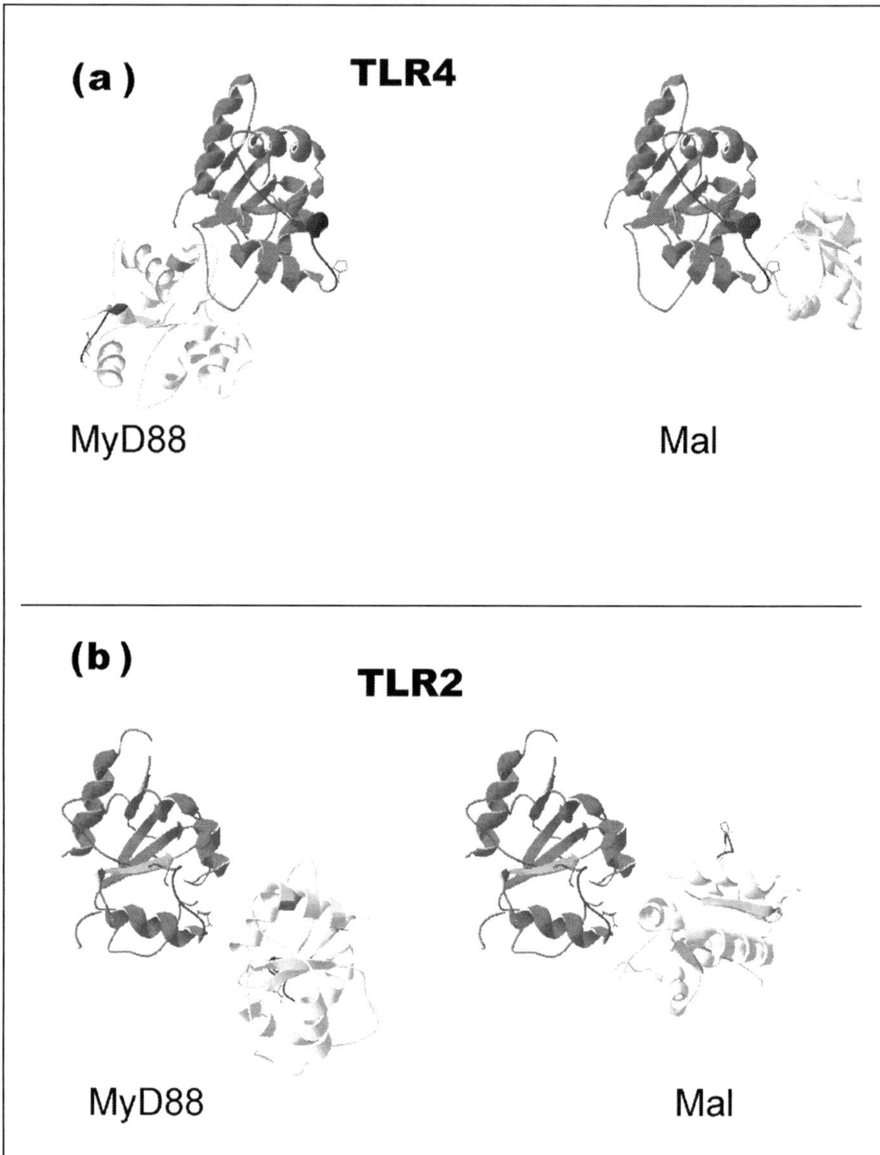

Figure 19. Model of the docking complexes formed between the TIR domains of hTLR2 and hTLR4 and the TIR domains of Mal and MyD88. Receptor TIR domains are shown in dark gray and adapter TIR domains in light gray. Structural features representing the conserved boxes of the TIR domains are shown in dark gray (box 1; box 2, BB loop; box 3, not present in Mal). The side chain of the semi-conserved proline residue in the BB loop is colored purple. Top panel: GRAMM docking of the MyD88 (left) and Mal TIR (right) domains to the TIR domain of hTLR4; bottom panel: docking of MyD88 (left) and Mal (right) TIR domains to the TIR domain of hTLR2. See ref. 155 for color reproduction.

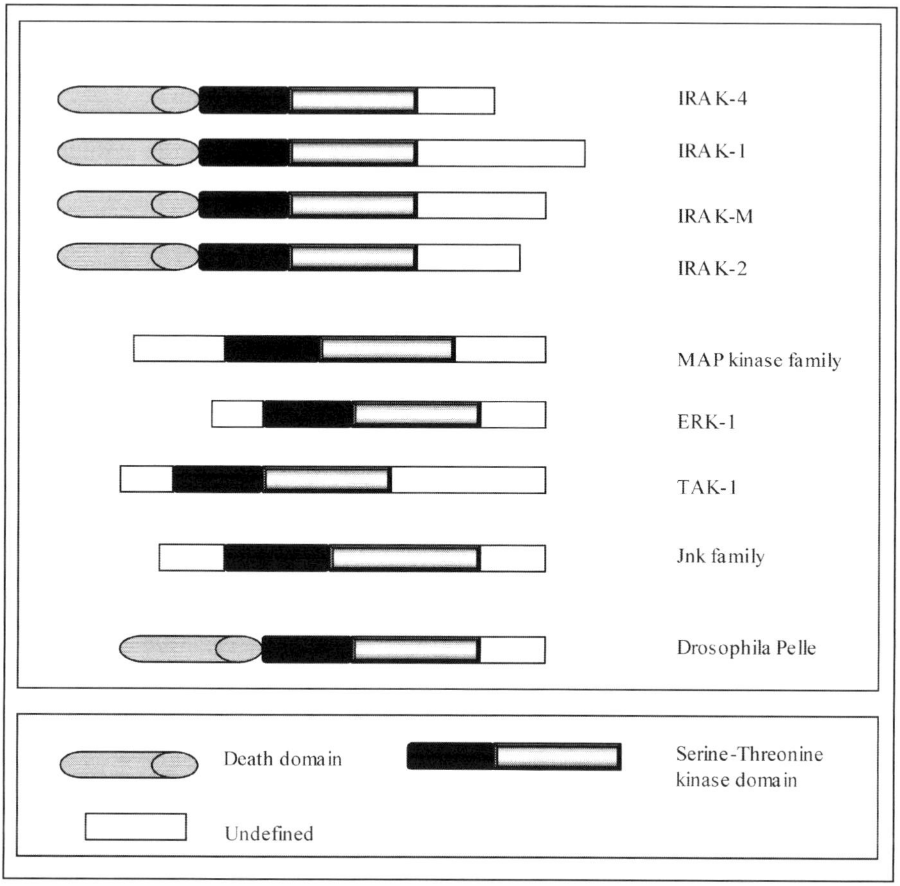

Figure 20. Representative kinase domain containing proteins in Toll-like receptor signaling.

Primary and Tertiary Structure Conservation in Protein Kinases

Shown in Figure 22 is a structure-based alignment of all known members of the IRAK superfamily to the cyclic AMP dependent kinase crystal structure. This alignment allows us to divide the kinase domain into 11 subdomains that are highly conserved in all kinases (labelled 1-11 in Fig. 22).

Since all protein kinases catalyze the same basic reaction, their tertiary structures are highly conserved. The kinase fold is composed of two lobes linked by a short polypeptide chain (Fig. 23). The N-terminal lobe (A) is made up of five anti-parallel β-strands and a single α-helix, called helix C. β strands 1 and 2 encompass the conserved signature phosphate anchor motif, GXGXY/FG. This phosphate anchor subdomain is conserved across all members of the IRAK family suggesting that they are all capable of binding ATP. The only anomaly lies with the conserved phenyalanine that usually caps the phosphate transfer site, which is replaced by an isoleucine in IRAK-M. Distortions in this region of the N-terminal lobe have been associated with inactive kinases. β strand 3 in subdomain 2 contains an invariant lysine that binds oxygens

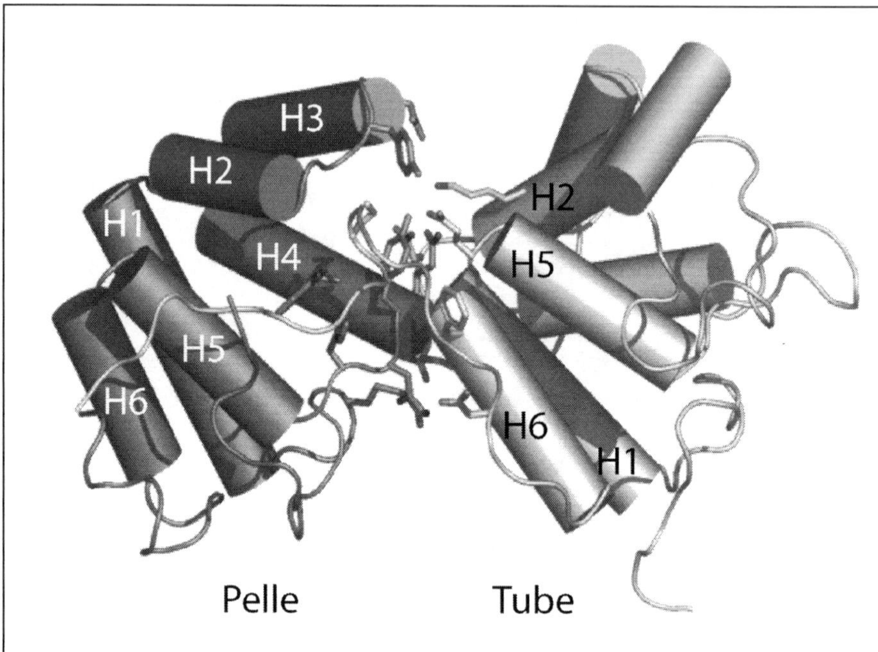

Figure 21. Crystal structure of the P2:T2 interface in the Tube/Pelle death domain complex.[109] Reprinted from Xiao T, Towb P, Wasserman SA et al. Three-dimensional structure of a complex between the death domains of Pell and Tube. Cell 1999; 99(5):545-555. ©2004, with permission from Elsevier.

in the β and γ phosphates of the ATP molecule. This lysine forms a stabilizing salt bridge with an invariant glutamic acid in helix C, thus orienting it properly for catalysis. Distortions that disrupt this salt bridge are also linked with kinase inactivity. The two lobes are connected by a flexible linker that stretches between β strand 5 and helix D (subdomain 5 in Fig. 22). The ATP molecule sits in a cleft between the two lobes and hydrogen bonds with linker residues.

The larger substrate binding C terminal lobe (B) is predominantly α-helical. The catalytic loop on subdomain 6b connects β strand 6 and 7. In most kinase structures the catalytic loop has a conserved tyrosine or histidine in position 1 and an arginine residue in position 2 that hydrogen bonds with conserved residues in subdomains 8 and 9. The arginine residue is conserved only in IRAK-4. Also found in the catalytic loop of IRAKs are the conserved catalytic aspartate and the metal binding asparagines, set five residues apart. Interestingly, IRAK-2 and IRAK-M have asparagine and serine residues respectively instead of the catalytic aspartate. Moreover, IRAK-M has a serine residue instead of the conserved ATP binding lysine ordinarily found next to the catalytic aspartate of serine/threonine kinases. These differences could provide an explanation for why IRAK-M and IRAK-2 are inactive kinases.

The activation loop between subdomains 7 and 8 contains a DFG motif that is conserved in all IRAKs save IRAK-2, where it is replaced by HPM. The aspartate in DFG forms part of the metal binding site. Also found in this region are the threonine and serine residues whose phosphorylation, in many protein kinases, induces a conformational change that results in full enzyme activation. For example, phosphorylation of S376 and T387 in the activation loop in IRAK-1, by itself or by IRAK-4, results in its activation and potential phosphorylation of IRAK-2 and IRAK-M.[112]

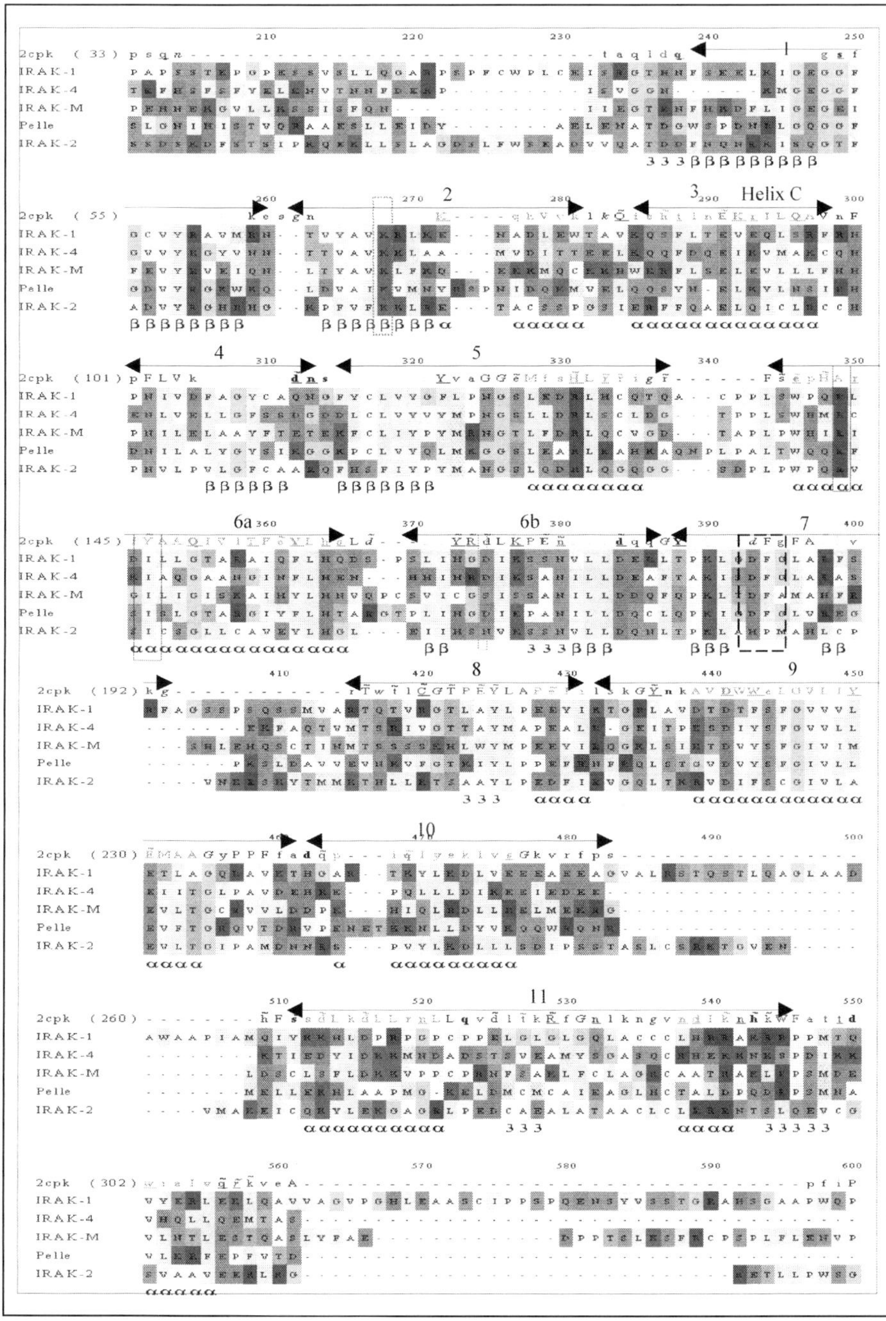

Figure 22. Structure based sequence alignment of IRAK family members with cyclic AMP dependent kinase.

Figure 23. Crystal structure of cyclic AMP-dependent protein kinase (cAPK) PDB: 2CPK

Conclusions

The last few years have seen significant advances in many areas of Toll receptor biology, genetics and biochemistry. At the same time we have only a rudimentary understanding of the structural basis for Toll signaling. Experimental structures have been determined only for some isolated components in the pathways, notably the TIR domains from TLR1 and 2 and the death domains of *Drosophila* Tube and Pelle.[101,109] This relative lack of success reflects the formidable difficulty involved in preparing the receptors and components of the pathway in a form that is suitable for high-resolution structural studies by X-ray crystallography and NMR. However advances in expression systems and protocols, protein purification methods and automated approaches to protein crystallization offer the prospect of significant progress. In particular, structural analysis of the protein complexes involved in activation of the receptors by pathogen associated molecular patterns and the nature of the post-receptor complexes will provide a powerful insight into the regulatory mechanisms of these pathways. This information will also facilitate structure-aided design of small molecule and protein therapeutics that can modulate TLR function in disease.

Developments in bioinformatics are also having an impact on these problems. Computational techniques for homology modeling now incorporate improved techniques for predicting loop conformations.[113] Programs for rigid-body docking of protein structures have also improved and now allow reliable prediction of heterotypic interactions by signaling adaptors.[114]

References

1. Akira S, Hemmi H. Recognition of pathogen-associated molecular patterns by TLR family. Immunol Lett 2003; 85(2):85-95.
2. Correia JD, Ulevitch RJ. MD-2 and TLR4 N-linked glycosylations are important for a functional lipopolysaccharide receptor. J Biol Chem 2002; 277(3):1845-1854.
3. He XLL, Bazan JF, McDermott G et al. Structure of the Nogo receptor ectodomain: A recognition module implicated in myelin inhibition. Neuron 2003; 38(2):177-185.

4. Evdokimov AG, Anderson DE, Routzahn KM et al. Unusual molecular architecture of the Yersinia pestis cytotoxin YopM: A leucine-rich repeat protein with the shortest repeating unit. J Mol Biol 2001; 312(4):807-821.
5. Price SR, Evens PR, Nagai K. Crystal structure of the spliceosomal U2B ''-U2A' protein complex bound to a fragment of U2 small nuclear RNA. Nature 1998; 394(6694):645-650.
6. Uff S, Clemetson JM, Harrison T et al. Crystal structure of the platelet glycoprotein Ib alpha N-terminal domain reveals an unmasking mechanism for receptor activation. J Biol Chem 2002; 277(38):35657-35663.
7. Kobe B, Deisenhofer J. A Structural Basis of the Interactions between Leucine-Rich Repeats and Protein Ligands. Nature 1995; 374(6518):183-186.
8. Huizinga EG, Tsuji S, Romijn RAP et al. Structures of glycoprotein Ib alpha and its complex with von Willebrand factor A1 domain. Science 2002; 297(5584):1176-1179.
9. Schubert WD, Urbanke C, Ziehm T et al. Structure of internalin, a major invasion protein of Listeria monocytogenes, in complex with its human receptor E-cadherin. Cell 2002; 111(6):825-836.
10. Yoder MD, Jurnak F. The Refined 3-Dimensional Structure of Pectate Lyase-C from Erwinia-Chrysanthemi at 2.2-Angstrom Resolution-Implications for an Enzymatic Mechanism. Plant Physiol 1995; 107(2):349-364.
11. Jones DT. Protein secondary structure prediction based on position-specific scoring matrices. J Mol Biol 1999; 292(2):195-202.
12. Buchanan SGS, Gay NJ. Structural and functional diversity in the leucine rich repeat family of proteins. Prog Biophys Mol Bio 1996; 65(1-2):1-44.
13. Kajava AV. Structural diversity of leucine-rich repeat proteins. J Mol Biol 1998; 277(3):519-527.
14. Kobe B, Deisenhofer J. The Leucine-Rich Repeat-a Versatile Binding Motif. Trends Biochem Sci 1994; 19(10):415-421.
15. Mitsuzawa H, Wada I, Sano H et al. Extracellular toll-like receptor 2 region containing Ser(40)-Ile(64) but not Cys(30)-Ser(39) is critical for the recognition of Staphylococcus aureus peptidoglycan. J Biol Chem 2001; 276(44):41350-41356.
16. Ozinsky A, Underhill DM, Fontenot JD et al. The repertoire for pattern recognition of pathogens by the innate immune system is defined by cooperation between Toll-like receptors. P Natl Acad Sci USA 2000; 97(25):13766-13771.
17. Schneider DS, Hudson KL, Lin TY et al. Dominant and Recessive Mutations Define Functional Domains of Toll, a Transmembrane Protein Required for Dorsal Ventral Polarity in the Drosophila Embryo. Gene Dev 1991; 5(5):797-807.
18. Burke D, Wilkes D, Blundell TL et al. Fibroblast growth factor receptors: lessons from the genes. Trends Biochem Sci 1998; 23(2):59-62.
19. Morisato D, Anderson KV. The Spätzle Gene Encodes a Component of the Extracellular Signaling Pathway Establishing the Dorsal-Ventral Pattern of the Drosophila Embryo. Cell 1994; 76(4):677-688.
20. DeLotto Y, Smith C, DeLotto R. Multiple isoforms of the Drosophila Spätzle protein are encoded by alternatively spliced maternal mRNAs in the precellular blastoderm embryo. Mol Gen Genet 2001; 264(5):643-652.
21. Weber A, Tauszig-Delamasure S, Hoffmann J et al. Binding of the Drosophila cytokine Spätzle to Toll is direct and establishes signaling. Nat Immunol 2003; 4:794-800.
22. Mizuguchi K, Parker JS, Blundell TL et al. Getting knotted: a model for the structure and activation of Spätzle. Trends Biochem Sci 1998; 23(7):239-242.
23. Bauer S, Kirschning CJ, Hacker H et al. Human TLR9 confers responsiveness to bacterial DNA via species-specific CpG motif recognition. P Natl Acad Sci USA 2001; 98(16):9237-9242.
24. Lien E, Means TK, Heine H et al. Toll-like receptor 4 imparts ligand-specific recognition of bacterial lipopolysaccharide. J Clin Invest 2000; 105(4):497-504.
25. Poltorak A, Ricciardi-Castagnoli P, Citterio S et al. Physical contact between lipopolysaccharide and Toll-like receptor 4 revealed by genetic complementation. P Natl Acad Sci USA 2000; 97(5):2163-2167.
26. Mizel SB, West AP, Hantgan RR. Identification of a sequence in human toll-like receptor 5 required for the binding of Gram-negative flagellin. J Biol Chem 2003; 278(26):23624-23629.

27. Lugtenberg B, Vanalphen L. Molecular Architecture and Functioning of the Outer-Membrane of Escherichia-Coli and Other Gram-Negative Bacteria. Biochim Biophys Acta 1983; 737(1):51-115.
28. Whitfield C. Biosynthesis of Lipopolysaccharide O-Antigens. Trends Microbiol 1995; 3(5):178-185.
29. Heinrichs DE, Yethon JA, Whitfield C. Molecular basis for structural diversity in the core regions of the lipopolysaccharides of Escherichia coli and Salmonella enterica. Mol Microbiol 1998; 30(2):221-232.
30. Raetz CRH, Ulevitch RJ, Wright SD et al. Gram-Negative Endotoxin-an Extraordinary Lipid with Profound Effects on Eukaryotic Signal Transduction. Faseb J 1991; 5(12):2652-2660.
31. Westphal O, Lüberitz O, Bister F. Über die Extraktion von Bakterien mit Phenol/Wasser. Z Naturforsch 1952; 7:148-155.
32. Galanos C, Luderitz O, Rietschel ET et al. Synthetic and Natural Escherichia-Coli Free Lipid-a Express Identical Endotoxic Activities. Eur J Biochem 1985; 148(1):1-5.
33. Morath S, Stadelmaier A, Geyer A et al. Synthetic lipoteichoic acid from Staphylococcus aureus is a potent stimulus of cytokine release. J Exp Med 2002; 195(12):1635-1640.
34. Muhlradt PF, Kiess M, Meyer H et al. Isolation, structure elucidation, and synthesis of a macrophage stimulatory lipopeptide from Mycoplasma fermentans acting at picomolar concentration. J Exp Med 1997; 185(11):1951-1958.
35. Aliprantis AO, Weiss DS, Radolf JD et al. Release of toll-like receptor-2-activating bacterial lipoproteins in Shigella flexneri culture supernatants. Infect Immun 2001; 69(10):6248-6255.
36. Takeuchi O, Kaufmann A, Grote K et al. Preferentially the R-stereoisomer of the mycoplasmal lipopeptide macrophage-activating lipopeptide-2 activates immune cells through a toll-like receptor 2-and MyD88-dependent signaling pathway. J Immunol 2000; 164(2):554-557.
37. Takeuchi O, Kawai T, Muhlradt PF et al. Discrimination of bacterial lipoproteins by Toll-like receptor 6. Int Immunol 2001; 13(7):933-940.
38. Samatey FA, Imada K, Nagashima S et al. Structure of the bacterial flagellar protofilament and implications for a switch for supercoiling. Nature 2001; 410(6826):331-337.
39. Eaves-Pyles TD, Wong HR, Odoms K et al. Salmonella flagellin-dependent proinflammatory responses are localized to the conserved amino and carboxyl regions of the protein. J Immunol 2001; 167(12):7009-7016.
40. Mimori-Kiyosue Y, Yamashita I, Fujiyoshi Y et al. Role of the outermost subdomain of Salmonella flagellin in the filament structure revealed by electron cryomicroscopy. J Mol Biol 1998; 284(2):521-530.
41. He XS, Rivkina M, Stocker BAD et al. Hypervariable Region-Iv of Salmonella Gene Flic(D) Encodes a Dominant Surface Epitope and a Stabilizing Factor for Functional Flagella. J Bacteriol 1994; 176(8):2406-2414.
42. Newton SMC, Kotb M, Poirier TP et al. Expression and Immunogenicity of a Streptococcal M-Protein Epitope Inserted in Salmonella Flagellin. Infect Immun 1991; 59(6):2158-2165.
43. Hemmi H, Kaisho T, Takeuchi O et al. Small anti-viral compounds activate immune cells via the TLR7 MyD88-dependent signaling pathway. Nat Immunol 2002; 3(2):196-200.
44. Lee J, Chuang TH, Redecke V et al. Molecular basis for the immunostimulatory activity of guanine nucleoside analogs: Activation of Toll-like receptor 7. P Natl Acad Sci USA 2003; 100(11):6646-6651.
45. Kawasaki K, Akashi S, Shimazu R et al. Mouse toll-like receptor 4 center dot MD-2 complex mediates lipopolysaccharide-mimetic signal transduction by taxol. J Biol Chem 2000; 275(4):2251-2254.
46. Byrd-Leifer CA, Block EF, Takeda K et al. The role of MyD88 and TLR4 in the LPS-mimetic activity of Taxol. Eur J Immunol 2001; 31(8):2448-2457.
47. Hacker G, Redecke V, Hacker H. Activation of the immune system by bacterial CpG-DNA. Immunology 2002; 105(3):245-251.
48. Yi AK, Peckham DW, Ashman RF et al. CpG DNA rescues B cells from apoptosis by activating NF kappa B and preventing mitochondrial membrane potential disruption via a chloroquine sensitive pathway. Int Immunol 1999; 11(12):2015-2024.
49. Tokunaga T, Yamamoto T, Yamamoto S. How BCG led to the discovery of immunostimulatory DNA. Jpn J Infect Dis 1999; 52(1):1-11.

50. Yamamoto S, Yamamoto T, Nojima Y et al. Discovery of immunostimulatory CpG-DNA and its application to tuberculosis vaccine development. Jpn J Infect Dis 2002;55(2):37-44.
51. Krieg AM, Hartmann G, Yi AK. Mechanism of action of CpG DNA. Curr Top Microbiol 2000; 247:1-21.
52. Krieg AM, Stein CA. Phosphorothioate Oligodeoxynucleotides-Antisense or Anti-Protein. Antisense Res Dev 1995; 5(4):241-241.
53. Van Uden JH, Tran CH, Carson DA et al. Type I interferon is required to mount an adaptive response to immunostimulatory DNA. Eur J Immunol 2001; 31(11):3281-3290.
54. Kandimalla ER, Bhagat L, Wang DQ et al. Divergent synthetic nucleotide motif recognition pattern: design and development of potent immunomodulatory oligodeoxyribonucleotide agents with distinct cytokine induction profiles. Nucleic Acids Res 2003; 31(9):2393-2400.
55. Imler JL, Hoffmann JA. Toll receptors in innate immunity. Trends Cell Biol 2001; 11(7):304-311.
56. Sato M, Sano H, Iwaki D et al. Direct binding of toll-like receptor 2 to zymosan, and zymosan-induced NF-kappa B activation and TNF-alpha secretion are down-regulated by lung collectin surfactant protein A. J Immunol 2003;171(1):417-425.
57. Gangloff M, Weber ANR, Gibbard RJ et al. Evolutionary relationships, but functional differences, between the Drosophila and human Toll-like receptor families. Biochem Soc T 2003; 31:659-663.
58. Triantafilou M, Triantafilou K. Lipopolysaccharide recognition: CD14, TLRs and the LPS-activation cluster. Trends Immunol 2002; 23(6):301-304.
59. Inohara N, Nunez G. NODs: Intracellular proteins involved in inflammation and apoptosis. Nat Rev Immunol 2003; 3(5):371-382.
60. Jack RS, Fan XL, Bernheiden M et al. Lipopolysaccharide-binding protein is required to combat a murine Gram-negative bacterial infection. Nature 1997; 389(6652):742-745.
61. Beamer LJ, Carroll SF, Eisenberg D. Crystal structure of human BPI and two bound phospholipids at 2.4 angstrom resolution. Science 1997; 276(5320):1861-1864.
62. Kleiger G, Beamer LJ, Grothe R et al. The 1.7 angstrom crystal structure of BPI: A study of how two dissimilar amino acid sequences can adopt the same fold. J Mol Biol 2000; 299(4):1019-1034.
63. Tobias PS, Soldau K, Gegner JA et al. Lipopolysaccharide-Binding Protein-Mediated Complexation of Lipopolysaccharide with Soluble Cd14. J Biol Chem 1995; 270(18):10482-10488.
64. Rock FL, Hardiman G, Timans JC et al. A family of human receptors structurally related to Drosophila Toll. P Natl Acad Sci USA 1998; 95(2):588-593.
65. Medzhitov R, PrestonHurlburt P, Janeway CA. A human homologue of the Drosophila Toll protein signals activation of adaptive immunity. Nature 1997; 388(6640):394-397.
66. Wright SD, Ramos RA, Tobias PS et al. Cd14, a Receptor for Complexes of Lipopolysaccharide (Lps) and Lps Binding-Protein. Science 1990; 249(4975):1431-1433.
67. Hailman E, Lichenstein HS, Wurfel MM et al. Lipopolysaccharide (Lps)-Binding Protein Accelerates the Binding of Lps to Cd14. J Exp Med 1994; 179(1):269-277.
68. Hailman E, Vasselon T, Kelley M et al. Stimulation of macrophages and neutrophils by complexes of lipopolysaccharide and soluble CD14. J Immunol 1996; 156(11):4384-4390.
69. Frey EA, Miller DS, Jahr TG et al. Soluble Cd14 Participates in the Response of Cells to Lipopolysaccharide. J Exp Med 1992; 176(6):1665-1671.
70. Pugin J, Schurermaly CC, Leturcq D et al. Lipopolysaccharide Activation of Human Endothelial and Epithelial-Cells Is Mediated by Lipopolysaccharide-Binding Protein and Soluble Cd14. P Natl Acad Sci USA 1993; 90(7):2744-2748.
71. Haziot A, Rong GW, Goyert SM. Recombinant Soluble Cd14 Mediates the Activation of Endothelial-Cells by Lps. J Immunol 1993; 150(8):A260-A260.
72. Schutt C, Schilling T, Grunwald U et al. Endotoxin-Neutralizing Capacity of Soluble Cd14. Res Immunol 1992; 143(1):71-78.
73. Stelter F, Loppnow H, Menzel R et al. Differential impact of substitution of amino acids 9-13 and 91-101 of human CD14 on soluble CD14-dependent activation of cells by lipopolysaccharide. J Immunol 1999; 163(11):6035-6044.
74. Mcginley MD, Narhi LO, Kelley MJ et al. Cd14-Physical-Properties and Identification of an Exposed Site That Is Protected by Lipopolysaccharide. J Biol Chem 1995; 270(10):5213-5218.
75. Shimazu R, Akashi S, Ogata H et al. MD-2, a molecule that confers lipopolysaccharide responsiveness on Toll-like receptor 4. J Exp Med 1999; 189(11):1777-1782.

76. Schromm AB, Lien E, Henneke P et al. Molecular genetic analysis of an endotoxin nonresponder mutant cell Line: A point mutation in a conserved region of MD-2 abolishes endotoxin-induced signaling. J Exp Med 2001; 194(1):79-88.
77. Viriyakosol S, Tobias PS, Kitchens RL et al. MD-2 binds to bacterial lipopolysaccharide. J Biol Chem 2001; 276(41):38044-38051.
78. Re F, Strominger JL. Monomeric recombinant MD-2 binds Toll-like receptor 4 tightly and confers lipopolysaccharide responsiveness. J Biol Chem 2002; 277(26):23427-23432.
79. Mullen GED, Kennedy MN, Visintin A et al. The role of disulfide bonds in the assembly and function of MD-2. P Natl Acad Sci USA 2003; 100(7):3919-3924.
80. Visintin A, Mazzoni A, Spitzer JA et al. Secreted MD-2 is a large polymeric protein that efficiently confers lipopolysaccharide sensitivity to Toll-like receptor 4. P Natl Acad Sci USA 2001; 98(21):12156-12161.
81. Inohara N, Nunez G. ML-a conserved domain involved in innate immunity and lipid metabolism. Trends Biochem Sci 2002; 27(5):219-221.
82. Akashi S, Nagai Y, Ogata H et al. Human MD-2 confers on mouse Toll-like receptor 4 species-specific lipopolysaccharide recognition. Int Immunol 2001; 13(12):1595-1599.
83. Wright CS, Li SC, Rastinejad F. Crystal structure of human GM2-activator protein with a novel beta-cup topology. J Mol Biol 2000; 304(3):411-422.
84. Hajjar AM, Ernst RK, Tsai JH et al. Human Toll-like receptor 4 recognizes host-specific LPS modifications. Nat Immunol 2002; 3(4):354-359.
85. Arbour NC, Lorenz E, Schutte BC et al. TLR4 mutations are associated with endotoxin hyporesponsiveness in humans. Nat Genet 2000; 25:187-191.
86. Koskinen R, Lamminmaki U, Tregaskes CA et al. Cloning and modeling of the first nonmammalian CD4. J Immunol 1999; 162(7):4115-4121.
87. Hoshino K, Takeuchi O, Kawai T et al. Cutting edge: Toll-like receptor 4 (TLR4)-deficient mice are hyporesponsive to lipopolysaccharide: Evidence for TLR4 as the Lps gene product. J Immunol 1999; 162(7):3749-3752.
88. Bin LH, Xu LG, Shu HB. TIRP, a novel Toll/interleukin-1 receptor (TIR) domain-containing adapter protein involved in TIR signaling. J Biol Chem 2003; 278(27):24526-24532.
89. Horng T, Barton GM, Flavell RA et al. The adaptor molecule TIRAP provides signalling specificity for Toll-like receptors. Nature 2002; 420(6913):329-333.
90. Fitzgerald KA, Palsson-McDermott EM, Bowie AG et al. Mal (MyD88-adapter-like) is required for Toll-like receptor-4 signal transduction. Nature 2001; 413(6851):78-83.
91. Medzhitov R, Preston-Hurlburt P, Kopp E et al. MyD88 is an adaptor protein in the hToll/IL-1 receptor family signaling pathways. Mol Cell 1998;2(2):253-258.
92. O'Neill LAJ, Fitzgerald KA, Bowie AG. The Toll-IL-1 receptor adaptor family grows to five members. Trends Immunol 2003; 24(6):287-290.
93. Gay NJ, Keith FJ. Drosophila Toll and Il-1 Receptor. Nature 1991; 351(6325):355-356.
94. Bowie A, Kiss-Toth E, Symons JA et al. A46R and A52R from vaccinia virus are antagonists of host IL-1 and toll-like receptor signaling. Proc Natl Acad Sci USA 2000; 97(18):10162-10167.
95. Harte MT, Haga IR, Maloney G et al. The poxvirus protein A52R targets toll-like receptor signaling complexes to suppress host defense. J Exp Med 2003; 197(3):343-351.
96. Vigers GPK, Caffes P, Evans RJ et al. X-Ray Structure of Interleukin-1 Receptor Antagonist at 2.0-Angstrom Resolution. J Biol Chem 1994; 269(17):12874-12879.
97. Slack JL, Schooley K, Bonnert TP et al. Identification of two major sites in the type I interleukin-1 receptor cytoplasmic region responsible for coupling to pro-inflammatory signaling pathways. J Biol Chem 2000; 275(7):4670-4678.
98. Poltorak A, He XL, Smirnova I et al. Defective LPS signaling in C3H/HeJ and C57BL/10ScCr mice: Mutations in TLR4 gene. Science 1998; 282(5396):2085-2088.
99. Heguy A, Baldari CT, Macchia G et al. Amino-Acids Conserved in Interleukin-1 Receptors (Il-1rs) and the Drosophila Toll Protein Are Essential for Il-1r Signal Transduction. J Biol Chem 1992; 267(4):2605-2609.
100. Huang WY, Valles S, Qwarnstrom EE. Translocation of the IL-1 receptor to focal adhesions is regulated through the C-terminal end of the cytoplasmic domain. Cell Biol Int 2001; 25(4):309-317.

101. Xu YW, Tao X, Shen BH et al. Structural basis for signal transduction by the Toll/interleukin-1 receptor domains. Nature 2000; 408(6808):111-115.
102. Tao X, Xu YW, Zheng Y et al. An extensively associated dimer in the structure of the C713S mutant of the TIR domain of human TLR2. Biochem Bioph Res Co 2002; 299(2):216-221.
103. Fesik SW. Insights into programmed cell death through structural biology. Cell 2000; 103(2):273-282.
104. Suzuki N, Suzuki S, Yeh WC. IRAK-4 as the central TIR signaling mediator in innate immunity. Trends Immunol 2002; 23(10):503-506.
105. Towb P, Galindo RL, Wasserman SA. Recruitment of Tube and Pelle to signaling sites at the surface of the Drosophila embryo. Development 1998; 125(13):2443-2450.
106. Schiffmann DA, White JHM, Cooper A et al. Formation and biochemical characterization of tube/pelle death domain complexes: Critical regulators of postreceptor signaling by the Drosophila Toll receptor. Biochemistry-Us 1999; 38(36):11722-11733.
107. Shen BH, Manley JL. Pelle kinase is activated by autophosphorylation during Toll signaling in Drosophila. Development 2002; 129(8):1925-1933.
108. Aravind L, Dixit VM, Koonin EV. The domains of death: evolution of the apoptosis machinery. Trends Biochem Sci 1999; 24(2):47-53.
109. Xiao T, Towb P, Wasserman SA et al. Three-dimensional structure of a complex between the death domains of Pelle and Tube. Cell 1999; 99(5):545-555.
110. Wesche H, Henzel WJ, Shillinglaw W et al. MyD88: An adapter that recruits IRAK to the IL-1 receptor complex. Immunity 1997; 7(6):837-847.
111. Janssens S, Beyaert R. Functional diversity and regulation of different interleukin-1 receptor-associated kinase (IRAK) family members. Mol Cell 2003; 11(2):293-302.
112. Li SY, Strelow A, Fontana EJ et al. IRAK-4: A novel member of the IRAK family with the properties of an IRAK-kinase. P Natl Acad Sci USA 2002; 99(8):5567-5572.
113. DePristo M, de Bakker P, Lovell S et al. Ab initio construction of polypeptide fragments: Efficient generation of consistent, representative ensembles. Proteins (Structure, Function and Genetics) 2003; 51:41-55
114. Tovchigrechko A, Wells CA, Vakser IA. Docking of protein models. Protein Sci 2002; 11(8):1888-1896.
115. Takeuchi O, Sato S, Horiuchi T et al. Cutting edge: Role of Toll-like receptor 1 in mediating immune response to microbial lipoproteins. J Immunol 2002; 169(1):10-14
116. Wyllie DH, Kiss-Toth E, Visintin A et al. Evidence for an accessory protein function for toll-like receptor 1 in anti-bacterial responses. J Immunol 2000; 165(12):7125-7132.
117. Bulut Y, Faure E, Thomas L et al. Cooperation of toll-like receptor 2 and 6 for cellular activation by soluble tuberculosis factor and Borrelia burgdorferi outer surface protein A lipoprotein: Role of Toll-interacting protein and IL-1 receptor signaling molecules in Toll-like receptor 2 signaling. J Immunol 2001; 167(2):987-994.
118. Aliprantis AO, Yang RB, Mark MR et al. Cell activation and apoptosis by bacterial lipoproteins through toll-like receptor-2. Science 1999; 285(5428):736-739.
119. Brightbill HD, Libraty DH, Krutzik SR et al. Host defense mechanisms triggered by microbial lipoproteins through toll-like receptors. Science 1999; 285(5428):732-736.
120. Lien E, Sellati TJ, Yoshimura A et al. Toll-like receptor 2 functions as a pattern recognition receptor for diverse bacterial products. J Biol Chem 1999; 274(47):33419-33425.
121. Hirschfeld M, Kirschning CJ, Schwandner R et al. Cutting edge: Inflammatory signaling by Borrelia burgdorferi lipoproteins is mediated by toll-like receptor 2. J Immunol 1999; 163(5):2382-2386.
122. Schwandner R, Dziarski R, Wesche H et al. Peptidoglycan- and lipoteichoic acid-induced cell activation is mediated by toll-like receptor 2. J Biol Chem 1999; 274(25):17406-17409.
123. Yoshimura A, Lien E, Ingalls RR et al. Cutting edge: Recognition of gram-positive bacterial cell wall components by the innate immune system occurs via toll-like receptor 2. J Immunol 1999; 163(1):1-5.
124. Iwaki D, Mitsuzawa H, Murakami S et al. The extracellular toll-like receptor 2 domain directly binds peptidoglycan derived from Staphylococcus aureus. J Biol Chem 2002; 277(27):24315-24320.

125. Lehner MD, Morath S, Michelsen KS et al. Induction of cross-tolerance by lipopolysaccharide and highly purified lipoteichoic acid via different toll-like receptors independent of paracrine mediators. J Immunol 2001; 166(8):5161-5167.
126. Means TK, Wang SY, Lien E et al. Human toll-like receptors mediate cellular activation by Mycobacterium tuberculosis. J Immunol 1999; 163(7):3920-3927.
127. Underhill DM, Ozinsky A, Smith KD et al. Toll-like receptor-2 mediates mycobacteria-induced proinflammatory signaling in macrophages. P Natl Acad Sci USA 1999; 96(25):14459-14463.
128. Hajjar AM, O'Mahony DS, Ozinsky A et al. Cutting edge: Functional interactions between toll-like receptor (TLR) 2 and TLR1 or TLR6 in response to phenol-soluble modulin. J Immunol 2001; 166(1):15-19.
129. Campos MA, Almeida IC, Takeuchi O et al. Activation of toll-like receptor-2 by glycosylphosphatidylinositol anchors from a protozoan parasite. J Immunol 2001; 167(1):416-423.
130. Opitz B, Schroder NWJ, Spreitzer I et al. Toll-like receptor-2 mediates Treponema glycolipid and lipoteichoic acid-induced NF-kappa B translocation. J Biol Chem 2001; 276(25):22041-22047.
131. Massari P, Henneke P, Ho Y et al. Cutting edge: Immune stimulation by neisserial porins is toll-like receptor 2 and MyD88 dependent. J Immunol 2002; 168(4):1533-1537.
132. Werts C, Tapping RI, Mathison JC et al. Leptospiral lipopolysaccharide activates cells through a TLR2-dependent mechanism. Nat Immunol 2001; 2(4):346-352.
133. Hirschfeld M, Weis JJ, Toshchakov V et al. Signaling by Toll-like receptor 2 and 4 agonists results in differential gene expression in murine macrophages. Infect Immun 2001; 69(3):1477-1482.
134. Asea A, Rehli M, Kabingu E et al. Novel signal transduction pathway utilized by extracellular HSP70-Role of Toll-like receptor (TLR) 2 and TLR4. J Biol Chem 2002; 277(17):15028-15034.
135. Vabulas RM, Wagner H, Schild H. Heat shock proteins as ligands of toll-like receptors. Toll-Like Receptor Family Members and Their Ligands 2002; 270:169-184.
136. Alexopoulou L, Holt AC, Medzhitov R et al. Recognition of double-stranded RNA and activation of NF-kappa B by Toll-like receptor 3. Nature 2001; 413(6857):732-738.
137. Jiang QQ, Akashi S, Miyake K et al. Cutting edge: Lipopolysaccharide induces physical proximity between CD14 and toll-like receptor 4 (TLR4) prior to nuclear translocation of NF-kappa B. J Immunol 2000; 165(7):3541-3544.
138. Qureshi ST, Lariviere L, Leveque G et al. Endotoxin-tolerant mice have mutations in toll-like receptor 4 (TLR4). J Exp Med 1999; 189(4):615-625.
139. Gomi K, Kawasaki K, Kawai Y et al. Toll-like receptor 4-MD-2 complex mediates the signal transduction induced by flavolipin, an amino acid-containing lipid unique to Flavobacterium meningosepticum. J Immunol 2002; 168(6):2939-2943.
140. Kurt-Jones EA, Popova L, Kwinn L et al. Pattern recognition receptors TLR4 and CD14 mediate response to respiratory syncytial virus. Nat Immunol 2000; 1(5):398-401.
141. Haynes LM, Moore DD, Kurt-Jones EA et al. Involvement of toll-like receptor 4 in innate immunity to respiratory syncytial virus. J Virol 2001; 75(22):10730-10737.
142. Rassa JC, Meyers JL, Zhang YM et al. Murine retroviruses activate B cells via interaction with toll-like receptor 4. P Natl Acad Sci USA 2002; 99(4):2281-2286.
143. Bulut Y, Faure E, Thomas L et al. Chlamydial heat shock protein 60 activates macrophages and endothelial cells through toll-like receptor 4 and MD2 in a MyD88-dependent pathway. J Immunol 2002; 168(3):1435-1440.
144. Sasu S, LaVerda D, Qureshi N et al. Chlamydia pneumoniae and chlamydial heat shock protein 60 stimulate proliferation of human vascular smooth muscle cells via Toll-like receptor 4 and p44/p42 mitogen-activated protein kinase activation. Circ Res 2001; 89(3):244-250.
145. Ohashi K, Burkart V, Flohe S et al. Heat shock protein 60 is a putative endogenous ligand of the toll-like receptor-4 complex. J Immunol 2000; 164(2):558-561.
146. Dybdahl B, Wahba A, Lien E et al. Inflammatory response after open heart surgery-Release of heat-shock protein 70 and signaling through toll-like receptor-4. Circulation 2002; 105(6):685-690.
147. Okamura Y, Watari M, Jerud ES et al. The extra domain A of fibronectin activates toll-like receptor 4. J Biol Chem 2001; 276(13):10229-10233.
148. Termeer C, Benedix F, Sleeman J et al. Oligosaccharides of hyaluronan activate dendritic cells via toll-like receptor 4. J Exp Med 2002; 195(1):99-111.

149. Johnson GB, Brunn GJ, Kodaira Y et al. Receptor-mediated monitoring of tissue well-being via detection of soluble heparan sulfate by toll-like receptor 4. J Immunol 2002; 168(10):5233-5239.
150. Smiley ST, King JA, Hancock WW. Fibrinogen stimulates macrophage chemokine secretion through toll-like receptor 4. J Immunol 2001; 167(5):2887-2894.
151. Hayashi F, Smith KD, Ozinsky A et al. The innate immune response to bacterial flagellin is mediated by Toll-like receptor 5. Nature 2001; 410(6832):1099-1103.
152. Hemmi H, Takeuchi O, Kawai T et al. A Toll-like receptor recognizes bacterial DNA. Nature 2000; 408(6813):740-745.
153. Leadbetter EA, Rifkin IR, Hohlbaum AM et al. Chromatin-IgG complexes activate B cells by dual engagement of IgM and Toll-like receptors. Nature 2002; 416(6881):603-607.
154. Mizuguchi K, Deane CM, Blundell TL et al. JOY: protein sequence-structure representation and analysis. Bioinformatics 1998; 14(7):617-623.
155. Dunne A, Ejdeback M, Ludidi PL et al. Structural complementarity of Toll/interleukin-1 receptor domains in Toll-like receptors and the adaptors Mal and MyD88. J Biol Chem 2003; 278:41443-41451.

CHAPTER 3

"Supramolecular" Activation Clusters in Innate Immunity

Martha Triantafilou and Kathy Triantafilou

Introduction

From birth every living organism must contend with an environment replete with infectious pathogens. Mammals and vertebrates have responded to this challenge by developing an intricate system of host defense that we collectively call the immune system. The immune system itself can be divided into two main components: innate and adaptive immunity.

The innate immune system is, in evolutionary terms, the more archaic of the two. It is the non-specific component of the immune system and is the first line of defense against invading pathogens. This defense is characterized by the production and activation of mediators and immune system cells that kill pathogens. The innate immune system uses a set of germline encoded receptors, called pattern recognition receptors (PRRs), to recognize specific molecular patterns or motifs called PAMPs (Pathogen-Associated Molecular Patterns) on invading pathogens.[1,2]

By contrast, the acquired immune response is present only in vertebrates and is pathogen specific. The main difference between the two wings of our immune system is that the acquired uses somatic gene rearrangement to produce specific antigen receptors. Two of the principle receptors involved are major histocompatibility (MHC) molecules and T-cell receptors. MHC molecules bind and present antigen to T-cells. T cell activation can only proceed if the T cell receptor interacts productively with the specific peptide-MHC complex being offered. In order to achieve this the T-cell requires the help of a plethora of co-receptors and accessory molecules. In the past few years, it has been shown that antigen recognition by the acquired immune system comprises an intricate multi-molecular choreography of receptors that takes place as the T-cell engages a peptide-MHC complex.[3,4] This interaction is dynamic and involves delicate contacts (termed immunological synapses) between the T-cell and antigen-presenting cell (which bears the MHC antigen complex). The successful formation of these contacts is dependent on the binding of the T cell receptor to a specific peptide-MHC complex, as well on the formation of different microdomains that contain clusters of different receptors.

Innate immune recognition was believed to be far simpler. PRRs were thought to bind and recognize conserved products of pathogens that were unique to the invading microorganism but not the host. Although this is true, it is becoming increasingly apparent that the model of a single PRR to recognize each microbial signature has been oversimplified. With the discovery of the Toll-like receptors as the main signal transducing molecules of the innate immune

Toll and Toll-Like Receptors: An Immunologic Perspective, edited by Tina Rich. ©2005 Eurekah.com and Kluwer Academic / Plenum Publishers.

system, a considerable research effort has now shown that PRRs are part of a multi-component sensor apparatus. In this chapter we'll attempt to critically examine recent advances in the field of innate recognition, with a primary focus on the receptors and microdomains that are involved in bacterial lipopolysaccharide (LPS) recognition.

Lipopolysaccharide Recognition

Lipopolysaccharide (LPS) or endotoxin constitutes the outer membrane of all Gram-negative bacteria and is one of the best known examples of a PAMP. LPS is perceived by the innate immune system as the molecular signature for Gram-negative bacterial infection and its recognition leads to a pro-inflammatory response. In some cases this recognition can lead to the uncontrolled production of pro-inflammatory mediators that can eventually lead to fatal sepsis syndrome in humans. Septic shock is the most common cause of death in intensive care units with the number of cases increasing since the 1930s. Consequently, there has been considerable interest in solving just how LPS is detected.

Structure of LPS

In determining how LPS stimulates the immune system we should first consider its structure. LPS from Gram-negative bacteria is generally composed of a polysaccharide attached to a lipid termed lipid A, hence the term lipopolysaccharide.[5] The hydrophilic part is composed of a polysaccharide core and O-antigen structures which can vary depending on the bacterial strain.[6,7] In contrast the lipid portion of LPS is highly conserved. Lipid A consists of a β-(1'$\rightarrow$6)-interlinked GlcpN-disaccharide, which is phosphorylated at positions 1 and 4', to which long chain fatty acids are attached. Biological activity in response to LPS is believed to be lipid A-dependent, although it has been shown that the polysaccharide portion enhances activity.[8]

Pattern Recognition Receptors Involved in LPS "Sensing"

Although LPS was chemically isolated more than 70 years ago,[9] the receptors responsible for its recognition proved more elusive. The seminal discovery of a serum protein that could bind LPS, termed lipopolysaccharide binding protein (LBP), came almost 50 years later and led to a new awareness of the molecular events that lead to LPS-recognition. Wright et al[10] then added another piece to the puzzle. They demonstrated that LBP potentiates the cellular response to LPS by aiding the transfer of LPS-LBP to a membrane receptor, CD14.

CD14

CD14, a 55 kDa glycosylphosphatidylinositol anchored protein expressed primarily on the surface of monocytes, was one of the first molecules to be identified as a pattern recognition receptor for LPS.[10] CD14-specific monoclonal antibodies inhibited LPS-induced TNF-α secretion[10-12] and transfection of CD14 into CD14-null cells enhanced LPS-induced activation.[13] The central role of CD14 in LPS recognition was further demonstrated by CD14-knockout mice that were 10,000-fold less sensitive to LPS compared to hemizygous littermates.[14]

CD14's role in LPS recognition was undoubted. However, it was less certain as to how a GPI-linked protein, not traversing the cell membrane, could transduce signals. Several reports hypothesized that CD14 must associate with a transmembrane receptor(s), termed the "LPS-signal transducer" to initiate LPS signaling.[15] This hypothesis was strengthened by several binding studies showing that CD14 specific mAbs could neither totally block FITC-LPS binding[16-18] especially at high LPS concentrations,[19] nor TNF-α production.[20] Collectively, these data suggested the existence of additional receptors.

Scavenger Receptors

Macrophages are able to recognize and internalize apoptotic and foreign cells and where appropriate, initiate inflammation and macrophage activation. Scavenger receptors are macrophage cell surface molecules associated with the endocytic uptake of lipoproteins. In addition to their uptake of low density lipoprotein, scavenger receptors on macrophages have also been identified as receptors for bacterial LPS.[21] Although scavenger receptors can bind LPS, in-vitro competition studies indicated that this binding was not responsible for the macrophage activation elicited by LPS. Interestingly, in-vivo, scavenger-receptor ligands were shown to greatly inhibit hepatic uptake of LPS in mice, suggesting an important role for these receptors in LPS clearance and detoxification.[21]

Integrins

The β_2 integrin family, especially complement receptors CR3* and CR4** have also been implicated in LPS recognition. They are transmembrane glycoproteins that are expressed on the surface of neutrophils, monocytes, macrophages and NK cells and are involved in numerous cell-cell interactions. Three members of the integrin family have been shown to mediate LPS signaling.[22,23] Additionally, CR3 was shown to be upregulated and to enhance the adhesive capacity of leukocytes after LPS stimulation.[24] More recently, CR3 has been implicated in cellular activation by LPS and Group B Streptococci,[25] and appears to mediate the activation of human neutrophils by LPS-coated erythrocytes.[26]

Although integrins are transmembrane receptors and were identified as LPS-receptors they did not seem to wholly solve the puzzle of LPS recognition. Their cytoplasmic domains were unnecessary for LPS-induced signal transduction.[27] Cells could be activated using LPS and a mutant CR3 protein, without a cytoplasmic domain.[28] Futhermore, CD11/CD18 integrins could enable LPS responsiveness independently of CD14.[28-30] Similarities between the signaling systems used by CD14 and CD11/CD18 integrins, such as the LBP dependent binding of whole Gram-negative bacteria, and the species specific effects of LPS analogues,[27] led to the idea that integrins, like CD14, may function by transferring LPS to a second receptor, the putative "LPS-transducer".

The Search for the LPS "Signal Transducer"

Although no molecule responsible for LPS-signaling was identified, LPS was known to trigger multiple signaling cascades. For example, the LPS induced production of TNF-α was shown to involve the activation of nuclear factor-κB (NF-κB).[31] LPS was also known to activate p38,[32] stress-activated protein kinase[33] and phosphatidylinositol-3 kinase.[34,35]

Several groups attempted to identify the "LPS-signal transducer" by isolating molecules that could directly bind LPS. Different approaches were used. Morrison et al were one of the first groups to identify a receptor molecule for LPS. They probed cells for membrane proteins that could recognize LPS labeled with sulfosuccinimidyl-2-(p-azidosalicyamino)-1,3-dithiopropionate (SASD). This led to the identification of an 80 kDa protein (with an approximate pI of 6.5) on a variety of LPS-responsive cell types.[36] Results with monoclonal antibodies raised against this protein, supported the notion that it may serve as a signaling-receptor for LPS. Dziarski and colleagues identified a 70 kDa protein by photochemical cross-linking with ASD-peptidoglycan (ASD-PG) from gram-positive bacteria and ASD-LPS.[37] Competitive binding experiments demonstrated that both PG and LPS bound to the same 70 kDa protein.

* also called Mac-1 and CD11b/CD18

** also called CD11c/CD18

A different approach was to mutagenize a murine macrophage-like cell line J774, then select for cells that were LPS resistant.[38] One such clone, termed LR-9, neither bound LPS nor responded to LPS stimulation. Presumably, LR-9 cells were resistant because they lacked the "LPS-signal transducer". These and parental cells were then challenged with ^{125}I-labelled-ASD-LPS with the result that LPS was cross-linked to two proteins (of 65 kDa and 55 kDa) expressed on the parental but not the LR-9 cells. Cross-linking a murine pre-B cell line (line 70Z/3) with ^{125}I-labelled-ASD-LPS failed to reveal the same 80 kDa protein that had been detected by Morrison. Instead two LPS binders were found, one of 18 kDa and the other, 25 kDa.[39] Ligand blotting has also been used to identify membrane proteins that bind LPS. Cellular proteins, immobilized on nitrocellulose, were incubated with radioactive LPS or anti-LPS antibodies. This approach, employed by Hampton et al to study lipid IVa binding,[40] revealed a 95 kDa LPS binder in cell membranes. The same approach, this time with erythrocytes, identified a 96 kD protein.[41] More recently, El-Samalouti et al has used immunoprecipitation to detect an 80 kDa protein that binds LPS.[42] Our group, using similar methods, has identified four new LPS binding proteins with the respective molecular weights of 80 kDa, 70 kDa, 55 kDa, and 40 kDa.[43] In summary, at least eight different membrane proteins have been identified as LPS binders, but till recently no conclusive role for them in mediating trans-membrane signaling had been described.

Toll-Like Receptors

Following the discovery of CD14 as an LPS receptor, nearly ten years elapsed before the next significant breakthrough. Work from the *Drosophila* field revealed that two members of the TLR family, dToll and 18-Wheeler, were responsible for the fly anti-fungal and anti-bacterial response.[30,44] The next step, coming from Medzditov and Janeway, was to discover the human homologue of Toll, "hToll".[45] This triggered a new field of research into the involvement of Toll-like receptors in innate immunity.

Toll-Like Receptor 4 (TLR4)

The work of Beutler and colleagues was seminal in elucidating the LPS "sensing" apparatus of the innate immune system. Beutler and co-workers demonstrated that two strains of mice that are hypo-responsive to LPS carry genetic defects in TLR4.[46] It was shown that C3H/HeJ mice had a mutation in the third exon of TLR4, whereas the second resistant strain, C57Bl/10ScCr, was homozygous for a null mutation of TLR4.[46-48] This was the first evidence that TLR4 was part of a multi-component LPS receptor, linking TLRs with innate immune recognition of LPS. The generation of TLR4-deficient mice proved its importance in LPS-signaling as these animals shared an identical phenotype with C3H/HeJ mice.[49]

Miyake and colleagues then demonstrated that another molecule, MD-2, physically associates with the extracellular domain of TLR4[48,50] and that this interaction was requisite for LPS recognition.[51,52] It is now widely accepted that TLR4 plays a central role in LPS signaling. This finding led to a break-through in our understanding of how innate immunity could be tuned to different pathogens and, largely through the work of Akira and colleagues, the involvement of several TLRs in microbial sensing has now been demonstrated.[53-55]

TLR2

TLR2 was initially believed to be involved in LPS recognition.[56,57] It was later revealed that these reports were flawed and that TLR4, and not TLR2, was responsible for LPS signal-transduction.[46,47,58] The contradictory results were attributed to contaminated LPS preparations.[59] It has now been demonstrated that TLR2 recognizes non-standard LPS, cell wall components from Gram-positive bacteria,[60] spirochetes,[61] mycobacteria[62] and yeast.[61-63] [64] TLR2 also detects *Listeria monocytogenes,*[65] lipoproteins from *Borrelia burgdorferi,*[66] and membrane lipopeptides from *Treponema pallidum, Mycoplasma fermentans* and *Mycobacterium avium.*[67]

TLR3

Alexopoulou and colleagues have recently discovered that TLR3 recognizes viral double stranded RNA[68] and that this recognition leads to the production of type I interferons. Not surprisingly, TLR3-deficient mice showed reduced responses to polyinosine-polycytidylic acid and reduced production of inflammatory cytokines.

TLR5

Hayashi and co-workers showed that mammalian TLR5 recognizes bacterial flagellin from both Gram-positive and Gram-negative bacteria.[69] Flagellin is a principal component of bacterial flagella and is a virulence factor that is recognized by the innate immune system. This would suggest that TLR5 has evolved in order to detect flagellated bacterial pathogens.

TLR7

The microbial specificity of TLR7 remains undetermined. However, TLR7 responds to imiquimod, a small anti-viral drug, leading some to speculate that TLR7 might be involved in viral recognition.*

TLR9

Mammalian DNA has a low frequency of CpG dinucleotides, most of which are methylated. In contrast, bacterial DNA possesses unmethylated CpG dinucleotides and has been found to have stimulatory effects on mammalian immune cells. Hemmi and colleagues have shown that TLR9 mediates immune responses to CpG DNA.[55] The importance of TLR9 in detecting bacterial DNA was demonstrated by TLR9 deficient mice that failed to respond to CpG DNA. In addition they were resistant to the lethal effects of CpG DNA without any elevation of their pro-inflammatory cytokines. This would suggest that the innate immune system has evolved a specific Toll-like receptor to distinguish bacterial from self-DNA.

Functional Interactions between Different TLRs

As mentioned above TLRs are a restricted family of receptors that recognize a panel of microbial pathogens. The question arises as to how a limited number of conserved molecules can recognize a wide repertoire of pathogen derived motifs. Recently Ozinsky and colleagues demonstrated that the recognition of PAMPs can be achieved by the combinatorial association of different TLRs.[64] They showed that peptidoglycan, a Gram-positive bacterial product, is recognized by TLR2 and TLR6. These two TLRs coordinate the activation of macrophages in response to Gram-positive bacteria, the yeast cell wall particle zymosan[64] and phenol-soluble modulin which is secreted by *Staphylococcus epidermis*.[63] Added variation is provided by the cytoplasmic domain of TLR2, which can heterodimerize with either TLR6 or TLR1 to induce cytokine production. These findings led the group to propose that some TLRs assemble as heteromeric and others as homomeric complexes (e.g., TLR4) and that these combinations define the microbial repertoire that can be recognized.[64] Subsequently, it has been found that multiple Toll-like receptors recognize Group B Streptococci (GBS). Henneke and colleagues demonstrated that CD14, TLR2 and TLR6 function as co-receptors for secreted microbial products derived from GBS, whereas cell wall components are recognized by TLRs distinct from TLR1, 2, 4, or 6.[70] More recently, cooperation between TLR1 and TLR2 has been shown to be essential for the recognition of native mycobacterial lipoprotein as well as several triacylated lipopeptides.[71]

* Since going to press murine TLR7 and human TLR8 have been shown to recognize ssRNA, both of viral and endogenous origin. (Heil F et al. Science 2004; 303:1526-1529. Diebold SS et al. Science 2004; 303:1529-1531.)

CD55

The 80 kDa LPS-binding protein that was co-precipitated with CD14 by El-Samalouti and colleagues was termed LMP-80.[42] This protein, expressed on human monocytes and endothelial cells, was eventually shown to be the decay accelerating factor DAF, or CD55.[72] Subsequently, it was shown that human CD55 could restore LPS signaling in LPS-hyporesponsive Chinese hamster ovary cells (CHO).[73] As CD55 is a GPI-anchored protein (like CD14) it was suggested that it must associate (perhaps transiently) with other signaling molecules to transduce LPS signals. Alternatively, it could reside in a multimeric LPS receptor complex. Recently Pfeiffer and colleagues have shown that CD55 is part of a complex of receptors that concentrates in lipid rafts following LPS stimulation.[74]

TREM-1

A new receptor of the immunoglobulin superfamily expressed on human neutrophils and monocytes has recently been identified termed triggering receptor expressed on myeloid cells (TREM)-1.[75] Bouchon and colleagues discovered that TREM-1 promotes cell activation through an associated molecule called DAP12. DAP12 has the ability to trigger the secretion of pro-inflammatory mediators[75] which prompted investigators to look at the role that TREM-1 plays in inflammation caused by microbial pathogens.

They found that TREM-1 expression was upregulated in response to Gram-positive bacteria, Gram-negative bacteria, and mycobacteria. Surprisingly, TREM-1 expression was significantly increased after incubation with extracellular bacteria, but not intracellular. Bouchon and colleagues went on to create a TREM-1-Ig Fc fusion protein that would compete with the unknown TREM-1 ligand. Administration of the fusion protein lowered serum levels of TNF and IL-1β. Most importantly they reported that the TREM-1 fusion protein could protect mice from death when given before or after LPS administration.[76] Few therapeutic interventions for septic shock can protect animals from death when administered more than 30 minutes after LPS.[77]

More recently the group have demonstrated that together, TREM-1 and Toll-like receptor mediated signals can synergise to increased the production of pro-inflammatory cytokines.[78] This adds TREM-1 to the list of factors involved in bacterial recognition. While the function of TREM molecules (TREM-1,-2 and -3), as well as DAP12 have just begun to be unveiled, emerging evidence suggests that these molecules play a crucial role in both innate and adaptive immune responses.[79]

RP105

RP105 (CD180) is a type I transmembrane protein with extracellular leucine repeats and a short cytoplasmic tail.[80] The multiple leucine-rich repeats are similar to those in Drosophila Toll, making RP105 the first mammalian molecule to be described with similarity to Toll. Expression of RP105 is restricted to mature B-cells and macrophages.[80] Ogata and colleagues demonstrated that RP105 regulates LPS signaling on B-cells[81] and showed that B-cells that lack RP105 were impaired in LPS-induced proliferation and antibody production.[81] RP105 has been shown to associate with an extracellular molecule, MD-1 and the RP105/MD-1 complex is strikingly similar to TLR4/MD-2. In the latter case, MD-2 is indispensable for LPS-induced cell activation. Likewise, MD-1 is indispensable for RP105 cell surface expression and B-cell responsiveness to LPS.[82] Thus the RP105/MD-1 complex constitutes an LPS-signaling complex on B-cells.

Platelet Activating Factor and ADAM10

ADAMs (a disintegrin and metalloprotease) are members of the metzincin superfamily of metalloproteases and Platelet-activating factor is a G protein-coupled receptor involved in responses to *Staphylococcus aureus* in epithelial cells. Bacterial LTA activates platelet activating factor which, in turn, activates ADAM10 and the epidermal growth factor receptor.[83] Interestingly, unlike responses in macrophages, epithelial responses to LTA do not seem to require TLR4.

Heat Shock Proteins

In our attempt to isolate LPS-binding receptors we identified four molecules,[43] two of which were heat shock proteins (HSPs), HSC70 and HSP90a.[58] HSPs are highly conserved proteins that are associated with tissue damage and stress. Although also considered as chaperones in antigen-presentation, HSPs have been found to exist on the cell surface[84-87] and are now classed as one of the groups of proteins that flag "danger" to the innate immune system.[88] HSPs have been shown to generate protective immunity, to activate antigen-specific T-cells[89,90] and to bind Taxol. The significance of the latter observation is that Taxol, a plant-derived anti-tumor agent, signals in a manner indistinguishable to that of bacterial LPS.[91] Recently HSPs have also been shown to interact with and activate TLRs. The work of Vabulas and co-workers has shown that different HSPs serve as endogenous stimuli for TLRs. HSP70 activates the innate immune system via the TLR signaling pathway.[92] Additionally, HSP60 as well as Gp96 signal via the TLR2/4 complex.[93,94] Furthermore they reported that HSP internalization is required in order to stimulate the TLR signaling pathway.

Interestingly recent studies have suggested that endotoxin (LPS) contamination of recombinant heat shock proteins might be responsible for the induction of cytokines observed when HSPs are administered[95] and for the interactions observed with the "LPS-sensing machinery".[96] Thus it is possible that HSPs complexed with LPS interact with the TLR pathway and induce CD14-dependent cytokine secretion.

In addition, HSP90 has been shown to bind directly to CpG oligonucleotides[97,98] and a specific inhibitor of HSP90, geldanamycin, prevents CpG driven cellular stimulation.[97] An interaction of HSP90 with TLR9, which is known to mediate CpG signaling, was also demonstrated.[98] Thus a new function for HSPs in immune responses is emerging. The evidence points towards a dual role for HSPs in bridging the innate and adaptive immune response. To do this HSPs may bind and present PAMPs to TLRs, at the same time binding and shuttling peptides to the antigen-presentation pathway. The expression of HSPs in lipid rafts[99,100] could facilitate both these functions.

Chemokine Receptor 4

Another molecule involved in the LPS-receptor cluster is chemokine receptor 4(CXCR4).[58] CXCR4, whose expression was originally thought to be restricted to leukocytes, is now known to be up-regulated on exposure to bacterial products[101] and facilitates the endothelial cell inflammatory response. Most importantly, CXCR4 is the receptor for HIV. Interestingly LPS is a potent inhibitor of HIV-1 replication, possibly by down-regulating the CXCR4 receptor.[102]

Activation Clusters

Recently our group has focused on unraveling the molecular events that leads to the innate recognition of bacterial LPS. Using non-invasive biophysical techniques, such as fluorescence recovery after photo-bleaching (FRAP), we investigated the lateral mobility of LPS and different LPS receptors on living cells, before and after LPS stimulation. We demonstrated that LPS initially binds CD14, is then released and transferred to an immobile receptor or complex of receptors.[103] Subsequently, we used affinity chromatography and peptide mass fingerprinting to identify a complex of four LPS binding receptors. This multimeric receptor complex is comprised of HSP70, HSP90, chemokine receptor 4 (CXCR4) and growth differentiation factor 5 (GDF5).[58] Using Fluorescence resonance energy transfer (FRET), we confirmed the association of all four proteins with LPS, finding that they formed a complex after LPS ligation. The functional significance of this complex was then demonstrated using antibody-inhibition experiments. These showed that incubation with antibodies against the four-identified receptors prior to LPS stimulation abrogated LPS-induced TNF-α secretion.[58]

Similar findings were presented by Pfeiffer and co-workers who recently demonstrated the existence of a diverse group of receptors forming an activation cluster following LPS stimulation.[74] Using FRET they observed clustering of CD11b/CD18, CD14, CD55, CD81, Fcγ-R CD16a, scavenger receptor CD36 and TLR4 following LPS stimulation. LTA induced a similar receptor cluster. It is worth noting that most of these receptors had already been implicated in LPS signaling (such as CD11b/CD18, CD14, scavenger receptor CD36, CD55 and TLR4). Interestingly the components of the activation cluster changed when cells were challenged with ceramide; the CD16a, TLR4 and CD81 components were absent. This would suggest that different combinational associations of receptors might be responsible for the wide range of microbial stimuli that the innate immune system can recognize. Our recent experiments would agree with this (Triantafilou et al unpublished observations). We have found that different activation clusters are formed following stimulation with different LPS analogues such as pentaacyl lipid A and lipid VIa. Interestingly TLR4 does not seem to be recruited into these clusters following stimulation by LPS antagonists. Heine et al have also identified CD55 as an LPS-binding receptor and active element of an LPS-receptor cluster.[73] The existence of activation-induced receptor clusters has also been shown by Perera et al[104] who demonstrated the interaction of CD14, CD11b/CD18 and TLR4 during LPS and taxol-inducible gene expression.

It is very likely that the activation clusters observed by each group[58,73,74,104] form by the same basic mechanism. In addition Seydel and colleagues have suggested that a Ca^{2+}-dependent K^+-channel is probably also associated with an LPS-induced activation cluster.[105] This becomes even more plausible given that activation of K^+-channels has been observed after exogenous addition of HSP70.[106]

Microdomains

The plasma membrane of mammalian cells was once believed to be homogeneous, but it is now clear that it is discontinuous and contains microdomains that are essential for cellular function. Lipid rafts are defined as dynamic assemblies of lipids and specific proteins in the biological membrane. One of the roles of these lipid rafts (or microdomains) is to recruit and concentrate molecules involved in cellular signaling.[107] The accumulation of receptors and their associated signal transduction machineries into these domains has the effect of enhancing signaling efficacy.[108]

The acquired immune response seems to employ similar microdomains in order to achieve T-cell activation.[109] Receptors involved in the innate immune response, such as CD14, have previously been reported to reside in such rafts.[110] We have found that the receptor molecules that are implicated in LPS driven cell activation such as HSP70, HSP90, CXCR4, GDF5 and TLR4 (in addition to CD14), are all present in microdomains following LPS stimulation.[99] Raft-disrupting drugs such as nystatin or MCD, inhibit LPS induced TNF-α secretion, suggesting that lipid raft integrity is essential for LPS signaling.

Pfeiffer et al have also reported receptor cluster formation in lipid rafts.[74] Using FRET they reported LPS induced clustering of diverse receptors, including CD14, TLR4 and integrins within microdomains. Based on these findings we suggest that bacterial recognition centers on the ligation of CD14 by bacterial components, then the recruitment of multiple signaling molecules to the site of CD14-LPS complexes in lipid rafts.

The Shape of LPS

Although LPS is a potent stimulus of the innate immune response, it is intriguing to note that several lipid As from nonenterobacteria such as *Rhodobacter sphaeroides* (RSLA) and *Rhodobacter capsulatus,* as well as precursors and analogues of toxic lipid A from *Escherichia coli* LPS (such as tetraacyldisaccharide lipid A precursor and pentaacyl LPS) have been shown to inhibit LPS activation.[111,112]

An important question is what is requisite for lipid A and LPS to be active or antagonistic with respect to endotoxin action. Schromm and colleagues demonstrated that biological activity is dependent on the shape of the lipid A component.[113] LPS that assumed a conical shape (such as that from *E. coli*) was found to be biologically active, whereas LPS with a cylindrical shape (such as the LPS from *Rhodobacter sphaeroides*) was inactive.[114] Netea and colleagues have suggested that the shape of LPS determines its interaction with Toll-like receptors and defines whether or not it can activate the TLR.[115] As examples they mention *P. gingivalis* and *Leptospiral* LPS molecules that stimulate via TLR2 [61] and *E. coli* LPS which acts through TLR4.[116] Interestingly *N.meningitidis* LPS which assumes an intermediate conformation seems to engage both TLR2 and TLR4 (Fig. 1).[117*]

Conclusions

Our current understanding of the innate immune system has leapt forward in the last twenty years. The discovery of CD14 as the LPS receptor, and the emergence of the TLRs as the "signal transducers" have changed our view of the single-receptor model of innate immune recognition. But what comes next? Although there is no doubt that TLRs are the dominant signaling molecules involved, the question remains: have we identified all the proteins that comprise the sensing apparatus? The lack of convincing evidence to show the direct engagement of PAMPs to TLRs, as well as the multiple signaling cascades that are triggered lead us to believe that additional receptors exist.

In this chapter we have detailed some of the molecules and pathways that are involved in the innate recognition of bacteria. We've shown the various routes and proteins that different cell types employ to respond to bacteria. How these proteins co-ordinate to achieve innate immune recognition remains to be thoroughly explored. Ozinsky and colleagues have suggested that the repertoire of innate recognition is defined by the combinational association of TLRs.[64] Based on our previous findings of a receptor complex within lipid rafts post LPS exposure,[58,99] we would extend this notion by suggesting that innate immune responses must also depend on different combinational associations of receptors.[118] Moreover, the immune response is regulated by the type of bacteria and the cell types involved. In addition to a "core cluster" of molecules (such as CD14, TLR4 and possibly HSPs), different receptor molecules would be recruited in a ligand dependent fashion. Compartmentalization of the membrane in the form of lipid rafts would provide a dynamic microenvironment for the clustering of these receptors. As Netea and colleagues[115] have suggested, the shape or conformation of LPS will surely play a role, not only in the recruitment of particular TLRs but also in the sequestering of receptor(s) within the cluster. This lipid driven tailoring of the TLR associated signaling complex may be an essential component in the immune recognition of microbes.

---

* Review discussing the link between LPS conformers and hydrophobic activatory surfaces (Seong and Matzinger. Nat Rev Immunol 2004; 4:469-478).

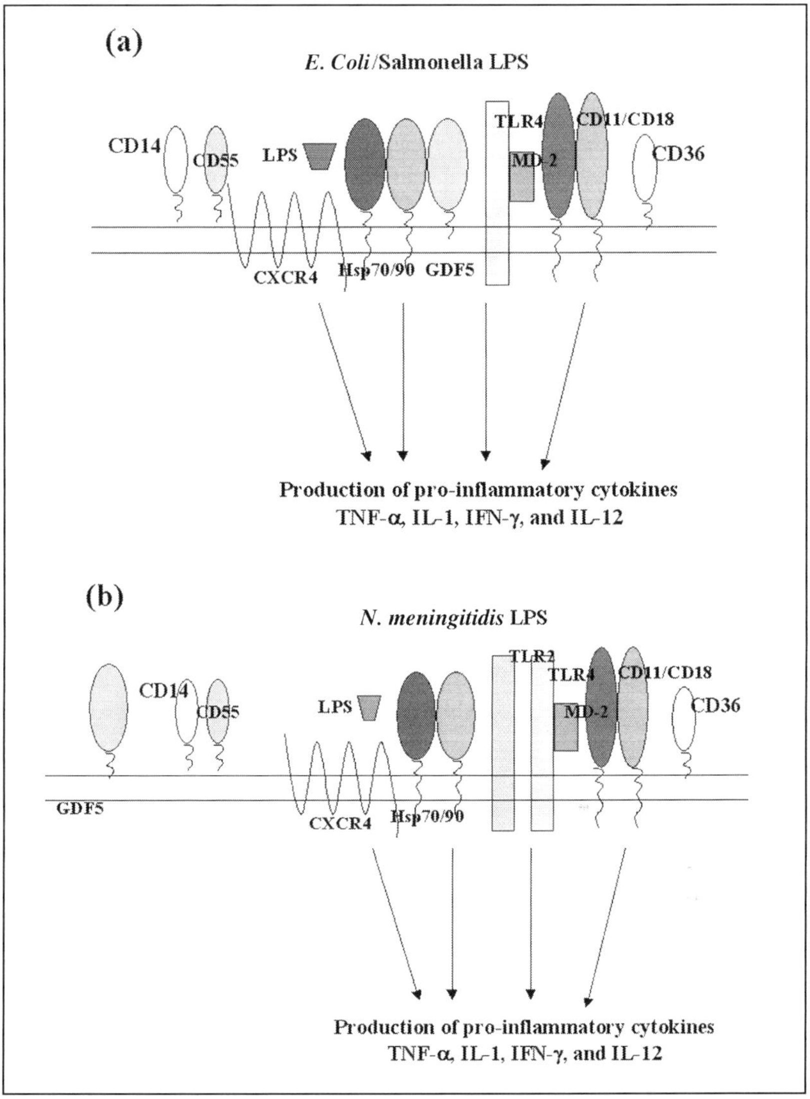

Figure 1. Hypothetical model of the formation of activation clusters in response to different stimuli. A) *E. coli* LPS (adopting a conical shape) induces a strong pro-inflammatory signal through the formation of an activation cluster containing a variety of receptors such as CD14, TLR4, HSPs, CXCR4, integrins etc. B) *N. meningitidis* with an intermediate shape induces the formation of a different activation cluster that contains HSPs, CXCR4, integrins, TLR2 and TLR4 and transduces a slightly different pro-inflammatory signal. Figure continued on next page.

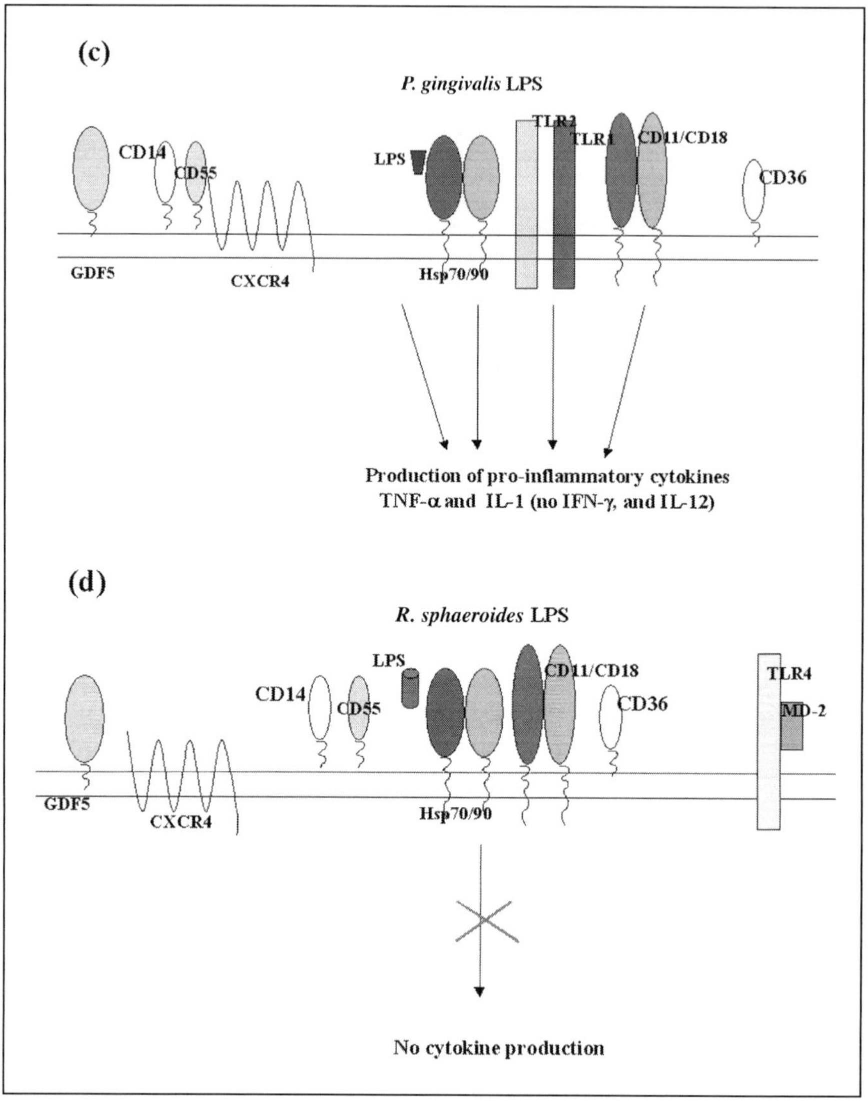

Figure 1. Continued. C) *P. gingivalis* LPS (conical) induces an activation cluster with HSPs, integrins, TLR2 and TLR1 and generates a weaker pro-inflammatory signal D) *R. sphaeroides* LPS (cylindrical) induces the formation of an activation cluster containing CD14, integrins, CD55, HSPs, but no TLRs and thus no pro-inflammatory signal.

There is no doubt that much more research needs to be done in order to complete our understanding of how the immune system "senses" pathogens and how the innate immune response activates the adaptive. Is it possible that heat shock proteins are the missing link between innate and adaptive immunity? Could they bind PAMPs for the innate wing and shuttle peptides to the acquired and does the combinational associations of receptors explain immune specificity? Lastly, how do TLRs associate with different accessory molecules and what is the role of transient supramolecular activation clusters in the non-specific branch of the immune system? Future work will, perhaps, find answers to these questions.

References

1. Medzhitov R, Janeway CA. An ancient system of host defense. Curr Opin Immunol 1998; 10:12-15.
2. Medzhitov R, Janeway CA. Decoding the patterns of self and nonself by the innate immune system. Science 2002; 296:298-300.
3. Grakoui A, Bromley SK, Sumen C et al. The immunological synapse: a molecular machine controlling T-cell activation. Science 1999; 285:221-227.
4. Monks CRF, Freiberg BA, Kupfer H et al. Three-dimensional segregation of supramolecular activation clusters in T-cells. Nature 1998; 395:82-86.
5. Shear MJ, Turner FC. Chemical treatment of tumors. V. Isolation of the hemorrhage-producing fraction from Serratia marcescens (Bacillus prodigiosus) culture filtrates. J Natl Cancer Inst 1943; 4:81-97.
6. Rietschel E, Brade H, Holst O et al. Bacterial endotoxin: chemical constitution, biological recognition, host response and immunological detoxification. Curr Top Microbiol Immunol 39-81.
7. Raetz C. Biochemistry of endotoxins. Ann Rev Biochem 1990; 59:129-170.
8. Haeffner-Cavaillon N, Caroff M, Cavaillon JM. Interleukin-1 induction by lipopolysaccharides: structural requirements of the 3-deoxy-D-manno-2-octulosonic acid (KDO). Mol Immunol 26:485-494.
9. Boivin A. Preparation of specific polysaccharides of bacteria. CR Seances Soc Biol Fil 490-504.
10. Wright SD, Ramos RA, Tobias PS et al. CD14, a receptor for complexes of lipopolysaccharide (LPS) and LPS binding-protein. Science 1990; 249:1431-1433.
11. Ulevitch RJ, Tobias PS. Recognition of endotoxin by cells leading to transmembrane Curr Opin Immunol 1994; 6:125-130.
12. Hailman E, Lichenstein HS, Wurfel MM et al. Lipopolysaccharide (LPS)-Binding protein accelerates the binding of LPS to CD14. J Exp Med 1994; 179:269-277.
13. Lee JD, Kato K, Tobias PS et al. Transfection of CD14 into 70Z/3 cells dramatically sensitivity to complexes of lipopolysaccharide (LPS) and LPS binding- protein. J Exp 175:1697-1705.
14. Haziot A, Ferrero E, Kontgen F et al. Resistance to endotoxin shock and reduced dissemination gram- negative bacteria in CD14-deficient mice. Immunity 1996; 4:407-414.
15. Gegner JA, Ulevitch RJ, Tobias PS. Lipopolysaccharide (LPS) Signal-transduction and clearance-roles for LPS binding-protein and membrane CD14. J Biol Chem 1995; 270:5319-5325.
16. Lynn WA, Liu Y, Golenbock DT. Neither CD14 nor serum is absolutely necessary forof mononuclear phagocytes by bacterial lipopolysaccharide. Infect Immun 1993; 61:4456-
17. Blondin C, Ledur A, Cholley B et al. Binding of LPS to normal human monocytes inis mediated by both sCD14 and mCD14 molecules. Tissue Antigens 1996; 48:445-450
18. Troelstra A, Antal-Szalmas P, de Graaf-Miltenburg LA et al. Saturable CD14-dependent of fluorescein-labelled lipopolysaccharide to human monocytes. Infect Immun 1997; 65:
19. Triantafilou M, Triantafilou K, Fernandez N. Rough and smooth forms of fluorescein- bacterial endotoxin exhibit CD14/LBP dependent and independent binding that is influenced endotoxin concentration. Eur J Biochem 2000; 267:2218-2226.
20. Gessani S, Testa U, Varano B et al. Enhanced production of LPS-induced cytokines during differentiation of human monocytes to macrophages. J Immunol 1993; 151:3758-3766.
21. Hampton RY, Golenbock, DT, Penman M et al. Recognition and plasma clearance of by scavenger receptors. Nature 1991; 352:342-344.
22. Wright SD, Jong MTC. Adhesion-promoting receptors on human macrophages recognize Escherichia coli by binding to lipopolysaccharide. J Exp Med 1996; 164:1876-1882.
23. Wright SD. Adhesion-promoting receptors on phagocytes. Agents Actions 1991; 35:
24. Wright SD, Ramos RA, Hermanowskivosatka A et al. Activation of the adhesive capacity on neutrophils by endotoxin-dependence on lipopolysaccharide binding-protein and CD14. Med 1991; 173:1280-1286.
25. Medvedev AE, Flo T, Ingalls RR et al. Involvement of CD14 and complement receptors CR4 in nuclear factor-kappa B activation and TNF production induced by lipopolysaccharide group B streptococcal cell walls. J Immunol 1998; 160:4536-4542.
26. Troelstra A, de Graaf-Miltenburg LAM, van Bommel T et al. Lipopolysaccharide-coated erythrocytes activate human neutrophils via CD14 while subsequent binding is through CD11b/ Immunol 1999; 162:4220-4225.

27. Ingalls R, Monks B, Savedra R et al. CD11/CD18 and CD14 share a common Lipid pathway. J Immunol 1998; 161:5413-5420.
28. Ingalls RR, Arnaout MA, Golenbock D. Outside-in signaling by lipopolysaccharide through tailless integrin. J Immunol 1997; 159:433-440.
29. Ingalls RR, Golenbock DT. CD11c/CD18, a transmembrane signaling receptor for lipopolysaccharide. J Exp Med 1995; 181:1472-1479.
30. Williams MJ, Rodriguez D, Kimbell A et al. The 18-wheeler mutation reveals complex antibacterial gene regulation in Drosophila host defense. EMBO J 1997; 16:6120-6125.
31. Shakhov AN, Collart MA, Vassalli P et al. kB-type enhancers are involved in lipopolysaccharide-mediated transcriptional activation of the tumor necrosis factor-a gene in primary macrophages. J Exp Med 1990; 171:35-47.
32. Han J, Lee JD, Bibbs L et al. A map kinase targeted by endotoxin and hyperosmolarity in mammalian-cells. Science 1994; 265:808-811.
33. Hambleton J, Weinstein SL, Lem L et al. Activation of c-Jun N-terminal kinase in bacterial lipopolysaccharide-stimulated macrophages. Proc Natl Acad Sci USA 1996; 93:2774-2778.
34. Herrera-Velit P and Reiner NE. Bacterial lipopolysaccharide induces the association and coordinate activation of p53/56lyn and phosphatidylinositol 3 kinase in human monocytes. J Immunol 1996; 156:1157-1165.
35. Herrera-Velit P, Knutson KL, Reiner NE. Phosphatidylinositol-3-kinase -dependent activation of protein kinase C in bacterial lipopolysaccharide-treated human monocytes. J Biol Chem 1997; 272:16445-16452.
36. Lei MG, Qureshi N, Morrison DC. Specific endotoxin lipopolysaccharide binding protein on murine splenocytes. J Immunol 1998; 141:996-1005.
37. Dziarski R. Peptidoglycan and lipopolysaccharide bind to the same binding site on lymphocytes. J Biol Chem 1991; 266:4719-4725.
38. Hara-Kuge S, Amano F, Nishijima M et al. Isolation of a lipopolysaccharide (LPS)-resistant mutant, with defective LPS binding, of cultured macrophage-like cells. J Biol Chem. 1990; 265:6606-6610.
39. Kirkland TN, Virea GD, Kuus-Reichel T et al. Identification of lipopolysaccharide binding proteins in 70Z/3 cells by photoaffinity crosslinking. J Biol Chem 1990; 265:9520-9525.
40. Hampton RY, Golenbock D, Raetz RHC. Lipid A binding sites in membranes of macrophage tumor cells. J Biol Chem 1988; 263:14802-14807.
41. Kirikae T, Kirikae F, Schade UF et al. Detection of lipopolysaccharide binding proteins on membranes of murine lymphocyte and macrophage-like cell lines. FEMS Microbiol and Immunol 1991; 76:327-336.
42. El-Samalouti VT, Schletter J, Brade H et al. Detection of lipopolysaccharide (LPS)-binding membrane proteins by immuno-coprecipitation with LPS and anti-LPS antibodies. Eur J Biochem 1997; 250:418-424.
43. Triantafilou K, Triantafilou M, Dedrick RL. Interactions of bacterial lipopolysaccharide and peptidoglycan with a 70kDa and an 80kDa protein on the cell surface of CD14 positive and CD14 negative cells. Hum Immunol 2001; 62:50-63.
44. Lemaitre B, Nicolas E, Michaut L et al. The dorsoventral regulatory gene cassette spatzel/Toll/cactus controls the potent antifungal response in Drosophila adults. Cell 1996; 86:973-975.
45. Medzhitov R, Preston-Hurlburt P, Janeway CA. A human homologue of the Drosophila Toll protein signals activation of adaptive immunity. Nature 1997; 388:394-397.
46. Poltorak A, He XL, Smirnova I et al. Defective LPS signaling in C3H/Hej and C57BL/10ScCr mice:Mutations in TLR4 gene. Science 1998; 282:2085-2088.
47. Qureshi ST, Lariviere L, Leveque G et al. Endotoxin-tolerant mice have mutations in toll-like receptor 4 (TLR4). J Exp Med 1999; 189:615-625.
48. Akashi S, Shimazu R, Ogata H et al. Cutting Edge:Cell Surface expression and Lipopolysaccharide signaling via the Toll-like receptor 4-MD-2 complex on mouse peritoneal macrophages. J Immunol 2000; 164:3471-3475.
49. Hoshino K, Takeuchi O, Kawai T et al. Cutting Edge:Toll-like receptor 4 (TLR4) deficient mice are hyporesponsive to lipopolysaccharide:Evidence for TLR4 as the LPS gene product. J Immunol 1999; 162:3749-3752.

50. Shimazu R, Akashi S, Ogata H et al. MD-2, a molecule that confers lipopolysaccharide responsiveness on Toll-like receptor 4. J Exp Med 1999; 189:1777-1782.
51. Akashi S, Ogata H, Kirikae F et al. Regulatory roles for CD14 and phosphatidylinositol in the signaling via Toll-like receptor 4-MD-2. Biochem Biophys Res Comm 2000; 268:172-177.
52. da Silva Correia J, Ulevitch RJ. MD-2 and TLR4 N-linked glycosylations are important for a functional lipopolysaccharide receptor. J Biol Chem 2002; 277:1845-1854.
53. Takeuchi O, Hoshino K, Kawai T et al. Differential roles of TLR2 and TLR4 in recognition of gram-negative and gram-positive bacterial cell wall components. Immunity 1999; 11:443-451.
54. Takeuchi O, Hoshino K, Akira S. TLR2-deficient and MyD88-deficient mice are highly susceptible to Staphylococcus aureus infection. J Immunol 2000; 165:5392-5396.
55. Hemni H, Takeuchi O, Kawai T. A Toll-like receptor recognizes bacterial DNA. Nature 2000; 408:740-745.
56. Yang RB, Mark MR, Gray A. Toll-like receptor-2 mediates lipopolysaccharide-included cellular signaling. Nature 1998; 395:284-288.
57. Kirschning CJ, Wesche H, Ayers TM. Human Toll-like receptor 2 confers responsiveness to bacterial ipopolyssacharide. J Exp Med 1998; 188:2091-2097.
58. Triantafilou K, Triantafilou M, Dedrick RL. A CD14-independent LPS receptor cluster. Nat Immunol 2001; 4:338-345.
59. Hirschfeld M, Ma Y, Weis JH et al. Repurification of lipopolysaccharide eliminates signaling through both human and murine toll-like receptor 2. J Immunol 2000; 165:618-622.
60. Yoshimura A, Lien E, Ingalls R et al. Cutting Edge:Recognition of Gram-positive bacterial cell wall components by the innate immune system occurs via Toll-like receptor 2. J Immunol 1999; 163:1-5.
61. Werts C, Tapping RI, Mathison JC et al. Leptospiral lipopolysaccharide activates cells through a TLR-2-dependent mechanism. Nat Immunol 2001; 2:346-352.
62. Means TK, Lien E, Yoshimura A et al. The CD14 ligands Lipoarabinomannan and lipopolysaccharide differ in their requirement for Toll-like receptors. J Immunol 1999; 163:6748-6755.
63. Hajjar AM, O'Mahony DS, Ozinsky A et al. Functional interactions between toll-like receptor (TLR) 2 and TLR1 or TLR6 in response to phenol-soluble modulin. J Immunol 2001; 166:15-19.
64. Ozinsky A, Underhill DM, Fontenot JD et al. The repertoire for pattern recognition of pathogens by the innate immune system is defined by cooperation between Toll-like receptors. Proc Natl Acad Sci USA 2000; 97:13766-13771.
65. Flo T, Halaas O, Lien E et al. Ryan L et al. Human Toll-like receptor 2 mediates monocyte activation by Listeria monocytogenes, but not by Group B Streptococci or lipopolysaccharide. J Immunol 2000; 164:2064-2069.
66. Hirschfeld M, Kirschning CJ, Schwandner R et al. Cutting Edge:Inflammatory signaling by Borrelia burgdorferi lipoproteins is mediated by Toll-like receptor 2. J Immunol 1999; 163:2382-2386.
67. Lien E, Sellati T, Yoshimura A et al. Toll-like receptor 2 functions as a pattern recognition receptor for diverse bacterial products. J Biol Chem 1999; 274:33419-33425.
68. Alexopoulou L, Holt AC, Medzhitov R et al. Recognition of double-stranded RNA and activation of NF-kappa B by Toll-like receptor 3. Nature 2001; 413:732-738.
69. Hayashi F, Smith KD, Ozinsky A et al. The innate immune response to bacterial flagellin is mediated by Toll-like receptor 5. Nature 2001; 410:1099-1103.
70. Henneke P, Takeuchi O, Van Strijp JA et al. Novel engagement of CD14 and multiple Toll-like receptors by Group B Streptococci. J Immunol 2001; 167:7069-7076.
71. Takeuchi O, Sato S, Horiuchi T et al. Cutting edge:role of Toll-like receptor 1 in mediating immune response to microbial lipoproteins. J Immunol 2002; 169:10-14. 2002.
72. El-Samalouti VT, Schletter J, Chyla I et al. Identification of the 80-kDa LPS-binding protein (LMP-80) as decay-accelerating factor (DAF, CD55). FEMS Immunol Med Microbiol 1999; 23:259-269.
73. Heine H, Ulmer AJ, El-Samalouti VT et al. Decay-accelerating factor (DAF/CD55) is a functional active element of the LPS receptor complex. J Endotoxin Res. 2001; 7:227-231.
74. Pfeiffer A, Bottcher A, Orso E et al. Lipopolysaccharide and ceramide docking to CD14 provokesligand-specific receptor clustering in rafts. Eur J Immunol 2001; 31:3153-3164.

75. Bouchon A, Dietrich J, Colonna M. Cutting Edge: Inflammatory responses can be triggered by TREM-1, a novel receptor expressed on neutrophils and monocytes. J Immunol 2000; 164:4991-4995.
76. Bouchon A, Facchetti F, Weigand MA et al. TREM-1 amplifies inflammation and is crucial mediator of septic shock. Nature 2001; 410:1103-1107.
77. Nathan C, Ding A. TREM-1:a new regulator of innate immunity in sepsis syndrome. Nat Med 2001; 7:530-532.
78. Bleharski JR, Kiessler V, Buonsanti C et al. A role for triggering receptor expressed on myeloid cells-1 in host defense during the early-induced and adaptive phases of the immune response. J Immunol 2003; 170:3812-3818.
79. Aoki N, Kimura S, Xing Z. Role of DAP12 in innate and adaptive immune responses. Curr Pharm Des 2003; 9:7-10.
80. Miyake K, Yamashita Y, Ogata M et al. RP105, a novel B cell surface molecule implicated in B cell activation, is a member of the leucine-rich repeat protein family. J Immunol 1995; 154:3333-3340.
81. Ogata H, Su I, Miyake K et al. The Toll-like Receptor Protein RP105 regulates lipopolysaccharide signaling in B cells. J Exp Med 2000; 192:23-29.
82. Nagai Y, Shimazu R, Ogata H et al. Requirement for MD-1 in cell surface expression of RP105/CD180 and B-cell responsiveness to lipopolysaccharide. Blood 2002; 99:1699-1705.
83. Lemjabbar H and Basbaum C. Platelet-activating factor receptor and ADAM10 mediate responses to Staphylococcus aureus in epithelial cells. Nat Med 2002; 8:41-46.
84. Kaur I, Voss SD, Gupta RS et al. Human peripheral gamma delta T cells recognize hsp 60 molecules on Daudi Burkitt's lymphoma cells. J Immunol 1993; 150:2046-2055.
85. Multhoff G, Otzler C, Jennen L et al. Heat shock protein 72 on tumour cells. J Immunol 1997; 158:4341-4350.
86. Tsuboi N, Ishikawa M, Tamura Y et al. Monoclonal antibody specifically reacting against 73-kiolodalton heat shock protein:possible expression on mammalian cell surface. Hybridoma 1994; 48:2798-2804.
87. Amura Y, Tsuboi N, Sato N et al. 70 kDa heat shock cognate protein is a transformation-associated antigen and a possible target for the host's anti-tumor immunity. J Immunol 1993; 151:5516-5520.
88. Todryk SM, Melcher AA, Dalgleish AG et al. Heat shock proteins refine the danger theory. Immunol 2000; 99:334-337.
89. Takashima S, Sato N, Kishi A et al. Involvement of peptide antigens in the cytotoxicity between 70 kDa heat shock cognate protein-like molecule and CD3+, CD4-,CD8-, TCRab- killer T-cells. J Immunol 1996; 157:3391-3395.
90. Breloer M, Fleischer B, von Bonin A. In vivo and in vitro activation of T cells after administration of Ag-negative heat shock proteins. J Immunol 1999; 162:3141-3147.
91. Byrd CA, Bornmann W, Erdjument-Bromage H et al. Heat shock 90 mediates macrophage activation by Taxol and bacterial lipopolysaccharide. Proc.Natl.Acad.Sci.USA 1999; 96:5645-5650.
92. Vabulas RM, Ahmad-Nejad P, Ghose S et al. Hsp70 as endogenous stimulus of the Toll/Interleukin-1 receptor signal pathway. J Biol Chem 2002; 277:15107-15112.
93. Vabulas RM, Ahmad-Nejad P, da Costa C et al. Endocytosed HSP60s use Toll-like receptor 2 (TLR2) and TLR4 to activate the Toll/Interleukin-1 receptor signaling pathway in innate immune cells. J Biol Chem 2001; 276:31332-31339.
94. Vabulas RM, Braedel S, Hilf N et al. The endoplasmic reticulum-resident heat shock protein Gp96 activates dendritic cells via the Toll-like receptor 2/4 pathway. J Biol Chem 2002; 277:20847-20853.
95. Asea A, Kraeft SK, Kurt-Jones EA et al. Hsp70 stimulates cytokine production through a CD14-dependent pathway, demonstrating its dual role as a chaperone and cytokine. Nat Med 2000; 6:435-442.
96. Gao B and Tsan MF. Endotoxin contamination in recombinant human heat shock protein 70 (Hsp70) preparation is responsible for the induction of tumor necrosis factor a release by murine macrophages. J Biol Chem 2003; 278:174-179.
97. Zhu FG and Pisetsky DS. Role of the heat shock protein 90 in immune response stimulation by bacterial DNA and synthetic oligonucleotides. Infect Immun 2001; 69:5546-5552.

98. Bandholtz L, Guo Y, Palmberg C et al. Hsp90 binds CpG oligonucleotides directly:implications for hsp90 as a missing link in CpG signaling and recognition. Cell Mol Life Sci 2003; 60:1-8.
99. Triantafilou M, Miyake K, Golenbock D et al. Mediators of the innate immune recognition of bacteria concentrate in lipid rafts and facilitate lipopolysaccharide-induced cell activation. J Cell Sci 2002; 115:2603-2611.
100. Broquet AH, Thomas G, Masliah J et al. Expression of the molecular chaperone hsp70 in detergent resistant microdomains correlates with its membrane delivery and release. J Biol Chem 2003; 278:21601-21606.
101. Moriuchi M, Moriuchi H, Turner W et al. Exposure to bacterial products renders macrophages highly susceptible to T-tropic HIV-1. J Clin Invest 1998; 102:1540-1550.
102. Juffermans NP, Paxton WA, Dekkers PE et al. Up-regulation of HIV coreceptors CXCR4 and CCR5 on CD4(+) T cells during human endotoxemia and after stimulation with mycobacterial antigens:the role of cytokines. Blood 2000; 96:2649-2654.
103. Triantafilou K, Triantafilou M, Ladha S et al. Fluorescence recovery after photobleaching reveals that lipopolysaccharide rapidly transfers from CD14 to heat shock proteins 70 and 90 on the cell membrane. J Cell Sci 2001; 114:2535-2545.
104. Perera PY, Mayadas TN, Takeuchi O et al. CD11b/CD18 acts in concert with CD14 and Toll-like receptor (TLR)4 to elicit full lipopolysaccharide and taxol-inducible gene expression. J Immunol 2001; 66:574-581.
105. Seydel U, Scheel O, Muller K et al. A K+ channel is involved in LPS signaling. J Endotox Res 2001; 7:243-247.
106. Negulyaev YA, Vedernikova EA, Kinev A et al. Exogenous heat shock protein 70 activates potassium channels in U937 cells. Biochem Biophys Acta 1996; 1282:156-162.
107. Pralle A, Keller P, Florin EL et al. Sphingolipid-cholesterol rafts diffuse as small entities in the plasma membrane of mammalian cells. J Cell Biol 2000; 148:997-1007.
108. Vereb G, Matko J, Vamosi G et al. Cholesterol-dependent clustering of IL-2Ra and its colocalization with HLA and CD48 on T lymphoma cells suggest their functional association with lipid rafts. Proc Natl Acad Sci USA 2000; 97:6013-6018.
109. Anderson HA., Hiltbold EM, Roche PA. Concentration of MHC class II molecules in lipid rafts facilitates antigen presentation. Nat Immunol 2000;1:156-162.
110. Wang PY, Kitchens R, Munford RS. Bacterial lipopolysaccharide binds to CD14 in low density domains of the monocyte-macrophage plasma membrane. J Inflamm 1996; 47:126-137.
111. Rose JR, Christ WJ, Bristol JR et al. Agonistic and antagonistic activities of bacterially derived Rhodobacter sphaeroides lipid A:comparison with activities of synthetic material of the proposed structure and analogs. Infect Immun 1995; 63:833-839.
112. Loppnow H, Libby P, Freudenberg MA et al. Cytokine induction by lipopolysaccharide (LPS) corresponds to the lethal toxicity and is inhibited by non-toxic Rhodobacter capsulatus LPS. Infect Immun 1990; 58:3743-3750.
113. Schromm A, Bradenburg K, Loppnow H et al. Biological activities of lipopolysaccharides are determined by the shape of their lipid A portion. Eur J Biochem 2000; 267:2008-2013.
114. Bradenburg K, Lindner B, Schromm A et al. Physiological characteristics of triacyl lipid A partial structure OM-174 in relation to biological activity. Eur J Biochem 2000; 267:3370-3377.
115. Netea MG, van Deuren M, Kullberg BJ et al. Does the shape of lipid A determine the interaction of LPS with Toll-like receptors? Trends Immunol 2002; 23:135-139
116. Hirschfeld M, Weis JJ, Toshchakov V et al. Signaling by toll-like receptor 2 and 4 agonists results in differential gene expression in murine macrophages. Infect Immun 2001; 69:1477-1482.
117. Wyllie DH, Kiss-Toth, E, Visintin A et al. Evidence for an accessory protein function for Toll-like receptor 1 in anti-bacterial responses. J Immunol 2000;165:7125-7132.
118. Triantafilou M, Triantafilou K. Lipopolysaccharide recognition:CD14, TLRs and the LPS-activation cluster. Trends Immunol 2002; 22:295-298.

Chapter 4

Interleukin-1 Receptor/Toll-Like Receptor Signaling

Harald Wajant, Peter Scheurich and Frank Henkler

Abstract

Upon stimulation Interleukin-1 receptor (IL-1R), the related IL-18R as well as the Toll-like receptors (TLRs), trigger signaling pathways that activate the transcription factors, nuclear factor κB (NF-κB), activator protein 1 (AP1) and interferon regulatory factor 3 (IRF-3). All members of the IL-1R/TLR family commonly induce NF-κB and AP1 whereas IRF-3 is targeted specifically by TLR3 and TLR4. While IRF-3 activation has only recently been recognized, NF-κB and AP1 activation, in particular by IL-1R, has been intensively studied. Indeed, most of the molecular components that transduce the IL-1R signal to activate NF-κB have been identified. There is considerable evidence that these proteins are also relevant to TLR signaling. Although there is, in some aspects, a very detailed understanding of how the various signaling proteins coordinate to achieve IL-1R driven NF-κB activation there are still gaps in the overall picture. The following chapter aims to give a comprehensive review on IL-1R/TLR signaling, focusing on the similarities and differences between these pathways in humans and in *Drosophila*. To facilitate a proper discussion of the molecular mechanisms involved in IL-1R/TLR signaling the relevant proteins will be introduced before IL-1R signaling in mammals and Toll signaling in flies are discussed in detail.

The "Hardware" of IL-1R/TLR Signaling

The IL-1R/TLR Family

The interleukin-1 receptor (IL-1R)/Toll-like receptor (TLR) family consists of two distinct subgroups of transmembrane receptors that express a conserved intracellular signaling domain called the Toll/interleukin-1 receptor (TIR) domain. TLRs share an extracellular domain comprising several leucine-rich repeats whereas IL-1R-related proteins contain immunoglobulin-like motifs in their extracellular domains. The TIR domain mediates protein-protein interactions with other TIR domain-containing proteins.[1] The TIR domain containing MyD88 adapters are one such group of proteins. In humans the IL-1R/TLR family comprises IL-1R, IL-1 receptor accessory protein (IL-1RAcp), IL-18R, IL-18RAcp, interleukin-1 receptor related protein (IL-1Rrp2), IL-1 receptor accessory protein-like (IL-1RAPL), T1/ST2 and TLR1 – TLR10. The fly family members include Toll, 18-Wheeler and Toll-3 – Toll-8.[2,3] All the TLRs described in mammals appear to have roles in innate immunity and IL-1R and IL-18R regulate both innate and adaptive immune responses. In *Drosophila* only Toll contributes

Toll and Toll-Like Receptors: An Immunologic Perspective, edited by Tina Rich. ©2005 Eurekah.com and Kluwer Academic / Plenum Publishers.

to innate immunity. For the other Toll proteins there are, as yet, no data to suggest anti-microbial activities.[2,3]

The MyD88 Adapter Protein Family

The MyD88 adapter proteins are involved in IL-1R/TLR signaling and are characterized by their TIR domain (Fig. 1). In mammals three TIR domain-containing adapters have been identified: Myeloid differentiation factor 88 (MyD88), MyD88-adapter-like (MAL) also called TIR domain-containing adapter protein (TIRAP) and TIR-containing adaptor molecule (TICAM)-1 also called TIR domain-containing adapter inducing IFN-beta (TRIF).[4-8] In flies there is a single equivalent protein called DmMyD88.[9] MyD88 and DmMyD88 consist of an amino-terminal death domain, a protein-protein-interaction motif originally described in the context of apoptotic signaling, and a carboxyl-terminal TIR domain. Both domains, separated by a short intermediate domain, drive homodimerization with other death domain and TIR domain containing proteins.[9,10] In contrast to MyD88 both MAL/TIRAP and TICAM-1/TRIF, lack death domains.[5-8]

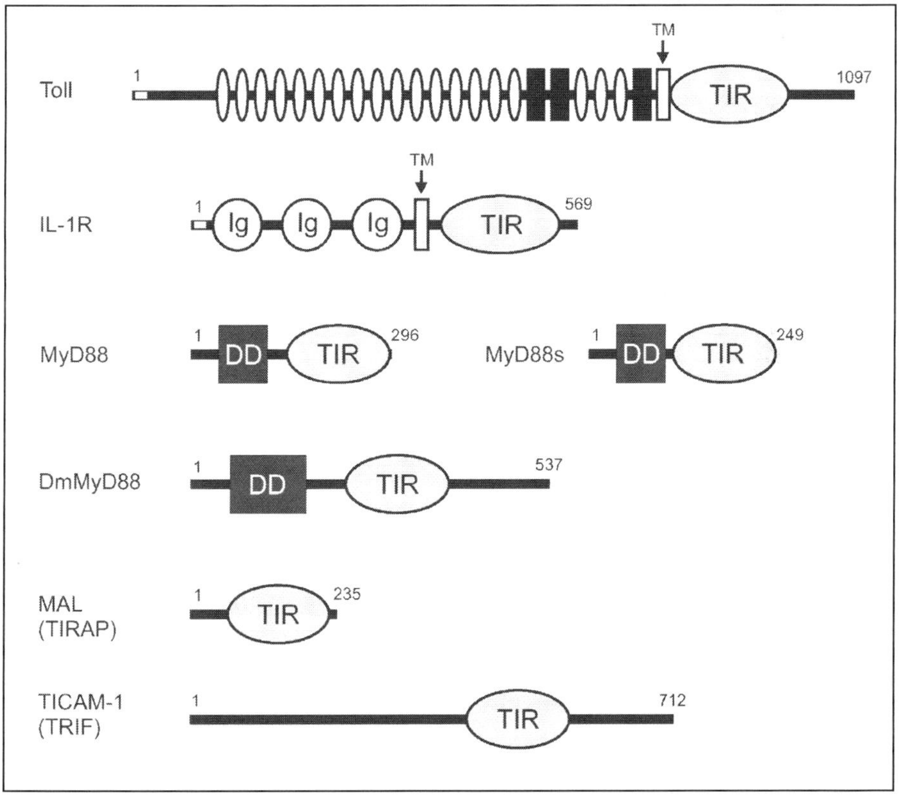

Figure 1. Domain architecture of TIR domain-containing proteins. IL-1R and Toll are representative of two subgroups of the IL-1R/TLR family that either have an extracellular domain with leucine-rich repeats (small open ellipses) or immunoglobulin-like motifs (Ig labeled circles). Toll/interleukin-1 receptor domains and death domains are labeled TIR and DD, respectively. Leader peptides and transmembrane domains of IL-1R and Toll are indicated by small gray boxes or open boxes labeled TM.

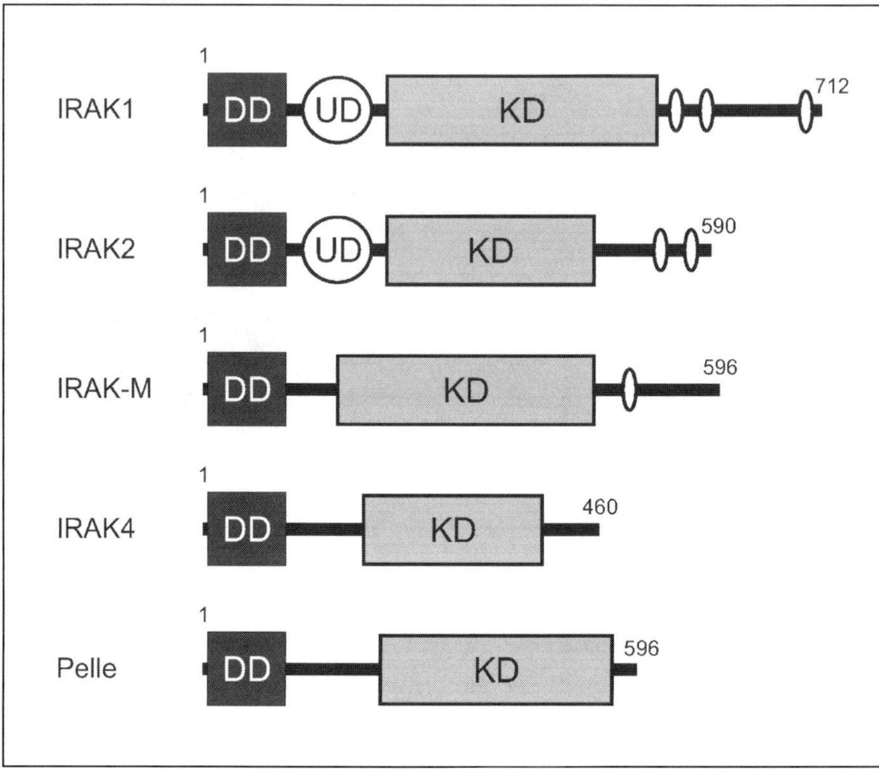

Figure 2. An undetermined domain found in IRAK1 and IRAK2, that splits NF-κB and JNK activation, at least by IRAK1, is labeled UD. Death domains and Kinase domains are labeled DD and KD respectively.

The Pelle/Interleukin-1 Receptor-Associated Kinase (IRAK) Family

The Pelle/IRAK family consists of a group of phylogenetically conserved serine-threonine kinases with an amino-terminal death domain and a central kinase domain (Fig. 2). Pelle was identified because of its role in Toll-mediated dorsal-ventral patterning in flies and IRAK1 was isolated as an IL-1R associated protein.[11,12] IRAK2, IRAK4 and IRAK-M were then identified in EST data base screens for IRAK1 related proteins.[13-16] Pelle, IRAK1 and IRAK4 all have a strong autophosphorylation capacity, unlike IRAK2 and IRAK-M.[14]

The TRAF Adapter Protein Family

The tumor necrosis factor (TNF) receptor-associated factor (TRAF) family of proteins comprises a group of phylogenetically conserved adapters. These adapters connect members of the IL-1R/TLR and TNF receptor superfamilies to signaling cascades which ultimately activate NF-κB and MAP kinases.[17] The TRAF proteins were named after their founder members which were identified on the basis of their interaction with TNF receptor 2.[18] The characteristic structural feature of the TRAF proteins is a carboxyl terminal domain of about 200 amino acids called the TRAF domain. This domain can be subdivided into a coiled coil amino-terminal TRAF-N domain and a highly conserved carboxyl-terminal TRAF-C domain (Fig. 3).[18] There are six TRAF proteins in humans and mice (TRAF1 – TRAF6), three in *Drosophila melanogaster* (DTRAF1 – DTRAF3) and one in *Caenorhabditis elegans* (CeTRAF).[17,19] The amino-termini of all the mammalian TRAFs, with the exception of TRAF1, contain a RING finger then five

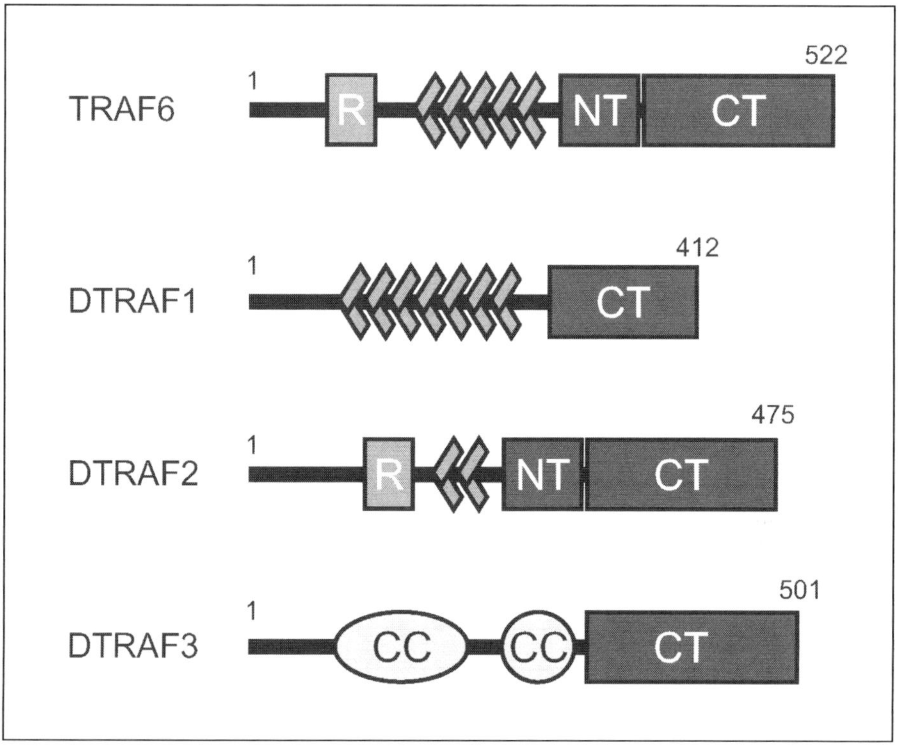

Figure 3. Domain architecture of TRAF proteins involved in IL-1R/TLR signaling. The highly conserved carboxyl-terminal domain and the less conserved amino-terminal domain of TRAF proteins are labeled. RING domains and regions adopting a coiled-coil structure are labeled R and CC. Zinc fingers are indicated by arrowheads.

or seven regularly spaced zinc fingers. The amino-terminal part of TRAF1 has a single zinc finger. Outside of this domain, TRAF1 shows no homology to any other protein domain of known function.[18] While the domain architecture of CeTRAF is similar to TRAF2–TRAF6, the domain composition at the amino-termini of fly TRAF proteins shows some variability (Fig. 3). DTRAF1 has seven zinc fingers and no RING domain, DTRAF2 has a RING domain followed by two zinc fingers and DTRAF3 is devoid of RING or zinc finger motifs.[17,19] Crystallographic analysis of the TRAF domains of TRAF2, TRAF3 and TRAF6 showed that TRAF proteins are organized as trimers with a three-fold symmetry.[20-23] The TRAF domains of three protomers adopt a mushroom-like structure with the TRAF-C domains as the cap and the elongated and intertwined coiled-coil domains as the stalk. TRAF proteins interact with "activated" trimeric TNF receptor complexes via their TRAF-C domain. Each TRAF-C domain binds a linear motif in the cytoplasmic portion of a single TNF receptor in the trimeric ligand-TNF receptor complex. Thermodynamic characterization of the interaction between a variety of monomeric receptor peptides and the TRAF domains of TRAF2 and TRAF6 resulted in dissociation constants between 40 and 2000 μM.[23,24] The low affinity of TRAF proteins to peptides derived from monomeric TNF receptors emphasizes the relevance of avidity for the recruitment of trimeric TRAF proteins to ligand-induced TNF receptor trimers or even clusters of receptor complexes. TRAF proteins may also be indirectly recruited to activated receptors. TRAF2 requires the adapter protein TNF receptor associated death domain

Table 1. TRAF6-associated proteins

Protein	Function	References
A20	Inhibition of NF-κB signaling	97,99,101
ASK1*	Disruption of TRAF6-TAK1 interaction and inhibition of IL-1-induced NF-κB activation	148
Cezanne*	Interferes with NF-κB activation	149
ECSIT*	Implicated in IL-1R/TLR-induced NF-κB activation by MEKK1 processing	150
IRAK1	Central role in IL-1R/TLR signaling, for details see text	-
IRAK2	Interacts with MAL/TIRAP and substitutes for IRAK1 in the IRAK1-MyD88 interplay	5,14
IRAK-M	Down-regulates IL-1R/TLR signaling	95
IRAK4	Central role in IL-1R/TLR signaling, for details see text	-
MEKK1	Implicated in IL-1-induced activation of JNK and NF-κB, but a crucial role has yet not been confirmed in knock out mice	47,51
NIK	Implicated in IL-1-induced activation of JNK and NF-κB, but a crucial role has yet not been confirmed in knock out mice	49,50
Pellino-1	Involved in the transfer of IRAK1, IRAK4 and TRAF6 from the IL-1R signaling complex to the membrane-associated TAK1 complex	40
c-Src	Implicated in TRAF6-dependent activation of Akt/PKB	151
SIMPL*	Enhancement of IRAK1-mediated activation of IKK2/IKKβ	152
T6BP*	Interacts with TRAF6 upon IL-1 stimulation, relevance for IL-1 signaling is unclear	153
TAB1	Involved in TRAF6- and TAK1-mediated activation of the IKK complex, for details see text	-
TAB2	Involved in TRAF6- and TAK1-mediated activation of the IKK complex, for details see text	-
TTRAP*	Interacts with TRAF6, relevance for IL-1 signaling is unclear	154
TRABID*	Interacts with TRAF6, relevance for IL-1 signaling is unclear	149
TIZ*	Interacts with TRAF6, relevance for IL-1 signaling is unclear	155
TIFA*	Constitutively associates with TRAF6 and with IRAK1 upon IL-1 stimulation	156

*these proteins are not discussed in the text: ASK1= apoptosis inducing kinase-1; Cezanne= cellular zinc finger anti-NF-kappa B; ECSIT= evolutionarily conserved signalling intermediate in Toll pathways; SIMPL= signalling molecule that associates with the mouse pelle-like kinase; T6BP= TRAF6 binding protein; TTRAP= TRAF and TNF receptor-associated protein; TRABID= TRAF-binding domain; TIZ= TRAF6-inhibitory zinc finger protein; TIFA= TRAF-interacting protein with a forkhead-associated (FHA) domain

protein (TRADD)[25] for recruitment to TNF-R1 and TRAF6 uses MyD88 to associate with members of the IL-1R/TLR family. While the TRAF-C domain is principally responsible for the direct or indirect interaction of TRAFs with activated receptor complexes, the TRAF-N domain allows binding to a wide variety of TRAF-interacting proteins (Table 1). These include kinases and regulators of TRAF-dependent signaling pathways.[17] The amino-terminal RING/zinc finger domain of TRAF proteins is typically responsible for triggering downstream signaling events. Deletion mutants of TRAF proteins that lack this domain act as dominant-negatives. As few proteins have been identified that bind to the RING/zinc finger domain, it has been suggested that its signaling function may be fulfilled by transient associations with proteins that bind to the N-TRAF domain.

The NF-κB Family of Transcription Factors

Like many other components of the IL-1R/TLR signaling pathway the transcription factors of the NF-κB family are phylogenetically conserved and have been described in distinct species such as man, fly and worms. NF-κB proteins are organized as homo- or heterodimers and are activated by a wide variety of stimuli, including physical stresses such as UV radiation and reactive oxygen species, by cytokines (e.g., IL1, TNF) and molecules associated with microbial pathogens. The structural hallmark of NF-κB proteins is the Rel homology domain (RHD) which comprises two immunoglobulin-like domains of about 300 amino acids.[26,27] The RHD mediates DNA binding, nuclear translocation, dimerization and interaction with members of the inhibitor of κB (I-κB) family which sequester NF-κB dimers in the cytoplasm and prevents their nuclear translocation. There are five NF-KB proteins in mammalian cells: RelA/p65, RelB, c-Rel, NF-κB1/p50 and NF-κB2/p52. The latter two are processed from the precursor proteins p105 (NF-κB1/p50) and p100 (NF-κB2/p52). While NF-κB1/p50 is cotranslationally produced by constitutive proteolysis of p105, NF-κB2/p52 is generally cleaved from p100 by induced proteolysis.[26,27] Nevertheless, both precursors are processed by the proteasome. RelA/p65, RelB, and c-Rel, but not p50 and p52, contain a transactivation domain in their carboxy termini.[26,27] Thus, homodimers of p50 and p52 can act as transcriptional repressors. There are also three NF-κB proteins in *Drosophila*: Dorsal, Dif and Relish. Similarly to NF-κB2/p52, the latter is also produced by inducible proteolysis, not by proteosome action, but by DREDD, a member of the caspase family of cysteine proteases.[28,29] I-κB proteins sequester NF-κBs in the cytoplasm by masking the nuclear localization sequence in their RHD domains. The I-κBs are defined by six or seven copies of a conserved helical protein domain, the ankyrin repeat. Remarkably, the p100, p105 and Relish NF-κB precursor proteins also contain six or seven ankyrin repeats in their carboxyl-terminal domains.[26,27] The mammalian I-κB protein family includes I-κBα, I-κBβ, I-κBε, I-κBγ, and Bcl3. The sole fly family member is Cactus.[26,27] Some of the I-κB proteins have an amino-terminal regulatory domain that allows their signal-induced proteasomal degradation, the key event in inducible NF-κB activation. I-κB protein degradation is triggered by serine phosphorylation-dependent ubiquitination. Phosphorylation of I-κBs is, in turn, mediated by the I-κB kinase (IKK) complex, which is activated by a wide variety of NF-κB inducers including IL-1, LPS and other pathogen-related substances. The IKK complex is composed of two related I-κB kinases, IKK1/IKKα and IKK2/IKKβ, the heat-shock protein 90 (Hsp90), its cochaperone Cdc37 and the catalytic scaffold protein NEMO/IKKγ. NEMO/IKKγ has a pivotal role in linking the IKK complex to diverse upstream pathways that trigger NF-κB activation.[26,27,30] Importantly, NF-κB-dependent transcription not only requires signal-induced I-κB degradation and nuclear translocation of NF-κB proteins, but also the modification of the NF-κB proteins by phosphorylation and perhaps acetylation. Phosphorylation and activation of NF-κB subunits can be achieved by pathways that are distinct from those that induce I-κB degradation, e.g., by the catalytic subunit of protein kinase A (PKA) or by casein kinase II.[26,27] However, there is evidence that some kinases, especially IKK2//IKKβ and atypical protein kinase ζ (PKCζ) contribute in a cell-type specific manner to I-κB degradation and NF-κB phoshorylation.

IL-1R Signaling

IL-1-Induced Activation of NF-κB

The initial event in IL-1R signaling is IL-1 binding-induced formation of a heteromeric complex of IL-1R and the IL-1 receptor accessory protein (IL-1RAcp) (Fig. 4).[31-33] This receptor complex recruits the cytosolic adapter proteins MyD88 and Tollip via homotypic interactions between the TIR domains of IL-1R, MyD88 and Tollip.[13,34,35] In resting cells, Tollip

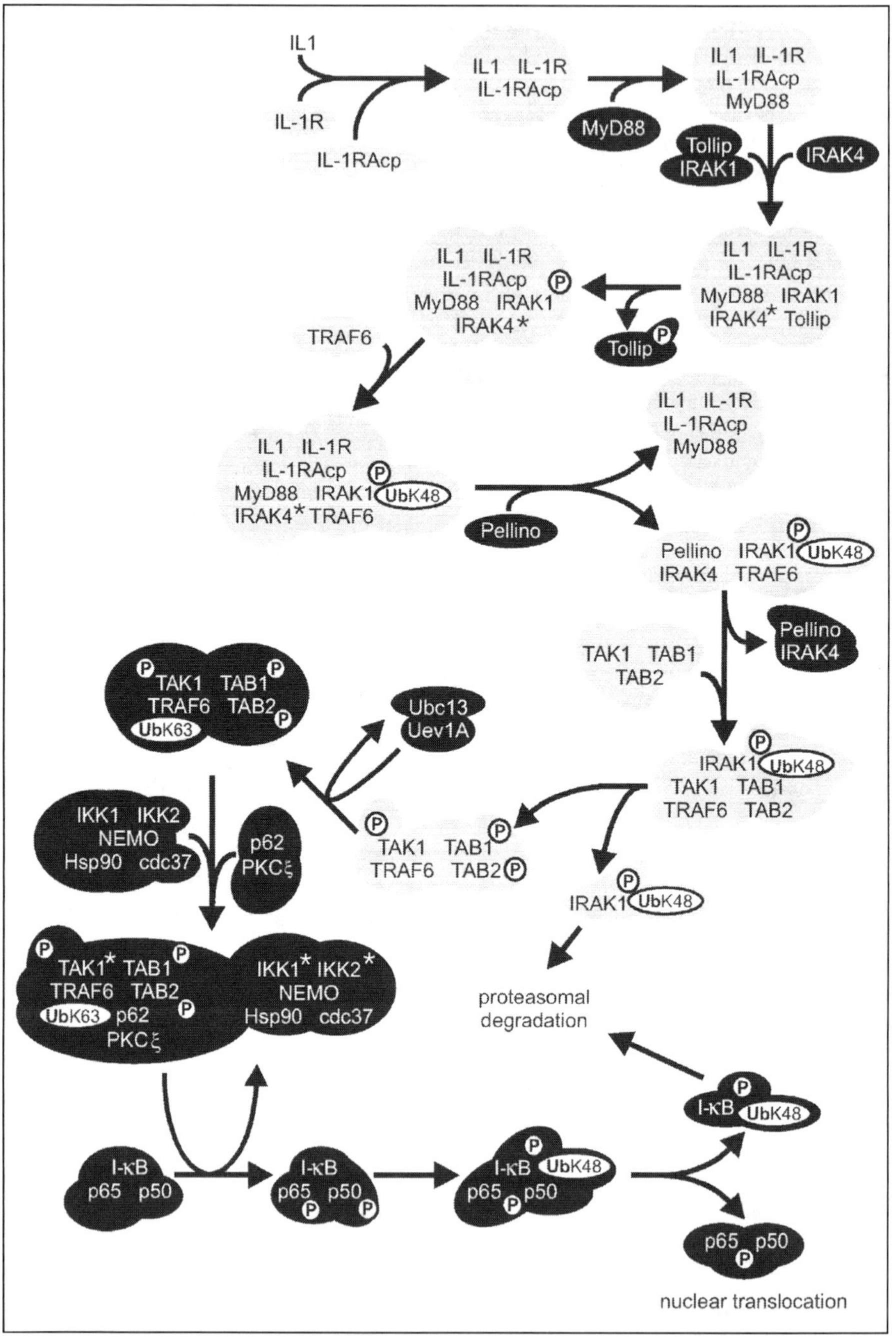

Figure 4. Model of IL-1-Induced NF-κB Activation. For details of the various steps, please see text. Membrane-associated protein complexes are in black fields, cytosolic complexes in gray fields. Phosphorylation of proteins is indicated by a circled P, ubiquitination by Ub and activated kinases are indicated by an asterisk.

interacts with nonphosphorylated IL-1 receptor-associated kinase-1 (IRAK1), most likely to silence this kinase in the absence of signaling competent IL-1R complexes.[35] As the Tollip-IRAK1 and (IL-1R)-(IL-1RAcp) complexes bind each other, IL-1R-bound MyD88 is also recruited (to IRAK1) to serve as an additional docking site for IRAK1 and IRAK4.[15,16] The MyD88-IRAK1/4 interaction relies on the homotypic association of death domains. By virtue of its own domain organization, MyD88 can act as an adapter for both TIR and death domain-containing proteins. In the context of IL-1R signaling, IRAK4 phosphorylates IRAK1, which is then autocatalytically hyperphosphorylated and competent for TRAF6 binding via its three TRAF6-binding motifs.[23,36,37] As a further consequence of hyperphosphorylation, IRAK1 is ubiquitinated and degraded by the proteasome, events which are not necessary for IL-1-induced NF-κB activation.[36,38] Besides undergoing autophosphorylation, IRAK1 also phosphorylates Tollip which then dissociates from the kinase.[39] Remarkably, the kinase-activity of IRAK1 is dispensable for IL-1 signaling, whereas IRAK4 needs catalytic integrity to fulfill its function.[15] There is evidence that the resultant TRAF6-containing IL-1R signaling complex interacts with the IRAK1-binding protein Pellino-1. This interaction leads to the release of an intermediate complex that contains Pellino1, IRAK1, IRAK4 and TRAF6. This complex transfers TRAF6 and IRAK1 to a membrane-associated ternary complex of TGFβ-activated kinase (TAK1), TAK1-binding protein 1 (TAB1) and TAK1-binding protein 2 (TAB2).[40] TAK1 was originally identified as a mitogen-activated protein kinase kinase kinase (MAP3K) involved in TGFβ signaling.[41] While TAB1 is a direct activator of TAK1, TAB2 acts as mediator to link TAK1 to cellular activators. [42,43] Formation of the IRAK1-TRAF6-TAB1-TAB2-TAK1 complex leads to the phosphorylation of TAK1, TAB1 and TAB2 by an unknown kinase.[37,38] Possible candidates are unlikely to be IRAK1 or TAK1 as TAK1, TAB1 and TAB2 can be phosphorylated in cells expressing a kinase-dead variant of IRAK1. Furthermore, TAK1 remains inactive in this complex. As TAK1, TAB1 and TAB2 are phosphorylated, the IRAK1-TRAF6-TAB1-TAB2-TAK1 complex dissociates to leave IRAK1 bound to the membrane whilst the TRAF6-TAB1-TAB2-TAK1 complex translocates to the cytosol.[37] Here, TAK1 activation takes place by cytosolic factors. Likely candidates include the dimeric ubiquitin-conjugating enzyme complex Ubc13 and Uev1A. Indeed, the Ubc13-Uev1A and the TAB1-TAB2-TAK1 complex have been biochemically purified as essential cofactors for TRAF6-mediated activation of the IKK complex.[44,45] Remarkably, TRAF6 acts as a ubiquitin ligase in concert with the Ubc13-Uev1A complex and a common ubiquitin-activating protein to catalyze its own Lys63-linked ubiquitination.[44,45] This type of polyubiquitination is distinct from the usual polyubiquitination via Lys48 which promotes proteasomal degradation. By an unknown mechanism, TRAF6 ubiquitination in the TRAF6-TAB1-TAB2-TAK1 complex induces the activation of TAK1. Once activated, TAK1 triggers IKK2/IKKβ activity by phosphorylation of S177 and S181. The crucial role of TAK1 for IL-1-induced IKK activation has been confirmed in knock down studies using TAK1-specific small interfering RNAs.[46] Besides TAK1, several other members of the MAP3K family have been implicated in IL-1 signaling and activate NF-κB when overexpressed.[47-49] These include mitogen-activated protein kinase/extracellular signal-regulated kinase kinase kinase-1 (MEKK1), MEKK2, MEKK3 and NF-κB inducing kinase (NIK). However, genetic studies have ruled out an essential role for MEKK1 and NIK in IL-1-induced NF-κB activation[50,51] though IL-1-induced NF-κB activation is impaired in MEKK3 deficient embryonic fibroblasts.[52] This suggests that MEKK3 and TAK1 either cooperate in IL-1-induced IKK activation or show cell specific redundancy. The latter possibility could explain the finding that embryonic fibroblasts derived from TAB2-deficient mice show normal IL-1-induced up-regulation of NF-κB target genes despite significantly reduced phosphorylation of TAB1 and TAK1.[53] However, the nonessential role of TAB2 deduced from this study could also be caused by a compensatory effect of the recently identified TAB2-related TAK1-binding protein 3 (TAB3).[54] Additional complexity in IL-1-induced NF-κB activation comes from studies that imply a role for the atypical protein kinase C family members PKCζ and PKCλ. These PKC isoforms interact with IKKs as well as with the ubiquitin binding protein p62.[55-57] Whether p62 interacts with ubiquitinated TRAF6 has yet to be clarified. The relevance of PKCζ for NF-κB activation is related to its ability to bind and phosphorylate p65 upon degradation of I-κB.[58] In accordance with this idea, PKCζ deficient embryonic fibroblasts show significantly reduced IL-1-dependent transcription of NF-κB target genes despite normal IKK activation and nuclear accumulation of NF-κB.[58] The central importance of the IKK complex for IL-1-induced NF-κB activation is evident from studies in NEMO/IKKγ deficient embryonic fibroblasts showing complete abrogation of NF-κB-dependent transcription upon treatment with IL-1.[59-61] NF-κB activity is also abolished in IKK1/IKKα and IKK2/IKKβ double-deficient mice emphasizing the crucial role of an intact IKK complex for this response.[62] However, a review of the literature shows that the relative contribution of the two IKKs to IL-1-induced NF-κB activation is far from clear. Reports that analyzed IL-1-induced NF-κB activity in IKK2/IKKβ deficient embryonic fibroblasts found a significant reduction in the DNA-binding capability of NF-κB as well as in NF-κB-dependent transcription.[59,63-65] Another study, this time using IKK1/IKKα -/- cells treated with IL-1, also reported a reduction in DNA-binding by NF-κB.[66] However, other experiments with IKK1/IKKα deficient cells could find no comparable effect.[65,67] Further, one of the latter two studies found normal transcription of IL-6 upon IL-1 treatment while in the other report this response was completely blocked.

IL-1-Induced Activation of c-Jun NH(2)-Terminal Kinase and p38 Mitogen-Activated Protein Kinase

c-Jun NH(2)-terminal kinase (JNK) and p38 mitogen-activated protein kinase (p38-MAPK) are activated by functionally distinct upstream kinases. While JNK activation occurs via MKK4 and MKK7, p38-MAPK activation is mediated by mitogen-activated protein kinase kinase-3 (MKK3) and MKK6.[68] However, which of the various MAP3Ks are relevant for IL-1-mediated activation of these intermediate kinases is unclear. Based on RNA interference studies, TAK1 has been implicated in LPS-induced JNK activation.[69] The similarity of the IL-1R and TLR signaling pathways leads us to speculate that TAK1 is also involved in IL-1-mediated JNK activation. However, in this case, TAK1 activation should occur by a pathway independent from those utilized in NF-κB signaling as IL-1-induced JNK activation is independent of the IRAK1-TRAF6 interaction.[70] With respect to IL-1-induced p38-MAPK activity there is evidence that a complex containing IRAK1, TRAF6, TAK1 and active Ras, links the IL-1R signaling complex to the p38-MAPK module.[71] As described above, IRAK1, TRAF6 and TAK1 are also part of the IKK activating complex utilized by IL-1. Thus, it appears possible that the IL-1R utilizes a common complex to signal both NF-κB and p38-MAPK but not JNK. Noteworthy, in context of TLR4 signaling, is a novel mechanism of p38-MAPK activation, which is independent of MKKs and MAP3Ks.[72] Future studies will evaluate whether IL-1 can also utilize this pathway, and if so, how it is related to the "classical" pathway of p38-MAPK activation.

Mechanisms of TLR Signaling

TLR-Induced Signaling in Mammalian Cells

Thus far, the IL-1R and mammalian TLRs interact with MyD88 in a mechanistically similar manner. Moreover, the negative regulators MyD88s and IRAK-M, which are discussed later, counter-regulate IL-1R and TLR signaling in a similar fashion. Thus, it seems feasible that these receptors use a common set of pathways downstream of the receptor-MyD88 complex to activate JNK, p38-MAPK and NF-κB. In support of this concept IL-1R, IL-18R, TLR2, TLR5, TLR7 and TLR9 signaling are all impaired in mice deficient for MyD88.[73-78] Moreover, mice deficient for IRAK4, IRAK1 or TRAF6 all show severe deficiencies in IL-1, IL-18 and LPS signaling, the latter mediated by TLR2 and TLR4.[16,79-83] However, while TRAF6 and MyD88 deficiency leads to a complete abrogation of IL-1-induced activation of JNK and NF-κB, the LPS response is slower and weaker.[73,75,82] Moreover, LPS-induced maturation of dendritic cells and the induction of IFN-regulated genes via TLR3 and TLR4 are normal in MyD88 deficient mice.[84,85] This discrepancy is explained by the use of a single MyD88 adapter in IL-1R signaling whereas some TLRs employ additional MyD88-related adapters called MAL/TIRAP and TICAM-1/TRIF (Fig. 5).[5-8] The MAL/TIRAP protein can interact with the TIR domain of TLR4 by homotypic interaction and acts upstream of TRAF6 and TAK1.[5,6] However, while MyD88 uses its death domain to interact with both IRAK1 and IRAK2, MAL/TIRAP selectively binds IRAK2 with its TIR domain.[5] Thus, MAL/TIRAP-IRAK2 signaling may partly compensate for the lack of MyD88-IRAK1 signaling in MyD88 deficient mice, explaining their delayed LPS-induced NF-κB response.[75] Interestingly, the LPS-triggered activation of JNK, p38-MAPK and ERKs in MAL/TIRAP deficient cells was delayed and weakened whereas the TLR2 ligand, mycoplasma-derived lipopeptide MALP-2, failed to induce any response. This points to an essential role for MAL/TIRAP in TLR2 signaling.[86] TIRAP has been found to be associated with protein kinase RNA-regulated (PKR), the PKR regulator p58 and protein activator of PKR (PACT).[6] Studies with PKR knock-out mice showed that PKR is an essential signaling intermediate of poly(IC)-, which signals via TLR3, and LPS-induced JNK and p38-MAPK activation.[87] Poly(IC) signaling is thought to be exclusively mediated by MyD88, suggesting that PKR is not only involved in TLR4-TIRAP signaling but also mediates MyD88 signaling by other TLRs. IL-1-induced activation of p38-MAPK, but not of JNK, is also impaired in PKR deficient embryonic fibroblasts. Remarkably, LPS as well as poly(IC) can still induce type I IFNs in MyD88 and MAL/TIRAP double-deficient mice, pointing to the existence of another adapter protein involved in IFNβ production. In this respect, a novel TIR domain-containing adapter protein, designated TIR-containing adapter molecule-1 (TICAM-1)[7] or TIR domain-containing adapter inducing IFNβ (TRIF)[8] has recently been identified. Except for a TIR domain, TICAM-1/TRIF displays no homology to MyD88 or MAL/TIRAP. Most remarkably down-regulation of TICAM-1/TRIF expression interferes with IFNβ production suggesting that this adapter is part of the MyD88-TIRAP independent pathway. This is consistent with the analysis of mice deficient in MyD88, MAL/TIRAP or both proteins.

An additional layer of complexity in TLR-mediated NF-κB activation became obvious from studies in mice deficient for poly(ADP-ribose) polymerase-1 (PARP-1), a nuclear DNA repair enzyme activated by DNA strand breaks.[88,89] PARP-1 $^{-/-}$ mice are resistant to LPS-induced endotoxic shock and exhibit a severely impaired NF-κB response.[90,91] While degradation of I-κB and translocation of NF-κB were similar in wild-type and PARP-1 deficient embryonic fibroblasts, DNA-binding and transcriptional activation were strongly reduced. There is evidence that PARP-1 binds directly to both, p65 and p50, but ambiguous data concerning the importance of the PARP-1 enzymatic activity for NF-κB activation is found in the literature.[89]

Recently, an unexpected mechanism for p38-MAPK activation that does not involve the prototypic kinase cascade has been identified. This involves p38-MAPK activation by autophosphorylation which correlates with the binding of TAB1 and TRAF6.[72] Remarkably, in ectopic expression experiments, TLR2 utilizes the "classical" kinase cascade to activate p38-MAPK whereas TLR4 and TLR9 employed, in the main, this novel TAB1-mediated auto-activation mechanism. Further studies will analyze how this novel pathway is linked to TLRs, IL-1R or its relative IL-18R and whether it cooperates with the classical pathway.

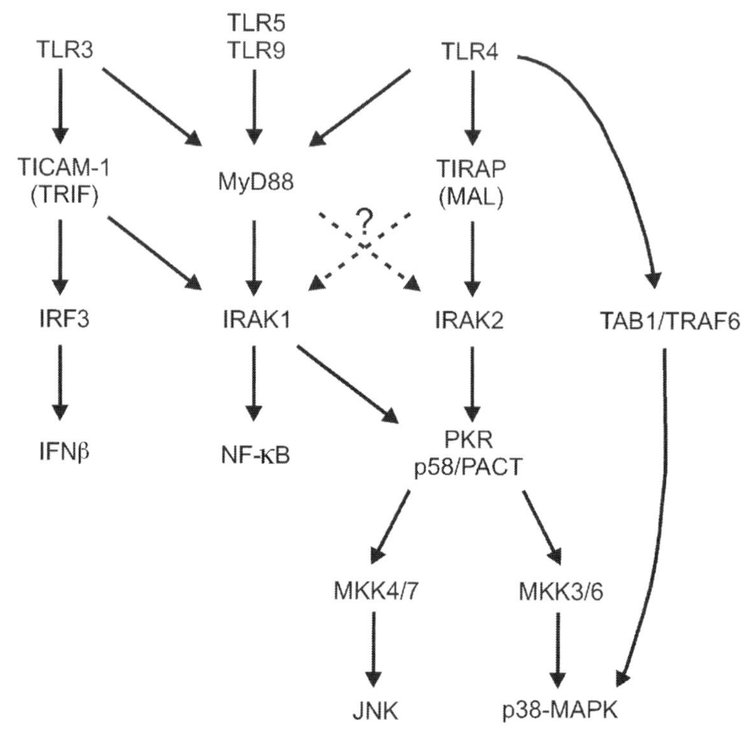

Figure 5. TLR3, TLR4, TLR5 and TLR9 use different TIR domain-containing adapter proteins for signaling. Note that TICAM-1/TRIF-mediated up-regulation of IFNβ via IRF-3 has also been suggested for TLR4 (not shown here).

Termination of IL-1R/TLR Signaling

Activation of IL-1R/TLR signaling, if left unchecked, can contribute to severe immunopathologies including Crohn's disease and septic shock.[92] One common mechanism to control IL-1R/TLR signaling is the transcriptional up-regulation of negative regulators that interfere with the formation of signaling competent IL-1R and/or TLR complexes. Examples include; A20 (Table 2), an NF-κB inducible zinc-finger protein,[93] MyD88s, a short splice variant of MyD88 that lacks the intermediate domain between the death and TIR domain,[94] and IRAK-M, a kinase-inactive member of the IRAK family.[95] These three proteins are induced by bacterial products and cytokines and inhibit IL-1R and TLR signaling. While MyD88s prevents IRAK4 recruitment to the receptor signaling complex,[96] IRAK-M inhibits the dissociation of IRAK1 and IRAK4 from MyD88, preventing the generation of the TRAF6-IRAK1-IRAK4-Pellino-1 complex.[95] A20 interacts with TRAF2, TRAF6 and IKKγ/NEMO to block TNF and IL-1-induced NF-κB activation upon ectopic expression.[97-99] However, A20 deficient mice show an impaired termination of TNF signaling but normal IL-1-induced NF-κB activation.[100] Although it is possible that the inhibitory action of IRAK-M and MyD88s on IL-1 signaling masks the effect of A20 deficiency, it cannot be ruled out that A20 regulates IL-1 and TNF signaling by distinct mechanisms. Indeed, there is evidence that A20 selectively interferes with TNF receptor-1 signaling.[101]

Table 2. Deficiencies in IL-1R/TLR signaling intermediates in mammals and flies

Mammalian Gene	Effect of Deficiency	Ref.	*Drosophila* Homologue	Effect of Deficiency	Ref.
A20	No effect on IL-1 signaling	100	-	-	-
IKK1/IKKα	In some studies reduced IL-1-induced NF-κB signaling	65-67	-	-	-
IKK2/IKKβ	No or reduced IL-1-induced NF-κB signaling	59,63-65	DmIKKβ	Normal Toll signaling, impaired IMD pathway	134
IRAK1	Reduced IL-1-mediated activation of JNK and NF-κB	79-81	Pelle	Impaired Toll signaling, normal IMD pathway	107
IRAK-M	Enhanced and sustained TLR signaling	95	-	-	
IRAK4	Impaired IL-1-mediated activation of JNK and NF-κB	16	-	-	-
MAL/TIRAP	Crucial role in MyD88-dependent TLR2 and TLR4 signaling	86	-	-	-
MEKK3	Reduced IL-1-mediated NF-κB activation	52	-	-	-
MyD88	Impaired IL-1-mediated activation of JNK and NF-κB	73-78	DmMyD88	Impaired Toll signaling, normal IMD pathway	9
NEMO/IKKγ	Impaired IL-1-mediated NF-κB activation	59-61	Kenny, DmIKKγ	Normal Toll signaling, impaired IMD pathway	133,134
P62		56	c-Ref(2)P	-	130
PKCζ	Impaired IL-1-mediated NF-κB activation (cell-type specific)	58	DaPKC	Impaired Toll signaling, normal IMD pathway	130
PKR	Impaired IL-1-mediated p38-MAPK activation	87	-	-	-
TAB2	No effect on IL-1 signaling	53	-	-	-
TRAF6	Impaired IL-1-mediated activation of JNK and NF-κB	82,83	DTRAF2		120
-	-	-	Tube	Impaired Toll signaling, normal IMD pathway	107
TAK1	-	45	DTAK1	Normal Toll signaling, impaired IMD pathway	141

The Toll Pathway in *Drosophila melanogaster*

In *Drosophila melanogaster* the Toll pathway was originally identified due to its crucial role in embryonic dorsoventral patterning.[102] During this process Toll is activated by a cleavage product of Spätzle, consisting of a disulfide-linked dimer of 106 amino acids that is released from a Spätzle precursor protein by a proteolytic cascade. This cascade comprises Gastrulation Defective, Snake, and Easter.[103-105] Later studies revealed that Toll signaling also played an important role in fly innate immunity. Toll signals induce the production of anti-microbial peptides (e.g., Drosomycin and Metchnikowin) that protect the fly against infection by fungi and Gram-positive bacteria. A loss-of-function mutant of Spätzle, but not Gastrulation Defective, Snake, or Easter, showed a vastly diminished Toll-mediated response to infection with both Gram-positive bacteria and fungi. This indicated that Toll's place in innate immune signaling was distinct from its role in the proteolytic cascade that drives dorsoventral patterning.[106,107] In further support of separate proteolytic cascades to regulate Spätzle-mediated Toll activation in immunity versus dorsoventral patterning, a serine protease inhibitor (Spn43Ac) that differentially blocks these processes has been identified.[106] Moreover, activation of Toll in innate immunity bifurcates upstream of Spätzle processing. While the soluble peptidoglycan recognition protein Semmelweis is essential for Spätzle-mediated Toll activation upon infection with Gram-positive bacteria, it is not involved in the Toll-dependent anti-fungal response.[108] As Toll activation in fly innate immunity is dependent on Spätzle processing, it appears that Toll itself does not directly recognize microbe-derived molecular patterns. In contrast, mammalian TLRs are activated without the requirement for activated proteases and can be incorporated into protein-complexes that directly bind microbial products.[109]

In dorsoventral patterning and immune activation of Toll, the *Drosophila* homologue of mammalian MyD88, designated DmMyD88 is recruited upstream of Pelle and Tube. Loss-of-function mutants of all three proteins abolish the Toll-mediated anti-microbial response. However, DmMyD88, Pelle and Tube are not involved in the Imd pathway which is a Toll-independent pathway responsible for defense against Gram-negative bacteria.[9,107] This second pathway is not triggered by any member of the IL-1R/TLR family, but is activated instead by PGRP-LC. PGRP-LC is a type II transmembrane protein containing three extracellular peptidoglycan-binding domains and an intracellular domain that lacks any homology to proteins of known function.[110,111]

Similar to its counterpart in mammalian cells, DmMyD88 binds Toll by homophilic interaction of TIR domains[9,112] then recruits Pelle. Nonphosphorylated Pelle can also interact directly with Toll via its carboxyl-terminal catalytic domain.[9,112,113] Pelle and Tube associate by homophilic interactions between their amino-terminal death domains in vitro. Pelle can also associate with Pellino via its kinase domain.[114-116] Pelle interacts with Tube preferentially in its nonphosphorylated form and phosphorylates itself, Toll and Tube in vitro.[113,114,117] However, only kinase-inactive mutants of Pelle increase colocalization of Tube and Toll and interfere with dorsal-ventral patterning, suggesting that Pelle-mediated protein phosphorylation triggers the release of a signaling-competent protein complex into the cytoplasm.[118] As no complexes between Pelle and Tube have been found in embryos[117,119] it has been suggested that Toll activated Pelle recruits Tube, phosphorylates Tube, then dissociates rapidly.[118] Thus, in respect to the functional importance of its kinase activity, Pelle resembles IRAK4 rather than IRAK1 in IL-1R/TLR signaling. Phosphorylation-dependent termination of the interaction of Pelle with the Toll signaling complex would mimic the situation described for IL 1R/TLR signaling, where the Pellino-TRAF6-IRAK1-IRAK4 complex dissociates from the IL-1R signaling complex upon IRAK1 phosphorylation. A further similarity to mammalian IL-1R/TLR signaling is evidence that a member of the TRAF protein family acts downstream of the receptor-signaling complex in *Drosophila*. DTRAF2, one of three fly TRAF proteins (DTRAF1, DTRAF2, DTRAF3), is sufficient to drive a Dorsal-regulated reporter gene upon overexpression.[120]

Dorsal is a *Drosophila* member of the NF-κB family that is activated by Toll signaling during embryonic patterning. Moreover, DTRAF2 weakly interacts with Pelle and both proteins synergistically activate Dorsal.[120] Full length DTRAF2 shows a weaker interaction with Pelle and Dorsal than its isolated RING/zinc finger domain. However, a deletion mutant of DTRAF2 that lacks the RING/zinc finger domain is still able to activate Dorsal.[120] This is unexpected given that the equivalent mammalian TRAF mutants typically have a dominant-negative influence on their intact counterparts. Indeed, TRAF6 binds to IRAK1 via its TRAF domain.[13] Moreover, IRAK1 interacts with the TRAF domain of TRAF6 via its amino-terminal death domain,[70] whereas Pelle seems to bind to DTRAF2 with its kinase domain.[120] However, as discussed above, IRAK4 rather than IRAK1 is the functional homologue of Pelle in mammalian cells. Thus, an interesting question remains whether, in the context of IL-1 signaling, the kinase domain of IRAK4 interacts with the RING/zinc finger domain of TRAF6.

Besides DTRAF2 another *Drosophila* TRAF protein, DTRAF1, has also been implicated in Toll signaling due to its capability to interact physically with Pelle. Remarkably, the binding of DTRAF1 to Pelle is based on the interaction of the TRAF domain of DTRAF1 and the death domain of Pelle.[121] Thus, the molecular mode of DTRAF1-Pelle interaction is similar to that of IRAK1 and TRAF6. Nevertheless, microinjection of mRNA encoding a dominant-negative mutant of DTRAF1 into embryos showed no effect on dorsal-ventral patterning, arguing against a role for DTRAF1 in Toll-mediated Dorsal activation.[121] However, it is possible that DTRAF1 has a role in Toll-mediated activation of the JNK pathway. For example, DTRAF1 was originally identified due to its binding to the MAP4K Misshapen. Misshapen is involved in the stimulation of the mitogen activated protein (MAP) kinase module that leads to JNK stimulation in *Drosophila*. Furthermore, ectopic expression of Pelle or Tube activates this pathway[122,123] and bifurcation of the *Drosophila* NF-κB and JNK signaling pathways at the level of the Pelle-TRAF interaction would mirror the divergence seen in the mammalian IL-1R/TLR signaling pathway.[70,124]

The *Drosophila* Toll signal leads to development-dependent activation of two members of the NF-κB family, namely Dorsal and Dif (Dorsal-related immunity factor). In adult flies the Toll-induced anti-fungal response relies on DIF alone whereas in larvae, Dorsal and DIF act redundantly.[125,126] Although the Dif protein is capable of restoring the defects in embryonic dorsal-ventral patterning seen in mothers with mutant Dorsal, this NF-κB protein is not normally used in this process.[127,128] Inhibition of Dorsal and Dif signaling relies on the I-κB protein Cactus.[129] Similarly to the mammalian system, signal-induced phosphorylation triggers the proteasomal degradation of Cactus, thereby allowing the nuclear import of these proteins. An additional similarity to mammalian IL-1R/TLR signaling is that the degradation of I-κB/Cactus is insufficient to trigger transcription of NF-κB-regulated genes, but requires a further phoshorylation-mediated activation of the transcriptional activity of Dorsal and Dif.[130] Moreover, there is evidence that nuclear import of free Dorsal or Dif also requires phosphorylation.[131] Nuclear translocation and transcriptional activation of Dorsal and DIF appear to be regulated by phosphorylation at distinct sites, mediated by distinct kinases. While DaPKC, the *Drosophila* homologue of the mammalian atypical PKCs, has been implicated in phosphorylation-dependent stimulation of Dif activity, the kinase(s) involved in nuclear translocation are still elusive.[130] Remarkably, DaPKC interacts with DTRAF2 via the adapter protein Ref(2)P, which is the *Drosophila* homologue of p62 and links PKCζ and TRAF6 in IL-1 signaling.[130,132] Up to this point the signaling mechanisms of *Drosophila* Toll and the mammalian IL-1/TLRs are strikingly similar in respect to the proteins involved and how they act. However, there are considerable differences in the kinase(s) that mediate the degradation of I-κB proteins. In mammalian cells, it seems that all stimuli inducing I-κB degradation utilize a single IKK complex consisting of IKK1/IKKα, IKK2/IKKβ and NEMO/IKKγ[26,27] An IKK complex composed of a NEMO/IKKγ homologue Kenny/DmIKKγ and a IKK2/IKKβ homologue DLAK/DmIKKβ also exists in *Drosophila*.[133-135] However, genetic studies indicate that the *Drosophila* IKK complex is not involved in Cactus degradation.[133,134] Instead the *Drosophila* IKK complex is required for signal-induced processing of the NF-κB precursor protein Relish. This is a crucial step in the Toll-independent Imd pathway.[28,134,136,137] By contrast, in mammalian cells the inducible processing of the p52 precursor protein p100 is performed by a NEMO/IKKγ-independent pathway that utilizes IKK1/IKKα.[138-140] With respect to TAK1 the situation is similar to the role of the IKK complex. In mammalian cells this MAP3K is involved in IL-1R/TLR–mediated NF-κB activation whereas the *Drosophila* homologue dTAK acts in the Imd pathway.[141]

TLR-Mediated Apoptosis

The induction of apoptosis by bacterial lipoproteins (BLPs) via TLR2 has recently been reported.[142,143] Downstream of TLR2 the apoptotic signal is transmitted by MyD88-mediated recruitment of the death domain-containing adaptor protein FADD (Fas-associated death domain protein), which has a crucial role in death receptor-induced apoptosis in mammalian cells.[144] Remarkably, it has also been shown that DmMyD88 interacts with dFADD, a homologue of mammalian FADD.[112,145] The apoptotic capacity of FADD relies on its interaction with the initiator casapse-8. In the context of death receptor signaling, several FADD molecules recruit several pro-caspase-8 molecules into multimeric death receptor complexes that lead to trans-processing and hence activation of caspase-8. Activated caspase-8 is then released from the death inducing signaling complex and triggers apoptosis by activating effector caspases, e.g., caspase-3. Likewise dFADD interacts and activates DREDD, a Drosophila homologue of caspase-8,[146] which has been implicated in Relish processing in the Imd pathway and in apoptosis-induction.[28,29,147] However, while genetic interference with dFADD expression convincingly demonstrated its essential role for Relish activation, a direct pro-apoptotic role for dFADD could not be confirmed. Future studies will reveal whether the Toll-DmMyD88-dFADD-DREDD axis deduced from ectopic expression data can act in vivo to induce apoptosis. Such a Toll-inducible apoptotic pathway could be difficult to detect as the concomitantly activated NF-κB pathway could have anti-apoptotic properties. Indeed, in mammalian cells a variety of anti-apoptotic genes are regulated by NF-κB and inhibition of NF-κB activity enhances TLR2-induced apoptosis.[143]

References

1. Silverman N, Maniatis T. NF-kappaB signaling pathways in mammalian and insect innate immunity. Genes Dev 2001; 15:2321-2342.
2. Barton GM, Medzhitov R. Toll-like receptors and their ligands. Curr Top Microbiol Immunol 2002; 270:81-92.
3. Dunne A, O'Neill LAJ. The Interleukin-1 Receptor/Toll-Like Receptor Superfamily: Signal Transduction During Inflammation and Host Defense. Sci STKE 2003; re3.
4. Lord KA, Hoffman-Liebermann B, Liebermann DA. Nucleotide sequence and expression of a cDNA encoding MyD88, a novel myeloid differentiation primary response gene induced by IL6. Oncogene 1990; 5:1095-1097.
5. Fitzgerald KA, Palsson-McDermott EM, Bowie AG et al. Mal (MyD88-adapter-like) is required for Toll-like receptor-4 signal transduction. Nature 2001; 413:78-83.
6. Horng T, Barton GM, Medzhitov R. TIRAP: An adapter molecule in the Toll signaling pathway. Nat Immunol 2001; 2:835-841.
7. Oshiumi H, Matsumoto M, Funami K et al. TICAM-1, an adaptor molecule that participates in Toll-like receptor 3-mediated interferon-beta induction. Nat Immunol 2003; 4:161-167.
8. Yamamoto M, Sato S, Mori K et al. Cutting edge: A novel Toll/IL-1 receptor domain-containing adapter that preferentially activates the IFN-beta promoter in the Toll-like receptor signaling. J Immunol 2002; 169:6668-6672.

9. Tauszig-Delamasure S, Bilak H, Capovilla M et al. Drosophila MyD88 is required for the response to fungal and Gram-positive bacterial infections. Nat Immunol 2002; 3:91-97.
10. urns K, Martinon F, Esslinger C et al. MyD88, an adapter protein involved in interleukin-1 signaling. J Biol Chem 1998; 273:12203-12209.
11. Shelton CA, Wasserman SA. Pelle encodes a protein kinase required to establish dorsoventral polarity in the Drosophila embryo. Cell 1993; 72:515-525.
12. Cao Z, Henzel WJ, Gao X. IRAK: A kinase associated with the interleukin-1 receptor. Science 1996; 271:1128-1131.
13. Muzio M, Ni J, Feng P et al. IRAK (Pelle) family member IRAK-2 and MyD88 as proximal mediators of IL-1 signaling. Science 1997; 278:1612-1615.
14. Wesche H, Gao X, Li X et al. IRAK-M is a novel member of the Pelle/interleukin-1 receptor-associated kinase (IRAK) family. J Biol Chem 1999; 274:19403-19410.
15. Li S, Strelow A, Fontana EJ et al. IRAK-4: A novel member of the IRAK family with the properties of an IRAK-kinase. Proc Natl Acad Sci USA 2002; 99:5567-5572.
16. Suzuki N, Suzuki S, Duncan GS et al. Severe impairment of interleukin-1 and Toll-like receptor signalling in mice lacking IRAK-4. Nature 2002; 416:750-756.
17. Wajant H, Henkler F, Scheurich P. The TNF-receptor-associated factor family: Scaffold molecules for cytokine receptors, kinases and their regulators. Cell Signal 2001; 13:389-400.
18. Rothe M, Wong SC, Henzel WJ et al. A novel family of putative signal transducers associated with the cytoplasmic domain of the 75 kDa tumor necrosis factor receptor. Cell 1994; 78:681-692.
19. Grech A, Quinn R, Srinivasan D et al. Complete structural characterisation of the mammalian and Drosophila TRAF genes: Implications for TRAF evolution and the role of RING finger splice variants. Mol Immunol 2000; 37:721-734.
20. Park YC, Burkitt V, Villa AR et al. Structural basis for self-association and receptor recognition of human TRAF2. Nature 1999; 398:533-538.
21. McWhirter SM, Pullen SS, Holton JM et al. Crystallographic analysis of CD40 recognition and signaling by human TRAF2. Proc Natl Acad Sci USA 1999; 96:8408-8413.
22. Ni CZ, Welsh K, Leo E et al. Molecular basis for CD40 signaling mediated by TRAF3. Proc Natl Acad Sci USA 2000; 97:10395-10399.
23. Ye H, Arron JR, Lamothe B et al. Distinct molecular mechanism for initiating TRAF6 signalling. Nature 2002; 418:443-447.
24. Ye H, Wu H. Thermodynamic characterization of the interaction between TRAF2 and tumor necrosis factor receptor peptides by isothermal titration calorimetry. Proc Natl Acad Sci USA 2000; 97:8961-8966.
25. Hsu H, Shu HB, Pan MG et al. TRADD-TRAF2 and TRADD-FADD interactions define two distinct TNF receptor 1 signal transduction pathways. Cell 1996; 84:299-308.
26. Karin M, Lin A. NF-kappaB at the crossroads of life and death. Nat Immunol 2002; 3:221-227.
27. Rothwarf DM, Karin M. The NF-kappa B activation pathway: A paradigm in information transfer from membrane to nucleus. Sci STKE 1999; 1999:RE1.
28. Stoven S, Ando I, Kadalayil L et al. Activation of the Drosophila NF-kappaB factor Relish by rapid endoproteolytic cleavage. EMBO Rep 2000; 1:347-352.
29. Leulier F, Rodriguez A, Khush RS et al. The Drosophila caspase Dredd is required to resist gram-negative bacterial infection. EMBO Rep 2000; 1:353-358.
30. Chen G, Cao P, Goeddel DV. TNF-induced recruitment and activation of the IKK complex require Cdc37 and Hsp90. Mol Cell 2002; 9:401-410.
31. Greenfeder SA, Nunes P, Kwee L et al. Molecular cloning and characterization of a second subunit of the interleukin 1 receptor complex. J Biol Chem 1995; 270:13757-13765.
32. Huang J, Gao X, Li S et al. Recruitment of IRAK to the interleukin 1 receptor complex requires interleukin 1 receptor accessory protein. Proc Natl Acad Sci USA 1997; 94:12829-12832.
33. Korherr C, Hofmeister R, Wesche H et al. A critical role for interleukin-1 receptor accessory protein in interleukin-1 signaling. Eur J Immunol 1997; 27:262-267.
34. Wesche H, Henzel WJ, Shillinglaw W et al. MyD88: An adapter that recruits IRAK to the IL-1 receptor complex. Immunity 1997; 7:837-847.
35. Burns K, Clatworthy J, Martin L et al. Tollip, a new component of the IL-1R pathway, links IRAK to the IL-1 receptor. Nat Cell Biol 2000; 2:346-351.

36. Yamin TT, Miller DK. The interleukin-1 receptor-associated kinase is degraded by proteasomes following its phosphorylation. J Biol Chem 1997; 272:21540-21547.
37. Jiang Z, Ninomiya-Tsuji J, Qian Y et al. Interleukin-1 (IL-1) receptor-associated kinase-dependent IL-1-induced signaling complexes phosphorylate TAK1 and TAB2 at the plasma membrane and activate TAK1 in the cytosol. Mol Cell Biol 2002; 22:7158-7167.
38. Qian Y, Commane M, Ninomiya-Tsuji J et al. IRAK-mediated translocation of TRAF6 and TAB2 in the interleukin-1- induced activation of NF-kappaB. J Biol Chem 2001; 276:41661-41667.
39. Zhang G, Ghosh S. Negative regulation of Toll-like receptor-mediated signaling by Tollip. J Biol Chem 2002; 277:7059-7065.
40. Jiang Z, Johnson HJ, Nie H et al. Pellino 1 is required for IL-1-mediated signaling through its interaction with IRAK4-IRAK-TRAF6. J Biol Chem 2002; 278:10952-10956
41. Yamaguchi K, Shirakabe K, Shibuya H et al. Identification of a member of the MAPKKK family as a potential mediator of TGF-beta signal transduction. Science 1995; 270:2008-2011.
42. Shibuya H, Yamaguchi K, Shirakabe K et al. TAB1: An activator of the TAK1 MAPKKK in TGF-beta signal transduction. Science 1996; 272:1179-1182.
43. Takaesu G, Kishida S, Hiyama A et al. TAB2, a novel adaptor protein, mediates activation of TAK1 MAPKKK by linking TAK1 to TRAF6 in the IL-1 signal transduction pathway. Mol Cell 2000; 5:649-658.
44. Deng L, Wang C, Spencer E et al. Activation of the IkappaB kinase complex by TRAF6 requires a dimeric ubiquitin-conjugating enzyme complex and a unique polyubiquitin chain. Cell 2000; 103:351-361.
45. Wang C, Deng L, Hong M et al. TAK1 is a ubiquitin-dependent kinase of MKK and IKK. Nature 2001; 412:346-351.
46. Takaesu G, Surabhi RM, Park KJ et al. TAK1 is critical for IkappaB Kinase-mediated activation of the NF-kappaB pathway. J Mol Biol 2003; 326:105-115.
47. Lee FS, Peters RT, Dang LC et al. MEKK1 activates both IkappaB kinase alpha and IkappaB kinase beta. Proc Natl Acad Sci USA 1998; 95:9319-9324.
48. Zhao Q, Lee FS. Mitogen-activated protein kinase/ERK kinase kinases 2 and 3 activate nuclear factor-kappaB through IkappaB kinase-alpha and IkappaB kinase-beta. J Biol Chem 1999; 274:8355-8358.
49. Malinin NL, Boldin MP, Kovalenko AV et al. MAP3K-related kinase involved in NF-kappaB induction by TNF, CD95 and IL-1. Nature 1997; 385:540-544.
50. Yin L, Wu L, Wesche H et al. Defective lymphotoxin-beta receptor-induced NF-kappaB transcriptional activity in NIK-deficient mice. Science 2001; 291:2162-2165.
51. Yujiri T, Ware M, Widmann C et al. MEK kinase 1 gene disruption alters cell migration and c-Jun NH2-terminal kinase regulation but does not cause a measurable defect in NF-kappa B activation. Proc Natl Acad Sci USA 2000; 97:7272-7277.
52. Yang J, Lin Y, Guo Z et al. The essential role of MEKK3 in TNF-induced NF-kappaB activation. Nat Immunol 2001; 2:620-624.
53. Sanjo H, Takeda K, Tsujimura T et al. TAB2 is essential for prevention of apoptosis in fetal liver but not for interleukin-1 signaling. Mol Cell Biol 2003; 23:1231-1238.
54. Munoz-Sanjuan I, Bell E, Altmann CR et al. Gene profiling during neural induction in Xenopus laevis: Regulation of BMP signaling by post-transcriptional mechanisms and TAB3, a novel TAK1-binding protein. Development 2002; 129:5529-5540.
55. Lallena MJ, Diaz-Meco MT, Bren G et al. Activation of IkappaB kinase beta by protein kinase C isoforms. Mol Cell Biol 1999; 19:2180-2188.
56. Sanz L, Sanchez P, Lallena MJ et al. The interaction of p62 with RIP links the atypical PKCs to NF-kappaB activation. EMBO J 1999; 18:3044-3053.
57. Vadlamudi RK, Joung I, Strominger JL et al. p62, a phosphotyrosine-independent ligand of the SH2 domain of p56lck, belongs to a new class of ubiquitin-binding proteins. J Biol Chem 1996; 271:20235-20237.
58. Leitges M, Sanz L, Martin P et al. Targeted disruption of the zetaPKC gene results in the impairment of the NF-kappaB pathway. Mol Cell 2001; 8:771-780.
59. Schmidt-Supprian M, Bloch W, Courtois G et al. NEMO/IKK gamma-deficient mice model incontinentia pigmenti. Mol Cell 2000; 5:981-992.

60. Rudolph D, Yeh WC, Wakeham A et al. Severe liver degeneration and lack of NF-kappaB activation in NEMO/IKKgamma-deficient mice. Genes Dev 2000; 14:854-862.
61. Makris C, Godfrey VL, Krahn-Senftleben G et al. Female mice heterozygous for IKK gamma/NEMO deficiencies develop a dermatopathy similar to the human X-linked disorder incontinentia pigmenti. Mol Cell 2000; 5:969-979.
62. Li Q, Estepa G, Memet S et al. Complete lack of NF-kappaB activity in IKK1 and IKK2 double-deficient mice: Additional defect in neurulation. Genes Dev 2000; 14:1729-1733.
63. Li Q, Van Antwerp D, Mercurio F et al. Severe liver degeneration in mice lacking the IkappaB kinase 2 gene. Science 1999; 284:321-325.
64. Tanaka M, Fuentes ME, Yamaguchi K et al. Embryonic lethality, liver degeneration, and impaired NF-kappaB activation in IKK-beta-deficient mice. Immunity 1999; 10:421-429.
65. Sizemore N, Lerner N, Dombrowski N et al. Distinct roles of the Ikappa B kinase alpha and beta subunits in liberating nuclear factor kappa B (NF-kappa B) from Ikappa B and in phosphorylating the p65 subunit of NF-kappaB. J Biol Chem 2002; 277:3863-3869.
66. Li Q, Lu Q, Hwang JY et al. IKK1-deficient mice exhibit abnormal development of skin and skeleton. Genes Dev 1999; 13:1322-1328.
67. Takeda K, Takeuchi O, Tsujimura T et al. Limb and skin abnormalities in mice lacking IKKalpha. Science 1999; 284:313-316.
68. Shaulian E, Karin M. AP-1 as a regulator of cell life and death. Nat Cell Biol 2002; 4:E131-E136.
69. Chen W, White MA, Cobb MH. Stimulus-specific requirements for MAP3 kinases in activating the JNK pathway. J Biol Chem 2002; 277:49105-49110.
70. Li X, Commane M, Jiang Z et al. IL-1-induced NFkappa B and c-Jun N-terminal kinase (JNK) activation diverge at IL-1 receptor-associated kinase (IRAK). Proc Natl Acad Sci USA 2001; 98:4461-4465.
71. McDermott EP, O'Neill LA. Ras participates in the activation of p38 MAPK by interleukin-1 by associating with IRAK, IRAK2, TRAF6, and TAK-1. J Biol Chem 2002; 277:7808-7815.
72. Ge B, Gram H, Di Padova F et al. MAPKK-independent activation of p38alpha mediated by TAB1-dependent autophosphorylation of p38alpha. Science 2002; 295:1291-1294.
73. Adachi O, Kawai T, Takeda K et al. Targeted disruption of the MyD88 gene results in loss of IL-1- and IL-18-mediated function. Immunity 1998; 9:143-150.
74. Hacker H, Vabulas RM, Takeuchi O et al. Immune cell activation by bacterial CpG-DNA through myeloid differentiation marker 88 and tumor necrosis factor receptor-associated factor (TRAF)6. J Exp Med 2000; 192:595-600.
75. Kawai T, Adachi O, Ogawa T et al. Unresponsiveness of MyD88-deficient mice to endotoxin. Immunity 1999; 11:115-122.
76. Hemmi H, Takeuchi O, Kawai T et al. A Toll-like receptor recognizes bacterial DNA. Nature 2000; 408:740-745.
77. Hayashi F, Smith KD, Ozinsky A et al. The innate immune response to bacterial flagellin is mediated by Toll-like receptor 5. Nature 2001; 410:1099-1103.
78. Hemmi H, Kaisho T, Takeuchi O et al. Small anti-viral compounds activate immune cells via the TLR7 MyD88-dependent signaling pathway. Nat Immunol 2002; 3:196-200
79. Kanakaraj P, Schafer PH, Cavender DE et al. Interleukin (IL)-1 receptor-associated kinase (IRAK) requirement for optimal induction of multiple IL-1 signaling pathways and IL-6 production. J Exp Med 1998; 187:2073-2079.
80. Kanakaraj P, Ngo K, Wu Y et al. Defective interleukin (IL)-18-mediated natural killer and T helper cell type 1 responses in IL-1 receptor-associated kinase (IRAK)-deficient mice. J Exp Med 1999; 189:1129-1138.
81. Thomas JA, Allen JL, Tsen M et al. Impaired cytokine signaling in mice lacking the IL-1 receptor-associated kinase. J Immunol 1999; 163:978-984.
82. Lomaga MA, Yeh WC, Sarosi I et al. TRAF6 deficiency results in osteopetrosis and defective interleukin-1, CD40, and LPS signaling. Genes Dev 1999; 13:1015-1024.
83. Naito A, Azuma S, Tanaka S et al. Severe osteopetrosis, defective interleukin-1 signalling and lymph node organogenesis in TRAF6-deficient mice. Genes Cells 1999; 4:353-362.
84. Kaisho T, Takeuchi O, Kawai T et al. Endotoxin-induced maturation of MyD88-deficient dendritic cells. J Immunol 2001; 166:5688-5694.

85. Alexopoulou L, Holt AC, Medzhitov R et al. Recognition of double-stranded RNA and activation of NF-kappaB by Toll- like receptor 3. Nature 2001; 413:732-738.
86. Yamamoto M, Sato S, Hemmi H et al. Essential role for TIRAP in activation of the signalling cascade shared by TLR2 and TLR4. Nature 2002; 420:324-329.
87. Goh KC, deVeer MJ, Williams BR. The protein kinase PKR is required for p38 MAPK activation and the innate immune response to bacterial endotoxin. EMBO J 2000; 19:4292-4297.
88. Burkle A. PARP-1: A regulator of genomic stability linked with mammalian longevity. Chembiochem 2001; 2:725-728.
89. Hassa PO, Hottiger MO. The functional role of poly(ADP-ribose)polymerase 1 as novel coactivator of NF-kappaB in inflammatory disorders. Cell Mol Life Sci 2002; 59:1534-1553.
90. Oliver FJ, Menissier-de Murcia J, Nacci C et al. Resistance to endotoxic shock as a consequence of defective NF-kappaB activation in poly (ADP-ribose) polymerase-1 deficient mice. EMBO J 1999; 18:4446-4454.
91. Hassa PO, Hottiger MO. A role of poly (ADP-ribose) polymerase in NF-kappaB transcriptional activation. Biol Chem 1999; 380:953-959.
92. Van Heel DA, McGovern DP, Jewell DP. Crohn's disease: Genetic susceptibility, bacteria, and innate immunity. Lancet 2001; 357:1902-1904.
93. Krikos A, Laherty CD, Dixit VM. Transcriptional activation of the tumor necrosis factor alpha-inducible zinc finger protein, A20, is mediated by kappa B elements. J Biol Chem 1992; 267:17971-17976.
94. Janssens S, Burns K, Tschopp J et al. Regulation of interleukin-1- and lipopolysaccharide-induced NF-kappaB activation by alternative splicing of MyD88. Curr Biol 2002; 12:467-471.
95. Kobayashi K, Hernandez LD, Galan JE et al. IRAK-M is a negative regulator of Toll-like receptor signaling. Cell 2002; 110:191-202.
96. Burns K, Janssens S, Brissoni B et al. Inhibition of Interleukin 1 Receptor/Toll-like receptor signaling through the alternatively spliced, short form of MyD88 Is due to its failure to recruit IRAK-4. J Exp Med 2003; 197:263-268.
97. Song HY, Rothe M, Goeddel DV. The tumor necrosis factor-inducible zinc finger protein A20 interacts with TRAF1/TRAF2 and inhibits NF-kappaB activation. Proc Natl Acad Sci USA 1996; 93:6721-6725.
98. Heyninck K, Beyaert R. The cytokine-inducible zinc finger protein A20 inhibits IL-1-induced NF-kappaB activation at the level of TRAF6. FEBS Lett 1999; 442:147-150.
99. Zhang SQ, Kovalenko A, Cantarella G et al. Recruitment of the IKK signalosome to the p55 TNF receptor: RIP and A20 bind to NEMO (IKKgamma) upon receptor stimulation. Immunity 2000; 12:301-311.
100. Lee EG, Boone DL, Chai S et al. Failure to regulate TNF-induced NF-kappaB and cell death responses in A20-deficient mice. Science 2000; 289:2350-2354.
101. He KL, Ting AT. A20 inhibits tumor necrosis factor (TNF) alpha-induced apoptosis by disrupting recruitment of TRADD and RIP to the TNF receptor 1 complex in Jurkat T cells. Mol Cell Biol 2002; 22:6034-6045.
102. Belvin MP, Anderson KV. A conserved signaling pathway: The Drosophila Toll-dorsal pathway. Annu Rev Cell Dev Biol 1996; 12:393-416.
103. Morisato D, Anderson KV. The spatzle gene encodes a component of the extracellular signaling pathway establishing the dorsal-ventral pattern of the Drosophila embryo. Cell 1994; 76:677-688.
104. DeLotto Y, DeLotto R. Proteolytic processing of the Drosophila Spatzle protein by easter generates a dimeric NGF-like molecule with ventralising activity. Mech Dev 1998; 72:141-148.
105. LeMosy EK, Hong CC, Hashimoto C. Signal transduction by a protease cascade. Trends Cell Biol 1999; 9:102-107.
106. Levashina EA, Langley E, Green C et al. Constitutive activation of Toll-mediated antifungal defense in serpin-deficient Drosophila. Science 1999; 285:1917-1919.
107. Lemaitre B, Nicolas E, Michaut L et al. The dorsoventral regulatory gene cassette spatzle/Toll/cactus controls the potent antifungal response in Drosophila adults. Cell 1996; 86:973-983.
108. Michel T, Reichhart JM, Hoffmann JA et al. Drosophila Toll is activated by Gram-positive bacteria through a circulating peptidoglycan recognition protein. Nature 2001; 414:756-759.

109. Akira S, Takeda K, Kaisho T. Toll-like receptors: critical proteins linking innate and acquired immunity. Nat Immunol 2001; 2:675-680.
110. Gottar M, Gobert V, Michel T et al. The Drosophila immune response against Gram-negative bacteria is mediated by a peptidoglycan recognition protein. Nature 2002; 416:640-644.
111. Ramet M, Manfruelli P, Pearson A et al. Functional genomic analysis of phagocytosis and identification of a Drosophila receptor for E. coli. Nature 2002; 416:644-648.
112. Horng T, Medzhitov R. Drosophila MyD88 is an adapter in the Toll signaling pathway. Proc Natl Acad Sci USA 2001; 98:12654-12658.
113. Shen B, Manley JL. Phosphorylation modulates direct interactions between the Toll receptor, Pelle kinase and Tube. Development 1998; 125:4719-4728.
114. Grosshans J, Bergmann A, Haffter P et al. Activation of the kinase Pelle by Tube in the dorsoventral signal transduction pathway of Drosophila embryo. Nature 1994; 372:563-566.
115. Galindo RL, Edwards DN, Gillespie SK et al. Interaction of the Pelle kinase with the membrane-associated protein tube is required for transduction of the dorsoventral signal in Drosophila embryos. Development 1995; 121:2209-2218.
116. Grosshans J, Schnorrer F, Nusslein-Volhard C. Oligomerisation of Tube and Pelle leads to nuclear localisation of dorsal. Mech Dev 1999; 81:127-138.
117. Edwards DN, Towb P, Wasserman SA. An activity-dependent network of interactions links the Rel protein Dorsal with its cytoplasmic regulators. Development 1997; 124:3855-3864.
118. Towb P, Bergmann A, Wasserman SA. The protein kinase Pelle mediates feedback regulation in the Drosophila Toll signaling pathway. Development 2001; 128:4729-4736.
119. Bergmann A, Stein D, Geisler R et al. A gradient of cytoplasmic Cactus degradation establishes the nuclear localization gradient of the dorsal morphogen in Drosophila. Mech Dev 1996; 60:109-123.
120. Shen B, Liu H, Skolnik EY et al. Physical and functional interactions between Drosophila TRAF2 and Pelle kinase contribute to Dorsal activation. Proc Natl Acad Sci USA 2001; 98:8596-8601.
121. Zapata JM, Matsuzawa S, Godzik A et al. The Drosophila tumor necrosis factor receptor-associated factor-1 (DTRAF1) interacts with Pelle and regulates NF-kappaB activity. J Biol Chem 2000; 275:12102-12107.
122. Liu H, Su YC, Becker E et al. A Drosophila TNF-receptor-associated factor (TRAF) binds the ste20 kinase Misshapen and activates Jun kinase. Curr Biol 1999; 9:101-104.
123. Bacher S, Grosshans J, Droge W et al. The Drosophila proteins Pelle and Tube induce JNK/AP-1 activity in mammalian cells. FEBS Lett 2001; 497:153-158.
124. Muzio M, Natoli G, Saccani S et al. The human Toll signaling pathway: Dvergence of nuclear factor kappaB and JNK/SAPK activation upstream of tumor necrosis factor receptor- associated factor 6 (TRAF6). J Exp Med 1998; 187:2097-2101.
125. Rutschmann S, Jung AC, Hetru C et al. The Rel protein DIF mediates the antifungal but not the antibacterial host defense in Drosophila. Immunity 2000; 12:569-580.
126. Manfruelli P, Reichhart JM, Steward R et al. A mosaic analysis in Drosophila fat body cells of the control of antimicrobial peptide genes by the Rel proteins Dorsal and DIF. EMBO J 1999; 18:3380-3391.
127. Stein D, Goltz JS, Jurcsak J et al. The Dorsal-related immunity factor (Dif) can define the dorsal-ventral axis of polarity in the Drosophila embryo. Development 1998; 125:2159-2169.
128. Ip YT, Reach M, Engstrom Y et al. Dif, a dorsal-related gene that mediates an immune response in Drosophila. Cell 1993; 75:753-763.
129. Wu LP, Anderson KV. Regulated nuclear import of Rel proteins in the Drosophila immune response. Nature 1998; 392:93-97.
130. Avila A, Silverman N, Diaz-Meco MT et al. The Drosophila atypical protein kinase C-ref(2)p complex constitutes a conserved module for signaling in the Toll pathway. Mol Cell Biol 2002; 22:8787-8795.
131. Drier EA, Govind S, Steward R. Cactus-independent regulation of Dorsal nuclear import by the ventral signal. Curr Biol 2000; 10:23-26.
132. Sanz L, Diaz-Meco MT, Nakano H et al. The atypical PKC-interacting protein p62 channels NF-kappaB activation by the IL-1-TRAF6 pathway. EMBO J 2000; 19:1576-1586.
133. Rutschmann S, Jung AC, Zhou R et al. Role of Drosophila IKK gamma in a Toll-independent antibacterial immune response. Nat Immunol 2000; 1:342-347.

134. Silverman N, Zhou R, Stoven S et al. A Drosophila IkappaB kinase complex required for Relish cleavage and antibacterial immunity. Genes Dev 2000; 14:2461-2471.
135. Kim YS, Han SJ, Ryu JH et al. Lipopolysaccharide-activated kinase, an essential component for the induction of the antimicrobial peptide genes in Drosophila melanogaster cells. J Biol Chem 2000; 275:2071-2079.
136. Hedengren M, Asling B, Dushay MS et al. Relish, a central factor in the control of humoral but not cellular immunity in Drosophila. Mol Cell 1999; 4:827-837.
137. Choe KM, Werner T, Stoven S et al. Requirement for a peptidoglycan recognition protein (PGRP) in Relish activation and antibacterial immune responses in Drosophila. Science 2002; 296:359-362.
138. Dejardin E, Droin NM, Delhase M et al. The lymphotoxin-beta receptor induces different patterns of gene expression via two NF-kappaB pathways. Immunity 2002; 17:525-535.
139. Claudio E, Brown K, Park S et al. BAFF-induced NEMO-independent processing of NF-kappa B2 in maturing B cells. Nat Immunol 2002; 3:958-965.
140. Senftleben U, Cao Y, Xiao G et al. Activation by IKKalpha of a second, evolutionary conserved, NF-kappa B signaling pathway. Science 2001; 293:1495-1499.
141. Vidal S, Khush RS, Leulier F et al. Mutations in the Drosophila dTAK1 gene reveal a conserved function for MAPKKKs in the control of rel/NF-kappaB-dependent innate immune responses. Genes Dev 2001; 15:1900-1912.
142. Aliprantis AO, Yang RB, Mark MR et al. Cell activation and apoptosis by bacterial lipoproteins through Toll- like receptor-2. Science 1999; 285:736-739.
143. Aliprantis AO, Yang RB, Weiss DS et al. The apoptotic signaling pathway activated by Toll-like receptor-2. EMBO J 2000; 19:3325-3336.
144. Schneider P, Tschopp J. Apoptosis induced by death receptors. Pharm Acta Helv 2000; 74:281-286.
145. Hu S, Yang X. dFADD, a novel death domain-containing adapter protein for the Drosophila caspase DREDD. J Biol Chem 2000; 275:30761-30764.
146. Chen P, Rodriguez A, Erskine R et al. Dredd, a novel effector of the apoptosis activators reaper, grim, and hid in Drosophila. Dev Biol 1998; 201:202-216.
147. Rodriguez A, Oliver H, Zou H et al. Dark is a Drosophila homologue of Apaf-1/CED-4 and functions in an evolutionarily conserved death pathway. Nat Cell Biol 1999; 1:272-279.
148. Mochida Y, Takeda K, Saitoh M et al. ASK1 inhibits interleukin-1-induced NF-kappaB activity through disruption of TRAF6-TAK1 interaction. J Biol Chem 2000; 275:32747-32752.
149. Evans PC, Taylor ER, Coadwell J et al. Isolation and characterization of two novel A20-like proteins. Biochem J 2001; 357:617-623.
150. Kopp E, Medzhitov R, Carothers J et al. ECSIT is an evolutionarily conserved intermediate in the Toll/IL-1 signal transduction pathway. Genes Dev 1999; 13:2059-2071.
151. Wong BR, Besser D, Kim N et al. TRANCE, a TNF family member, activates Akt/PKB through a signaling complex involving TRAF6 and c-Src. Mol Cell 1999; 4:1041-1049.
152. Vig E, Green M, Liu Y et al. SIMPL is a tumor necrosis factor-specific regulator of nuclear factor-kappaB activity. J Biol Chem 2001; 276:7859-7866.
153. Ling L, Goeddel DV. T6BP, a TRAF6-interacting protein involved in IL-1 signaling. Proc Natl Acad Sci USA 2000; 97:9567-9572.
154. Pype S, Declercq W, Ibrahimi A et al. TTRAP, a novel protein that associates with CD40, tumor necrosis factor (TNF) receptor-75 and TNF receptor-associated factors (TRAFs), and that inhibits nuclear factor-kappa B activation. J Biol Chem 2000; 275:18586-18593.
155. Shin JN, Kim I, Lee JS et al. A novel zinc finger protein that inhibits osteoclastogenesis and the function of tumor necrosis factor receptor-associated factor 6. J Biol Chem 2002; 277:8346-8353.
156. Takatsuna H, Kato H, Gohda J et al. Identification of TIFA as an adapter protein that links TRAF6 to IRAK-1 in IL-1 receptor signaling. J Biol Chem 2003; 278:12144-12150.

Chapter 5

Virus Induced Signaling to Initiate the Interferon Mediated Anti-Viral Host Response

Claudia Wietek and Luke A.J. O'Neill

Abstract

Signaling cascades leading to the activation of an anti-viral host response have been subject to intense investigations over the last decade. Consequently our understanding of the initial switches that launch this response have increased greatly. The establishment of an anti- viral state hinges on the co-ordinated production of type I interferons. These cytokines link the innate and adaptive anti-viral response. The expression of "early phase" interferons, comprising IFNβ, murine IFNα4 or human IFNα1, are controlled by Toll-like receptor-induced signaling cascades that activate the latent transcription factors NF-κB and interferon regulatory factor (IRF)-3. In turn, the early IFNs induce signals to promote the expression and activation of transcription factors. These go on to induce the expression of a variety of cytokines and chemokines that propel cells of the adaptive immune system towards an anti-viral response.

Introduction

The central task of our immune system is to distinguish between self and non-self, allowing us to recognize and eliminate invading pathogens. Distinct pathogenic threats are sensed by specific receptors, the Toll- like receptors (TLRs), which induce signaling cascades to activate transcription factors. These regulators activate the expression of a defined subset of cytokines and chemokines. The signals induced by viruses result primarily in the production of interferons. These are crucial players in establishing an anti-viral state and eventually, a cell-mediated immune response. Interferons (IFNs) are a heterogeneous family of multifunctional secreted proteins. Based on their molecular properties IFNs can be divided into type I IFNs (comprising 13 closely related IFNαs and a single IFNβ) and the type II IFN, IFNγ. The production of interferons is controlled by several transcription factors, most notably members of the interferon- regulated- factor (IRF) family, NF-κB, and signal transducers and activators of transcription (STATs). The activation of transcription factors occurs in two phases. Firstly, viral products induce the activation of IRF-3 and NF-κB leading to IFNβ expression. Autocrine induction by IFNβ then activates the transcription factors that orchestrate the expression of "second phase" anti-viral genes. Some viruses have evolved mechanisms to evade these anti-viral

Toll and Toll-Like Receptors: An Immunologic Perspective, edited by Tina Rich. ©2005 Eurekah.com and Kluwer Academic / Plenum Publishers.

measures and increase their virulence. Their stratagems include the production of proteins that block the signaling cascades as well as the interference of the transcription factors themselves.

Initiation of the Host Immune Response

Upon infection, cells of the innate immune system such as macrophages, natural killer cells and dendritic cells (DCs), build an immediate defense barrier. Exposure to a foreign stimulus activates these cells to phagocytose and kill the invading pathogens. This innate immunity, an evolutionarily old branch of the immune system, has long been perceived as a rather non-specific and general response to pathogens.[1] Specificity in the host response was assumed to be conferred by adaptive immunity, which is only found in vertebrates.[2] Dendritic cells are crucial players in launching this delayed, secondary response. The function of DCs is not to kill the pathogen but rather to present pathogen- associated antigens to T cells, polarizing them towards a Th1 or Th2 phenotype.[3] Importantly, DCs are activated by pathogen-specific signals, the so-called pathogen-associated molecular patterns (PAMPs). PAMPs are usually discrete structures that are essential for microbial survival and which cannot be eliminated without compromising the microbe. They do not usually occur in the host itself. Different PAMPs will induce different sets of cytokines by DCs such that T cells become either Th1 or Th2 cells.[4]

Bacterial and viral infections typically induce Th1 cell-mediated immune responses. These are characterized by increased production of interferon gamma. In contrast, the humoral Th2 response is provoked by parasites and is accountable for allergic reactions. Hence, recognition of the type of pathogen and infection by the innate immune system determines the type of the adaptive immune response. A key question has been to identify the PAMP receptors and TLRs now appear to be the most likely candidates.

Toll-Like Receptors

TLRs are characterized by a cytoplasmic TIR (Toll/IL-1 receptor) domain and an extra-cellular portion that contains leucine rich repeats.[5] The extra-cellular portion distinguishes this family from the TIR bearing IL1 receptor family, which express extracellular immunoglobulin domains. The TIR domain is evolutionary well conserved and is found in plants, insects and mammals.[5] Strikingly, it is involved in the functional anti pathogen response in all these organisms. The first human homolog of *Drosophila* Toll, TLR4, was found in 1998 and its importance in the activation of adaptive immunity quickly acknowledged.[6,7] TLR4 turned out to be the receptor for lipopolysaccharide (LPS), the major component of the outer membrane of Gram negative bacteria. Positional cloning in the LPS- hypo-responsive C3H/HeJ mouse strain revealed that these mice bore a point mutation in the *tlr4* gene, which introduced a proline to histidine amino acid change. This mutation was in the BB-loop, a region crucial for the functional integrity of the TIR domain.[7] At the time of writing, ten members of the human TLR family have been described and specific ligands for most of them identified. The principle receptors for bacterial products are TLR2 (lipoprotein), TLR4 (LPS), TLR5 (flagellin) and TLR9 (bacterial DNA) (Fig. 1). Interestingly, TLR4 also recognizes viral components such as the F protein of Respiratory Syncytial Virus (RSV),[8,9] as well as host stress proteins (Hsp60 and Hsp70) that are released during tissue injury or infection.[10,11] TLR3 recognizes the artificial ligand polyIC, which mimics viral double-stranded RNA.[12] Although their natural ligands have not yet been identified, TLR7 and TLR8 both recognize the anti-viral imidazoquinoline (R-848). The R-848 ligand induces synthesis of IFNα and other cytokines in various cell types[13,14] which would suggest that the function of its receptor might be in viral host defense.*

* Since going to press murin TLR7 and human TLR8 have been shown to recognize ssRNA, both of viral and endogenous origin (Heil F et al. Science 2004; 1526-1529. Diebold SS et al. Science 2004; 303:1529-1531.)

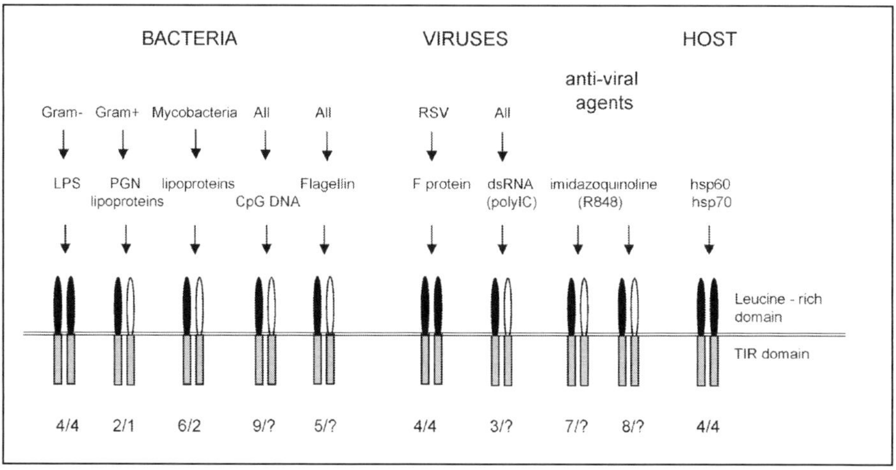

Figure 1. Toll-like receptors and their ligands. Selected TLRs and their virus derived or bacterial ligands are presented. An extracellular leucine-rich domain and a TIR domain in the cytosolic portion typify TLRs. TLR specific PAMPs bind to dimerized or multimeric receptors to stimulate signaling. TLR4 transduces signals as a homodimer whereas TLR2 can associate with TLR6 or TLR1. The partners for TLR3, TLR5, TLR7, TLR8 and TLR9 have not yet been determined. TLRs sense common bacterial and viral PAMPs such as bacterial CpG DNA (the ligand for TLR9) or dsRNA (perceived by TLR3). Additionally, more specific PAMPs are recognized by other TLRs to tailor the cellular response. TLR4 detects LPS, a Gram- negative specific molecular structure whilst TLR2 senses either Gram positive derived lipoproteins (when associated with TLR1) or mycobacterial lipoproteins (with TLR6 as its partner). TLR4 also senses F protein, which occurs in a Respiratory Syncytial Virus (RSV) PAMP.

Upon stimulation with their corresponding ligands, TLRs recruit TIR-adapter molecules to their cytoplasmic TIR domains. This initiates signals that activate transcription factors, most importantly NF-κB,[5] which then induce an array of genes, most notably the cytokines such as TNF, IL-1, IL-12 and IL-6. Some of these polarize T cells towards a Th1 response. The most common cytoplasmic TIR-adapter is myeloid differentiation factor 88 (MyD88).[15] Its loss prevents signaling from most of the TLRs with the exception of TLRs 3 and 4.[12,16,17] TLR3 and TLR4 both initiate "MyD88-independent"-pathways through a recently described adapter termed TIR domain-containing adapter inducing IFNβ (TRIF) or TIR-containing adapter molecule-1 (TICAM-1).[18-21] This alternative TRIF mediated pathway results in delayed activation of NF-κB and most importantly, in the activation of the interferon responsive element-3 (IRF-3) and IFNβ-induced gene expression (Fig. 2).[17,20,21] While TLR3 presumably only signals through TRIF, TLR4 also recruits MyD88 and at least one other adapter termed MyD88 adapter- like (Mal) or TIR adapter protein (TIRAP).[22,23] The phenotypes of Mal and MyD88 single or double knockout mice are essentially convergent in that LPS induced IRF-3 activation and dendritic cell maturation are unaffected but NF-κB activation is delayed. This would suggest that these adapters are components of the same TLR4 pathway that induces rapid activation of NF-κB. Interestingly, the loss of either of these adapters abolishes TLR2 induced NF-κB activation, indicating that both are recruited to TLR2 but (in contrast to TLR4 signaling) are probably not functionally redundant.[24,25] Two further TIR

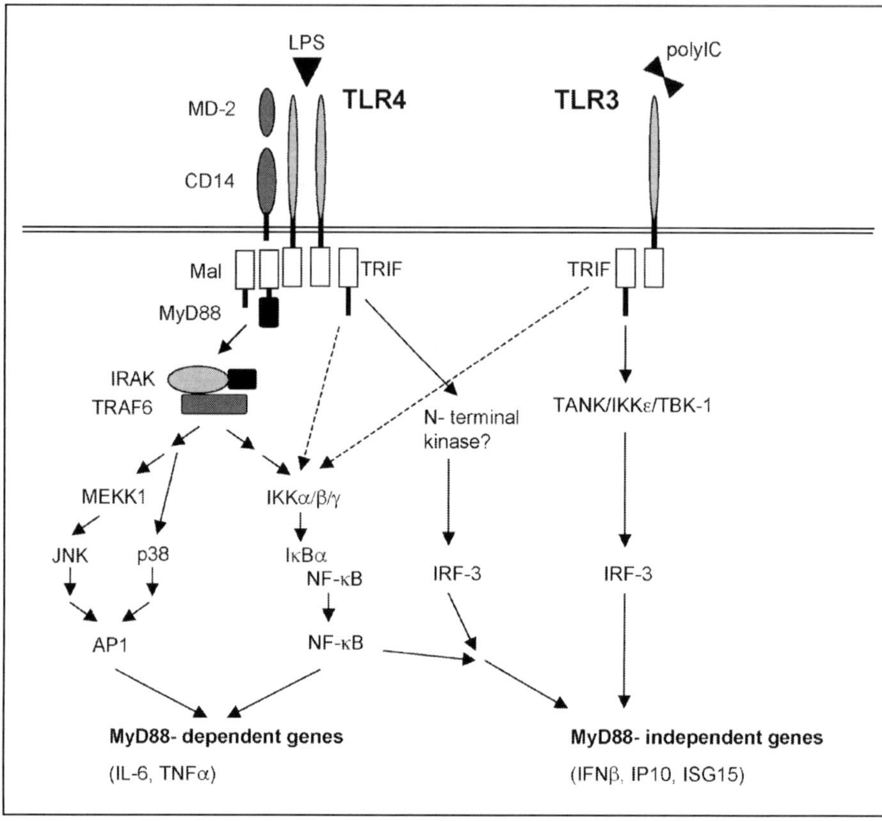

Figure 2. TLR4 and TLR3 induced signaling cascades. Upon LPS stimulation TLR4 together with its co-receptors CD14 and MD-2, stimulate signaling cascades mediated by the adapters MyD88, Mal and TRIF that interact via their TIR domains with the receptor. Mal and MyD88 interact with IRAKs, and phosphorylated IRAKs dissociate from the receptor complex and signal via TRAF6 to activate the transcription factors AP-1 and NF-κB. Alternatively, TLR4 signals via TRIF to activate IRF-3, presumably using an N-terminal IRF-3 kinase, and NF-κB. TRIF dependent activation of NF-κB is delayed compared to MyD88/Mal induced NF-κB activation, indicated with hatched arrows. Both, TLR4 and TLR3 induced TRIF dependent pathways result in activation of early phase anti-viral genes. TLR3 also signals through TRIF to activate IRF-3 and NF-κB. IRF-3 activation, however, differs from the TLR4 induced pathway since TRIF presumably directly activates the TANK/IKKε TBK-1 complex to phosphorylate IRF-3 at its C-terminus. White boxes in the receptors and adapters indicate TIR domains and black boxes signify death domains in MyD88 and IRAK.

adapters have been found, TIR domain containing adapter protein (TIRP),[26] also called TRIF-related adapter molecule (TRAM) and a protein called sterile alpha and TIR motif containing (SARM).[27] Whilst TRAM has been implicated in IL-1 signaling the role of SARM is still unknown.* Whether either of these compensates for the loss of Mal and MyD88 in TLR4 signaling has yet to be determined.

* During publication of this chapter the TIR adapter TRAM has been further characterized as a specific TLR4 adapter, which acts upstream of TRIF to activate IRF-3 and NFκB.[133]

Double Stranded RNA As the Principle Viral PAMP

Double stranded RNA (dsRNA) produced by replicating virus, or in the form of the viral genome, is a key stimulus in the activation of transcription factors that induce type I interferons (IFNs). The production of these cytokines is a crucial event in the initial host response to viral infection. Viral dsRNA activates quiescent precursor forms of the dsRNA dependent protein kinase R (PKR) and the 2'-5' oligoadenylate synthetases (OASs).[28-31] OASs catalyze the synthesis of ATP oligomers consisting of three to five adenosines linked by 2'-5' bonds. These unstable molecules bind and activate endoribonuclease (RNase) L, which catalyses the cleavage of single-stranded RNAs including mRNA and 28S ribosomal RNA. This results in translational inhibition.[31,32] PKR is a multifunctional serine/threonine kinase, which regulates transcription and translation upon viral infection and also mediates the apoptotic death of infected cells.[33,34] In its active form PKR is a dimer and the N-terminal regulatory domain of each PKR molecule is used to bind a single molecule of dsRNA. Mutual trans-phosphorylation at several serine and threonine residues renders the dimer active.[35] One of its substrates is the IκB kinase (IKK) 2, which phosphorylates IκBα, the inhibitor of NF-κB.[36,37] IκBα binds NF-κB in the cytoplasm, preventing its translocation to the nucleus. Once phosphorylated (at Ser 32 and Ser 36) IκBα becomes ubiquitinated by an E3 ubiquitin ligase and is targeted for proteasomal degradation. NF-κB is released and translocates to activate its target genes.[38,39] PKR is also involved in the regulation of other transcription factors such as the signal transducer and activator of transcription (STAT)-1[40,41] and interferon regulatory factor (IRF)-1.[42] Another important function of PKR is to inhibit translation in virus infected cells. PKR phosphorylates the α subunit of the eukaryotic translation initiation factor (eIF)-2 which then prevents the recycling of initiation factors at the pre-initiation complex.[33,43] In addition to its function as a sensor for dsRNA, PKR has also been implicated in TLR4 signaling by its interaction with the adapter Mal/TIRAP.[23] However, PKR is not entirely accountable for the establishment of a full anti-viral state since PKR knockout mice still show significant resistance to viral infections and cells from these animals still respond to polyIC.[29] As mentioned before, studies with TLR3 knockout mice indicated its possible function as a dsRNA sensor. TLR3 deficient mice showed reduced responses to polyIC and a decreased production of inflammatory cytokines, suggesting a role for TLR3 in viral recognition.[12] However, a recent publication suggests that TLR3 may only recognize extracellular dsRNA (viral particles for example) leaving PKR to respond to intracellular dsRNA.[44]

Interferon Regulatory Factors (IRFs)

As stated earlier, a crucial signal activated by TLRs is the transcription factor NF-κB. The process leading to its activation has been dealt with in several reviews.[5,39] Less well understood is the impact of TLRs on other transcription factors. Attention has recently been drawn to interferon regulatory factors (IRFs) and in particular IRF-3, because of their crucial role in initiating the host response to viral infection mediated by type I IFNs. IRF-3 is the target of the so-called "MyD88- independent pathway" induced by TLR4 and TLR3 signals.

IRF-1 was identified as a protein that binds to the virus inducible elements in the IFNβ promoter.[45] A second related factor termed IRF-2 was found in a cross- hybridization screen.[46] Mutational analysis identified a DNA-binding domain (DBD) in the N-termini of these transcription activators. Further crystallographic studies of this domain, bound to DNA, revealed that a conserved tryptophan repeat (five tryptophans spaced in 10 to 18 amino acids) mediates binding of a novel helix-turn-helix motif to a tandem repeat of the GAAA core sequence.[46,47] Based on the homology of their N-terminal DBDs, seven more human IRFs were discovered. These comprise IRF-3, IRF-4 (Pip, LSIRF, ICSAT, MUM1), IRF-5, IRF-6, IRF-7, IRF-8 (ICSBP) and IRF-9 (p48, ISGFγ). Additional viral members of this family (vIRFs) were found in the genome of human herpes virus 8 (HHV-8)[48-50] and will be discussed later. IRFs bind to DNA motifs termed Interferon Regulatory Elements (IRF-E). IRF-Es are present in the Positive Regulatory Domains (PRD I and III) of the IFNβ promoter where they are bound by IRF-1 and IRF-2. Similarly, the Interferon Consensus Sequence (ICS) of the MHC class I promoter is bound by IRF-8 and the Interferon Stimulated Response Element (ISRE), found in most IFN inducible genes, is recognized by IRF-3, IRF-7 and the trimeric (ISGF) 3 complex. IRFs can act as transcriptional activators (IRF-1, IRF-3, IRF-9), repressors (IRF-8) or both (IRF-2, IRF-4, IRF-7).[51]

IRF-3 and IRF-7

The structurally related activators IRF-3 and IRF-7 are crucial for the expression of type I IFNs and IFN inducible genes.[52-54] Adjacent to their N-terminal DBDs these proteins bear a nuclear export signal (NES) followed by an IRF activation domain (IAD) and a C-terminal activation domain which comprises a serine/threonine cluster for phosphorylation.[55,56] Both factors are expressed in a variety of cells, but while constitutive expression of IRF-3 remains unaltered by viral infection or stimulation with IFNs, transcription of IRF-7 is induced by type I IFNs.[57,58] The two proteins bind to the ISRE consensus sequence ($^{A}/_{G}$NGAAANNGAAACT) but have distinct preferences for specific residues in this motif.[59] Activation of IRF-3 is the initial cellular response to stimulation with viral products resulting in the production of low amounts of IFNβ, murine IFNα4 or human IFNα1 and the so-called early phase genes of the biphasic IFN feedback loop (discussed later). IFNβ and IFNα activate the trimeric transcription factor, interferon stimulated gene factor (ISGF)-3, which in turn induces the transcription of late phase genes including IRF-7. IRF-7 itself acts on the IFNβ and IFNα promoters, mainly in cooperation with IRF-3, to enhance IFN production and expression of IFN induced genes.[60] IRF-3, however, mainly affects transcription of IFNβ whereas IRF-7 regulates expression of both IFNβ and the IFNαs.[61,62]

IRF-9

IRF-9, previously termed p48 and ISGF-3γ, is a DNA binding subunit of the ISGF-3 complex, which also comprises the signal transducers and activators of transcription (STAT)1 and STAT2.[63,64] IRF-9 knockout mice are phenotypically similar to mice deficient in the type I IFN receptor IFNAR in that they lack production of IFNαs and IRF-7 but are only slightly impaired in IFNβ expression.[65] These observations are consistent with the finding that IFNβ knockout mice are impaired in IFNα production.[66] Depletion of IRF-3 in IRF-3 knockout mice significantly diminished the production of type I IFN after viral infection and IFNα/β transcription was entirely abolished in IRF-3 IRF-9 double knockout mice, which do not produce IRF-7. Normal IFNα production can only be restored when both IRF-3 and IRF-7 are coexpressed in the double knockout cells.[62] These studies identified IRF-3 and IRF-7 as key regulators in the virus induced IFN response, with IRF-3 initiating the response through the induction of IFNβ and early ISRE dependent genes. IRF-7 and IRF-3 then amplify the anti-viral response by further upregulation of type I IFN and the expression of additional IFN inducible genes.

IRF-3 Activation

Regulation of IRF-3 and the pathways leading to its activation have been subject to detailed investigations. IRF-3 in its inactive form is constitutively present in a quiescent cytoplasmic pool. Upon stimulation with viral products such as dsRNA IRF-3 is phosphorylated in its C-terminus which exposes an interferon association domain (IAD) that enables dimerization (Fig. 3).[67-69] While other stimuli like LPS, UV-light or osmoactive substances also activate IRF-3 they do not induce C-terminal phosphorylation but rather phosphorylation of the

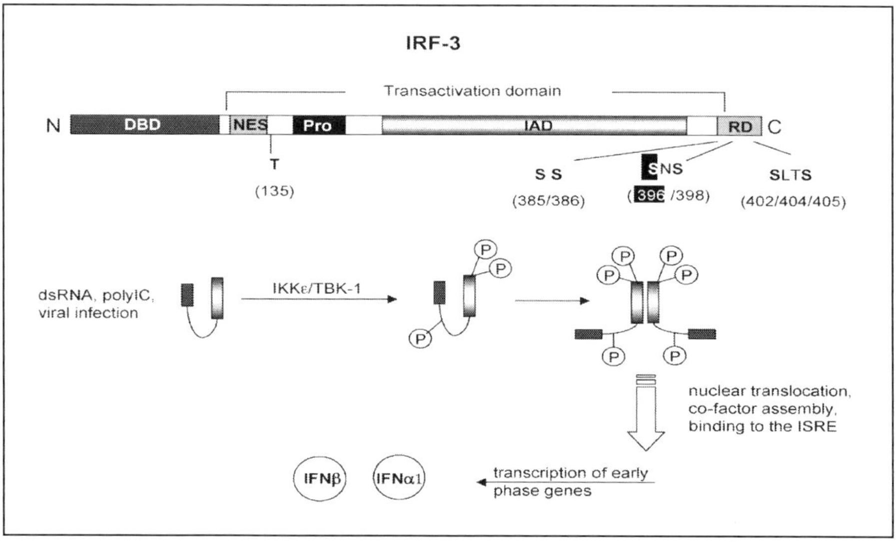

Figure 3. Virus induced activation of IRF-3. Schematic presentation of IRF-3: The N-terminal binding domain (DBD), the nuclear export sequence (NES), a proline- rich region (Pro), the IRF association domain (IAD) and the response domain (RD) are indicated in gray boxes. Threonine 135 in the N- terminus is targeted by stress-induced phosphorylation. Phosphorylation of serine 396 in the C-terminal serine/threonine cluster (by the IKKε TBK-1 complex) triggers dimerization of IRF-3, which subsequently translocates to the nucleus to associate with its co-regulators CBP or p300 and bind to the ISRE. This ISRE activation complex induces transcription of the early phase genes, most importantly the type I interferons IFNβ and IFNα1.

N-terminus.[70] Though the nature of this phosphorylation is uncertain, it is presumably insufficient to activate IRF-3.[52] Mutational analysis led to the identification of serine 396 in a C-terminal 6 serine/ 1 threonine phosphorylation cluster as the minimal phospho-acceptor site to activate IRF-3.[71] This residue is not involved in the activation induced by TLR4 signaling, hinting at different stimuli- dependent activation mechanisms.[70,71] Recent studies have identified two IKK isoforms (IKKε and TANK- binding kinase (TBK)-1) as components of the virus activated kinase (VAK) complex which activates IRF-3 and IRF-7.[72,73] In support of these findings, an inactive mutant of IKKε blocked the C-terminal phosphorylation of IRF-3 and IRF-7 induced expression from the IFNα4 promoter.[72]

In its activated and dimeric form, IRF-3 translocates to the nucleus and binds the paralog co-activators CBP (CREB- binding protein) or p300.[74] CBP/p300 prevent nuclear export of the activated transcription factor and are essential for the binding of IRF-3 to the ISRE.[75] As discussed earlier, signaling through either TLR3 (stimulated by polyIC) or TLR4 (the sensor for LPS and the F protein of RSV) induces IRF-3 activation and an IFN regulated anti-viral, or in the case of LPS, anti bacterial state. Interestingly the two TLRs employ distinct signaling pathways and activation mechanisms for IRF-3. While the polyIC/TLR3 pathway induces phosphorylation of IRF-3 by the IKKε TBK-1 complex, signaling through TLR4 does not lead to C-terminal phosphorylation.[70] Also, IKKε deficient mouse embryonic macrophages (MEFs) are not impaired in IRF-3 activation following stimulation with the TLR4 ligand LPS.[76] TLR4 signaling to IRF-3 is blocked by NF-κB inhibitors and p65 deficient cells are significantly impaired in TLR4 mediated activation of the ISRE. However, depletion of NF-κB has no effect on the TLR3 response. Furthermore, IRF-3 and NF-κB are found in the same TLR4 stimulated activation complex at the ISRE.[77] The reason for these fundamental differences in

IRF-3 activation is uncertain. However, it appears likely that differently activated forms of IRF-3 are bound in diverse activation complexes and that these either control similar ISRE sites or activate identical sites differentially.

The IFN Loop

The anti-viral state of the host cell is initiated by activated transcription factors such as IRF-3 and NF-κB. These induce the expression of early or primary response genes and, subsequently enhanced in an autocrine/paracrine loop activated by IFNβ, result in the expression of late or secondary response genes (Fig. 4). Since IRF-3 and NF-κB are both activated from a latent pool, the activation of the early response to viral stimuli does not require de novo protein synthesis. Early response genes include chemotactic chemokines such as RANTES and IP-10, the interferon stimulated gene (ISG) 15 (which is a structural homologue of ubiquitin), IFIT1 (which is thought to be involved in protein processing) and most importantly IFNβ and murine IFNα4 or human IFNα1.[76,78] Secreted IFNβ feeds back on the IFNα/β receptor (IFNAR) which consists of two major subunits, IFNAR1 and IFNAR2. These associate with the "Janus" tyrosine kinases (Jak) Tyk2 and Jak1 respectively.[79-81] IFNAR1 and IFNAR2 dimerize upon binding of type I IFNs to allow the trans-phosphorylation of the two kinases.[81] Activated Tyk2 phosphorylates STAT2, which is also bound to IFNAR2 and IFNAR1 at crucial tyrosine residues. The SH2 domain of STAT2 then docks at the phosphotyrosine of IFNAR1 whereas STAT1 binds to the phosphorylated tyrosine residue of STAT2 via its SH2 domain.[80,82-84] STAT1 is subsequently phosphorylated [85] and the activated STAT1/STAT2 dimer dissociates

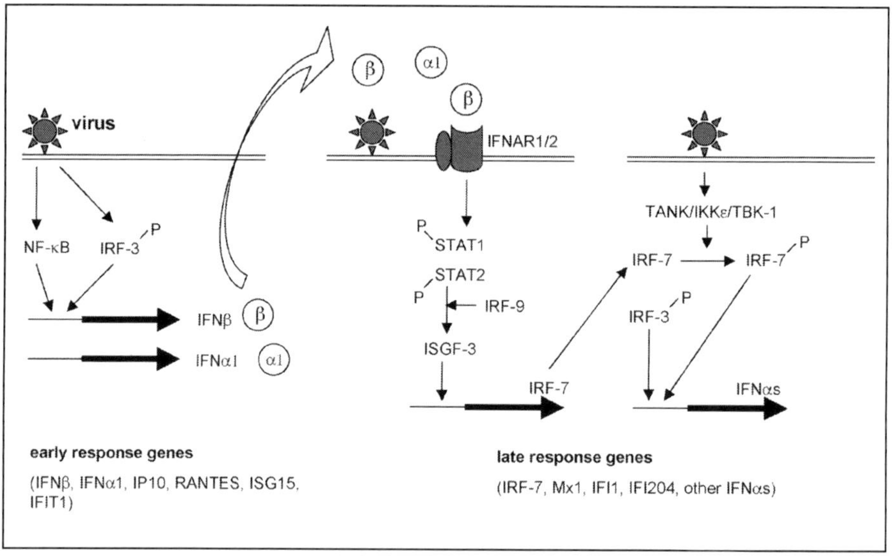

Figure 4. The interferon loop. Virus derived PAMPs induce activation of the quiescent cytosolic transcription factors NF-κB and IRF-3, which co-operatively activate transcription of early phase genes to launch the host response against the virus. IFNβ and IFNα1, which are produced as a result of this initial signaling event, feed back on their heterodimeric receptor IFNAR1/2 in a paracrine loop to activate STAT1 and STAT2. The two STATs form a trimeric complex with IRF-9 termed ISGF-3, which binds to the ISRE and induces expression of IRF-7. IRF-7 is structurally and functionally similar to IRF-3. Phosphorylation of its C-terminus by the virus activated IKKε-TBK-1 complex renders it competent to homo- or heterodimerize with IRF-3. The dimer, in association with co-activators, induces the transcription of further late phase ISRE controlled genes including IFNαs.

from the receptor complex, translocates to the nucleus, and binds IRF-9 to form the trimeric transcriptional activator ISGF-3.[86] ISGF-3 binds to the ISRE (consensus $^{A}/_{G}$NGAAANNGAAACT) and activates transcription of IFNβ induced secondary response genes. As opposed to the primary response, transcription of these genes can be blocked with either IFNβ specific inhibitory antibodies or the translational inhibitor cycloheximide.[76,87] Moreover, primary response genes are co-regulated by NF-κB and IRF-3 whilst the NF-κB site is redundant for expression of secondary response genes.[59,88] Activated STAT1 is also able to homodimerize and bind to the GAS (IFNγ activated site) to induce transcription. This STAT1 homodimer termed GAF (IFNγ activated factor) is a crucial transcriptional activator in IFNγ signaling and is activated as a result of a tyrosine phosphorylation cascade that initiates at the IFNγ receptor (IFNGR). However, GAF is less important in type I IFN signaling (reviewed in ref. 89). ISGF-3 both directly and indirectly activates transcription of the late phase genes via IRF-7 and IRF-5, both of which are induced by ISGF-3. The late phase genes include IFNαs, the growth inhibiting chemokine IFI204 and the GTPases Mx1 and IFI1. ISGF-3 also upregulates the basal expression levels of PKR and OAS.[76,90,91] Additionally, the expression of early response genes such as IFNβ and IFNα1 are up-regulated as a result of these autocrine IFNβ effects.[92] The second wave of anti-viral protein thus amplifies the initial response, tailoring it according to the pathogenic threat by using different combinations of activated transcription factors. The positive feedback loop through IFNAR is essential for the successful establishment of the anti-viral state following TLR4 or TLR3 signaling.[93] TLR3 is generally a stronger inducer of the anti-viral host response than TLR4. This is, in part, due to the fact that TLR3 amplifies its own expression via IFNβ. In contrast, TLR4 expression is unaffected by IFNβ signaling.[76] Differences in activation and complex formation of the initiating transcriptional activator IRF-3 may also contribute to a more potent TLR3 anti-viral response.[72,76,77]

The interferon loop is eventually attenuated via different mechanisms. Negative regulators of the IRF family (IRF-2 and IRF-8) have been shown to down-regulate expression of ISRE dependent genes and hence the IFN response.[46,94] Moreover, the suppressors of cytokine signaling (SOCS) proteins have been implicated as important negative regulators for IFN signaling by interfering with Jaks and STATs at the receptor complex.[95-97] Expression of these negative regulators is induced by various cytokines including IFNγ. Interestingly, bacterial TLR ligands, namely LPS, lipoteichoic acid (LTA) and CpG DNA, can also stimulate the production of SOCS proteins.[98-100] In addition to their function as inhibitors of IFN signaling, SOCS have recently been implicated in TLR signal transduction and are thought to be involved in LPS tolerance.[101,102]

Interactions at IFN Promoters

The most investigated ISRE containing promoter is the IFNβ promoter which serves as the prototype for combinatorial interactions of transcriptional activators.[103-105] Cooperative assembly of NF-kB, members of the IRF family (IRF-3, IRF-7 and IRF-1), and the ATF-2/c-Jun heterodimer at the overlapping positive regulatory domains, PRDII, PRDI-III, and PRD IV (in the enhancer site -110 to -45bp upstream of the transcriptional start site) induce transcription of IFNβ.[105,106] Interestingly, the positioning of the enhancer relative to the transcriptional start site is not decisive for induction of transcription. Assembly of the IFNβ enhancesome is assisted by a small flexible protein with DNA binding and multiple protein/protein interaction domains. This protein, called the high mobility group protein HMG I(Y),[104,107,108] binds four sites in the enhancer to unwind the DNA and help recruit activators.[109] HMG I(Y) also chaperones the interaction between transcription factors at the enhancer.

Other than the IFNβ promoter, the IFNα promoters do not contain an NF-κB binding site in their virus responsive enhancers but rather express several AANNGAAAs to bind IRFs.[59,110] The main regulators for IFNα transcription are IRF-7, IRF-3 and IRF-5. Relative levels of these IRFs, in particular the ratio between IRF-7 and IRF-3, determines the differential expression of cell type specific IFNα subtypes. IRF-3 and IRF-7 are closely related and are both activated by phosphorylation, which enables them to homo- or heterodimerise to activate their target promoters. The composition of IRFs varies between different IFNα enhanceosomes.[111] IRF-3 and IRF-7 presumably heterodimerise to induce the IFNA2 promoter. An additional member of the IRF family, IRF-5, has been found to up-regulate expression of IFNA8 and was found in the IFNA1 enhanceosome in cells infected with Newcastle disease virus (NDV). Interestingly, IRF-5 seems to be phosphorylated and activated only upon infection with NDV but is unresponsive to other viral stimuli.[112] Human IFNα1 and murine IFNα4 are the only IFNα genes which do not require ongoing protein synthesis for their expression and are hence early response genes like IFNβ.[78] The varying composition of IFN enhanceosomes indicates that stimulus- specific activation of IRFs and the differential binding capacities of the IFN enhancer target sequences orchestrate the expression profile of type I IFN subtypes. Further investigations of the interactions at the IFNα enhanceosome are necessary to understand specificity of IFNα transactivation and virus specific host responses.

Viral Evasion Strategies

During co-evolution with their hosts viruses have developed various strategies to attenuate host defense mechanisms and evade clearance by the host cell. TLR and IFN induced signaling cascades and the interferons themselves are the main targets to block the anti-viral response. Vaccinia virus for example encodes two small TIR domain containing proteins termed A46R and A52R, which interfere with IL-1 and TLR signaling.[113] A52 has severe effects on IL-1, IL-18, TLR4 and TLR3 signaling. A more detailed study demonstrated that A52R associates with the interleukin 1 receptor associated kinase 2 (IRAK2) and tumor necrosis factor receptor associated factor 6 (TRAF6). Both are components of the NF-κB activation pathway, so we can presume that A52R disrupts TLR stimulated signaling.[114] Also, deletion of A52R in the vaccinia genome results in attenuation of viral infection in a mouse model.[114] A variety of viruses interfere with IFN production and signaling. IRFs are a common target for viral evasive mechanisms. For example, E6 of human papillomavirus type 16 (HPV-16) interacts with and neutralizes IRF-3, the inducer of the IFNβ promoter.[115] The Ebola virus VP35 protein and the NS3/4A serine protease of hepatitis C virus (HCV) both prevent the phosphorylation of IRF-3.[116,117] The HPV-16 E7 protein binds the IRF-9 subunit of ISGF-3 and prevents DNA binding of the complex.[118] The K9 gene product of human herpes virus 8 (HHV-8) also targets IRF-1.[115] Interestingly, inhibition of either of these IRFs is not sufficient to block the establishment of an anti- viral state, suggesting that their function is, in part, redundant. HHV-8 additionally encodes a viral IRF-1 homologue, termed vIRF-1, which inhibits host IRF-1 and IFN signaling and is presumably involved in oncogenic transformation.[119,120] vIRF-1 also inhibits recruitment of the CBP/p300 coactivators to IRF-3.[121] Three further vIRFs have been found in the HHV-8 genome, namely vIRF-2, vIRF-3 and vIRF-4.[49,50] The viral homolog of IRF-4, vIRF-3, has also been shown to act as a dominant negative form of both IRF-3 and IRF-7 and inhibits virus mediated transcriptional activity of the murine IFNα4 promoter.[49] Soluble IFN receptor homologues are encoded by various poxviruses that bind secreted IFNs and prevent them from signaling.[122-124] Moreover, the IFN induced Jak/STAT signaling pathway can be blocked at different stages. The T antigen of murine polyomavirus (MPyV) associates with Jak1 to prevent type 1 and type 2 IFN signaling.[125] STATs are also a common target for several viruses such as paramyxovirus simian virus 5 (SV5), which mediates the proteasomal degradation of STAT1, and human parainfluenza virus 2 (hPIV2), which targets STAT2.[126,127] A variety of viruses also bind dsRNA to prevent the activation of antiviral enzymes like PKR, OAS and the RNAseL system or to inhibit these enzymes directly.[128-132]

Conclusions

TLRs are the first defense barrier against pathogens and launch signaling cascades to initiate an adequate response to infections. Viral stimuli, perceived via TLR3 and TLR4, essentially induce activation of a family of transcriptional regulators, the IRFs. IRFs control the expression of interferons, which link innate to adaptive immunity. Differences in activation of IRFs and the differential expression of their various subtypes controls the temporal and stimulus dependent expression of a defined subset of target genes. Interferons themselves are subject to the control of enhanceosomes, which consist of a variety of transcriptional activators and co-regulators, that induce transcription. The ISRE is the main activating element targeted by virus induced IFN signaling. Despite the fact that this binding site is recognized by a variety of IRFs and the trimeric ISGF-3 complex, preferences for specific variations of the ISRE consensus sequence trigger the formation of disparate complexes at the respective promoters. We are only beginning to understand the capacity that these multi-protein complexes provide for a tailored host response to the infecting virus. In order to overcome clearance by the host, viruses have also come up with different mechanisms to disturb these regulatory anti-viral networks. A better understanding of these signaling events and the viral counter-measures is of great pharmaceutical interest since it will reveal targets for therapies against specific viruses.

References

1. Hoffmann JA, Kafatos FC, Janeway CA et al. Phylogenetic perspectives in innate immunity. Science 1999; 284:1313-1318.
2. Fearon DT, Locksley RM. The instructive role of innate immunity in the acquired immune response. Science 1996; 272:50-53.
3. Banchereau J, Steinman RM. Dendritic cells and the control of immunity. Nature 1998; 392:245-252.
4. Akira S, Takeda K, Kaisho T. Toll-like receptors: Critical proteins linking innate and acquired immunity. Nat Immunol 2001; 2:675-80.
5. Dunne A, O'Neill LA. The interleukin-1 receptor/Toll-like receptor superfamily: Signal transduction during inflammation and host defense. Sci STKE 2003; (171):re3.
6. Rock FL, Hardiman G, Timans JC et al. Family of human receptors structurally related to Drosophila Toll. Proc Natl Acad Sci USA 1998; 95:588-593.
7. Poltorak A, He X, Smirnova I et al. Defective LPS signaling in C3H/HeJ and C57BL/10ScCr mice: Mutations in Tlr4 gene. Science 1998; 282:2085-2088.
8. Kurt-Jones EA, Popova L, Kwinn L et al. Pattern recognition receptors TLR4 and CD14 mediate response to respiratory syncytial virus. Nat Immunol 2000; 1:398-401.
9. Haynes LM, Moore DD, Kurt-Jones EA et al. Involvement of toll-like receptor 4 in innate immunity to respiratory syncytial virus. J Virol 2001; 75:10730-10737.
10. Ohashi K, Burkart V, Flohe S et al. Cutting edge: Heat shock protein 60 is a putative endogenous ligand of the toll-like receptor-4 complex. J Immunol 2000; 164:558-561.
11. Vabulas RM, Ahmad-Nejad P, Ghose S et al. HSP70 as endogenous stimulus of the Toll/interleukin-1 receptor signal pathway. J Biol Chem 2002; 277:15107-15112.
12. Alexopoulou L, Holt AC, Medzhitov R et al. Recognition of double-stranded RNA and activation of NF-kappaB by Toll- like receptor 3. Nature 2001; 413:732-738.
13. Hemmi H, Kaisho T, Takeuchi O et al. Small anti-viral compounds activate immune cells via the TLR7 MyD88- dependent signaling pathway. Nat Immunol 2002; 3:196-200.
14. Jurk M, Heil F, Vollmer J et al. Human TLR7 or TLR8 independently confer responsiveness to the antiviral compound R-848. Nat Immunol 2002; 3:499.
15. Burns K, Martinon F, Esslinger C et al. MyD88, an adapter protein involved in interleukin-1 signaling. J Biol Chem 1998; 273:12203-12209.
16. Kawai T, Adachi O, Ogawa T et al. Unresponsiveness of MyD88-deficient mice to endotoxin. Immunity 1999; 11:115-122.

17. Kawai T, Takeuchi O, Fujita T et al. Lipopolysaccharide stimulates the MyD88-independent pathway and results in activation of IFN-regulatory factor 3 and the expression of a subset of lipopolysaccharide-inducible genes. J Immunol 2001; 167:5887-5894.
18. Oshiumi H, Matsumoto M, Funami K et al. TICAM-1, an adapter molecule that participates in Toll-like receptor 3- mediated interferon-beta induction. Nat Immunol 2003; 4:161-167.
19. Yamamoto M, Sato S, Mori K et al. Cutting edge: A novel Toll/IL-1 receptor domain-containing adapter that preferentially activates the IFN-beta promoter in the Toll-like receptor signaling. J Immunol 2002; 169:6668-6672.
20. Yamamoto M, Sato S, Hemmi H et al. Role of Adapter TRIF in the MyD88-Independent Toll-Like Receptor Signaling Pathway. Science 2003; 301:640-643.
21. Hoebe K, Du X, Georgel P et al. Identification of Lps2 as a key transducer of MyD88-independent TIR signalling. Nature 2003; 424:743-748.
22. Fitzgerald KA, Palsson-McDermott EM, Bowie AG et al. Mal (MyD88-adapter-like) is required for Toll-like receptor-4 signal transduction. Nature 2001; 413:78-83.
23. Horng T, Barton GM, Medzhitov R. TIRAP: An adapter molecule in the Toll signaling pathway. Nat Immunol 2001; 2:835-841.
24. Horng T, Barton GM, Flavell RA et al. The adapter molecule TIRAP provides signalling specificity for Toll- like receptors. Nature 2002; 420:329-333.
25. Yamamoto M, Sato S, Hemmi H et al. Essential role for TIRAP in activation of the signalling cascade shared by TLR2 and TLR4. Nature 2002; 420:324-329.
26. Bin LH, Xu LG, Shu HB. TIRP: A novel TIR domain-containing adapter protein involved in Toll/interleukin-1 receptor signaling. J Biol Chem 2003; 278:24526-24532
27. O'Neill LA, Fitzgerald KA, Bowie AG. The Toll-IL-1 receptor adapter family grows to five members. Trends Immunol 2003; 24:286-90.
28. Maran A, Maitra RK, Kumar A et al. Blockage of NF-kappa B signaling by selective ablation of an mRNA target by 2-5A antisense chimeras. Science 1994; 265:789-792.
29. Yang YL, Reis LF, Pavlovic J et al. Deficient signaling in mice devoid of double-stranded RNA-dependent protein kinase. Embo J 1995; 14:6095-6106.
30. Kerr IM, Brown RE. pppA2'p5'A2'p5'A: An inhibitor of protein synthesis synthesized with an enzyme fraction from interferon-treated cells. Proc Natl Acad Sci USA 1978; 75:256-260.
31. Silverman RH, Cirinio NM. RNA decay by the interferon- regulated 2-5A system as a host defense against viruses. New York: Wiley-Liss Inc, 1997.
32. Silverman RH. 2-5A- dependent RNAseL: A regulated endoribonuclease in the interferon system. New York: Academic Press, 1997.
33. Williams BR. Signal integration via PKR. Sci STKE 2001; re2.
34. Dagon Y, Dovrat S, Vilchik S et al. Double-stranded RNA-dependent protein kinase, PKR, down-regulates CDC2/cyclin B1 and induces apoptosis in non-transformed but not in v- mos transformed cells. Oncogene 2001; 20:8045-8056.
35. Robertson HD, Mathews MB. The regulation of the protein kinase PKR by RNA. Biochimie 1996; 78:909-914.
36. Chu WM, Ostertag D, Li ZW et al. JNK2 and IKKbeta are required for activating the innate response to viral infection. Immunity 1999; 11:721-731.
37. Zamanian-Daryoush M, Mogensen TH, DiDonato JA et al. NF-kappaB activation by double-stranded-RNA-activated protein kinase (PKR) is mediated through NF-kappaB-inducing kinase and IkappaB kinase. Mol Cell Biol 2000; 20:1278-1290.
38. Israel A. The IKK complex: An integrator of all signals that activate NF-kappaB? Trends Cell Biol 2000; 10:129-133.
39. Perkins ND. Achieving transcriptional specificity with NF-kappa B. Int J Biochem Cell Biol 1997; 29:1433-1448.
40. Wong AH, Tam NW, Yang YL et al. Physical association between STAT1 and the interferon-inducible protein kinase PKR and implications for interferon and double-stranded RNA signaling pathways. Embo J 1997; 16:1291-1304.
41. Ramana CV, Grammatikakis N, Chernov M et al. Regulation of c-myc expression by IFN-gamma through Stat1-dependent and -independent pathways. Embo J 2000; 19:263-272.

42. Kumar A, Yang YL, Flati V et al. Deficient cytokine signaling in mouse embryo fibroblasts with a targeted deletion in the PKR gene: Role of IRF-1 and NF-kappaB. Embo J 1997; 16:406-416.
43. Clemens MJ, Elia A. The double-stranded RNA-dependent protein kinase PKR: Structure and function. J Interferon Cytokine Res 1997; 17:503-524.
44. Diebold SS, Montoya M, Unger H et al. Viral infection switches non-plasmacytoid dendritic cells into high interferon producers. Nature 2003; 424:324-328.
45. Miyamoto M, Fujita T, Kimura Y et al. Regulated expression of a gene encoding a nuclear factor, IRF-1, that specifically binds to IFN-beta gene regulatory elements. Cell 1988; 54:903-913.
46. Harada H, Fujita T, Miyamoto M et al. Structurally similar but functionally distinct factors, IRF-1 and IRF- 2, bind to the same regulatory elements of IFN and IFN-inducible genes. Cell 1989; 58:729-739.
47. Fujii Y, Shimizu T, Kusumoto M et al. Crystal structure of an IRF-DNA complex reveals novel DNA recognition and cooperative binding to a tandem repeat of core sequences. Embo J 1999; 18:5028-5041.
48. Moore PS, Boshoff, C Weiss RA et al. Molecular mimicry of human cytokine and cytokine response pathway genes by KSHV. Science 1996; 274:1739-1744.
49. Lubyova B, Pitha PM. Characterization of a novel human herpesvirus 8-encoded protein, vIRF-3, that shows homology to viral and cellular interferon regulatory factors. J Virol 2000; 74:8194-8201.
50. Cunningham C, Barnard S, Blackbourn DJ et al. Transcription mapping of human herpesvirus 8 genes encoding viral interferon regulatory factors. J Gen Virol 2003; 84:1471-1483.
51. Taniguchi T, Ogasawara K, Takaoka A et al. IRF family of transcription factors as regulators of host defense. Annu Rev Immunol 2001; 19:623-655.
52. Servant MJ, Grandvaux N, Hiscott J. Multiple signaling pathways leading to the activation of interferon regulatory factor 3. Biochem Pharmacol 2002; 64:985-992.
53. Servant MJ, Tenoever B, Lin R. Overlapping and distinct mechanisms regulating IRF-3 and IRF-7 function. J Interferon Cytokine Res 2002; 22:49-58.
54. Levy DE, Marie I, Smith E et al. Enhancement and diversification of IFN induction by IRF-7-mediated positive feedback. J Interferon Cytokine Res 2002; 22:87-93.
55. Mamane Y, Heylbroeck C, Genin P et al. Interferon regulatory factors: The next generation. Gene 1999; 237:1-14.
56. Lin R, Mamane Y, Hiscott J. Multiple regulatory domains control IRF-7 activity in response to virus infection. J Biol Chem 2000; 275:34320-34327.
57. Sato M, Hata N, Asagiri M et al. Positive feedback regulation of type I IFN genes by the IFN-inducible transcription factor IRF-7. FEBS Lett 1998; 441:106-110.
58. Marie I, Durbin JE, Levy DE. Differential viral induction of distinct interferon-alpha genes by positive feedback through interferon regulatory factor-7. Embo J 1998; 17:6660-6669.
59. Lin R, Genin P, Mamane Y et al. Selective DNA binding and association with the CREB binding protein coactivator contribute to differential activation of alpha/beta interferon genes by interferon regulatory factors 3 and 7. Mol Cell Biol 2000; 20:6342-6353.
60. Au WC, Moore PA, Lowther W et al. Identification of a member of the interferon regulatory factor family that binds to the interferon-stimulated response element and activates expression of interferon-induced genes, Proc Natl Acad Sci USA 1995; 92:11657-11661.
61. Sato M, Tanaka N, Hata N et al. Involvement of the IRF family transcription factor IRF-3 in virus- induced activation of the IFN-beta gene. FEBS Lett 1998; 425:112-116.
62. Sato M, Suemori H, Hata N et al. Distinct and essential roles of transcription factors IRF-3 and IRF-7 in response to viruses for IFN-alpha/beta gene induction. Immunity 2000; 13:539-548.
63. Bluyssen AR, Durbin JE, Levy DE. ISGF3 gamma p48, a specificity switch for interferon activated transcription factors. Cytokine Growth Factor Rev 1996; 7:11-7.
64. Darnell Jr JE, Kerr IM, Stark GR. Jak-STAT pathways and transcriptional activation in response to IFNs and other extracellular signaling proteins. Science 1994; 264:1415-1421.
65. Harada H, Matsumoto M, Sato M et al. Regulation of IFN-alpha/beta genes: Evidence for a dual function of the transcription factor complex ISGF3 in the production and action of IFN- alpha/ beta. Genes Cells 1996; 1:995-1005.
66. Erlandsson L, Blumenthal R, Eloranta ML et al. Interferon-beta is required for interferon alpha production in mouse fibroblasts. Curr Biol 1998; 8:223-226.

67. Lin R, Heylbroeck C, Pitha PM et al. Virus-dependent phosphorylation of the IRF-3 transcription factor regulates nuclear translocation, transactivation potential, and proteasome-mediated degradation. Mol Cell Biol 1998; 18:2986-2996.
68. Lin R, Mamane Y, Hiscott J. Structural and functional analysis of interferon regulatory factor 3: Localization of the transactivation and autoinhibitory domains. Mol Cell Biol 1999; 19:2465-2474.
69. Yoneyama M, Suhara W, Fukuhara Y et al. Direct triggering of the type I interferon system by virus infection: Activation of a transcription factor complex containing IRF-3 and CBP/p300. Embo J 1998; 17:1087-1095.
70. Servant MJ, ten Oever B, LePage C et al. Identification of distinct signaling pathways leading to the phosphorylation of interferon regulatory factor 3. J Biol Chem 2001; 276:355-363.
71. Servant MJ, Grandvaux N, tenOever BR et al. Identification of the minimal phosphoacceptor site required for in vivo activation of interferon regulatory factor 3 in response to virus and double-stranded RNA. J Biol Chem 2003; 278:9441-9447.
72. Sharma S, tenOever BR, Grandvaux N et al. Triggering the interferon antiviral response through an IKK-related pathway. Science 2003; 300:1148-1151.
73. Fitzgerald KA, McWhirter SM, Faia KL et al. IKKepsilon and TBK1 are essential components of the IRF3 signaling pathway. Nat Immunol 2003; 4:491-496.
74. Suhara W, Yoneyama M, Iwamura T et al. Analyses of virus-induced homomeric and heteromeric protein associations between IRF-3 and coactivator CBP/p300. J Biochem (Tokyo) 2000; 128:301-307.
75. Suhara W, Yoneyama M, Kitabayashi I et al. Direct involvement of CREB-binding protein/p300 in sequence-specific DNA binding of virus-activated interferon regulatory factor-3 holocomplex. J Biol Chem 2002; 277:22304-22313.
76. Doyle S, Vaidya S, O'Connell R et al. IRF3 mediates a TLR3/TLR4-specific antiviral gene program. Immunity 2002; 17:251-263.
77. Wietek C, Miggin SM, Jefferies CA et al. IRF-3- mediated activation of the interferon- sensitive response element by TLR4 but not TLR3 requires the p65 subunit of NF-κB. J Biol Chem 2003; 278:50923-50931.
78. Civas A, Island ML, Genin P et al. Regulation of virus-induced interferon-A genes. Biochimie 2002; 84:643-654.
79. Mogensen KE, Lewerenz M, Reboul J et al. The type I interferon receptor: Structure, function, and evolution of a family business. J Interferon Cytokine Res 1999; 19:1069-1098.
80. Colamonici OR, Uyttendaele H, Domanski P et al. p135tyk2, an interferon-alpha-activated tyrosine kinase, is physically associated with an interferon-alpha receptor. J Biol Chem 1994; 269:3518-3522.
81. Novick D, Cohen B, Rubinstein M. The human interferon alpha/beta receptor: Characterization and molecular cloning. Cell 1994; 77:391-400.
82. Yan H, Krishnan K, Lim JT et al. Molecular characterization of an alpha interferon receptor 1 subunit (IFNaR1) domain required for TYK2 binding and signal transduction. Mol Cell Biol 1996; 16:2074-2082.
83. Leung S, Qureshi SA, Kerr IM et al. Role of STAT2 in the alpha interferon signaling pathway. Mol Cell Biol 1995; 15:1312-1317.
84. Qureshi SA, Leung S, Kerr IM et al. Function of Stat2 protein in transcriptional activation by alpha interferon. Mol Cell Biol 1996; 16:288-293.
85. Shuai K, Ziemiecki A, Wilks AF et al. Polypeptide signalling to the nucleus through tyrosine phosphorylation of Jak and Stat proteins. Nature 1993; 366:580-583.
86. Veals SA, Schindler C, Leonard D et al. Subunit of an alpha-interferon-responsive transcription factor is related to interferon regulatory factor and Myb families of DNA-binding proteins. Mol Cell Biol 1992; 12:3315-3324.
87. Toshchakov V, Jones BW, Perera PY et al. TLR4, but not TLR2, mediates IFN-beta-induced STAT1alpha/beta-dependent gene expression in macrophages. Nat Immunol 2002; 3:392-398.
88. Yeow WS, Au WC, Lowther WJ et al. Downregulation of IRF-3 levels by ribozyme modulates the profile of IFNA subtypes expressed in infected human cells. J Virol 2001; 75:3021-3027.
89. Stark GR, Kerr IM, Williams BR et al. How cells respond to interferons. Annu Rev Biochem 1998; 67:227-264.

90. Ward SV, Samuel CE. Regulation of the interferon-inducible PKR kinase gene: The KCS element is a constitutive promoter element that functions in concert with the interferon-stimulated response element. Virology 2002; 296:136-146.
91. Rebouillat D, Hovnanian A, David G et al. Characterization of the gene encoding the 100-kDa form of human 2',5' oligoadenylate synthetase. Genomics 2000; 70:232-240.
92. Ohmori Y, Hamilton TA. Requirement for STAT1 in LPS-induced gene expression in macrophages. J Leukoc Biol 2001; 69:598-604.
93. Doyle SE, O'Connell R, Vaidya SA et al. Toll-like receptor 3 mediates a more potent antiviral response than toll-like receptor 4. J Immunol 2003; 170:3565-3571.
94. Nelson N, Marks MS, Driggers PH et al. Interferon consensus sequence-binding protein, a member of the interferon regulatory factor family, suppresses interferon-induced gene transcription. Mol Cell Biol 1993; 13:588-599.
95. Endo TA, Masuhara M, Yokouchi M et al. A new protein containing an SH2 domain that inhibits JAK kinases. Nature 1997; 387:921-924.
96. Naka T, Narazaki M, Hirata M et al. Structure and function of a new STAT-induced STAT inhibitor. Nature 1997; 387:924-929.
97. Starr R, Willson TA, Viney EM et al. A family of cytokine-inducible inhibitors of signalling. Nature 1997; 387:917-921.
98. Boisclair YR, Wang J, Shi J et al. Role of the suppressor of cytokine signaling-3 in mediating the inhibitory effects of interleukin-1beta on the growth hormone-dependent transcription of the acid-labile subunit gene in liver cells. J Biol Chem 2000; 275:3841-3847.
99. Dalpke AH, Opper S, Zimmermann S et al. Suppressors of cytokine signaling (SOCS)-1 and SOCS-3 are induced by CpG-DNA and modulate cytokine responses in APCs. J Immunol 2001; 166:7082-7089.
100. Stoiber D, Kovarik P, Cohney S et al. Lipopolysaccharide induces in macrophages the synthesis of the suppressor of cytokine signaling 3 and suppresses signal transduction in response to the activating factor IFN-gamma. J Immunol 1999; 163:2640-2647.
101. Kinjyo I, Hanada T, Inagaki-Ohara K et al. SOCS1/JAB is a negative regulator of LPS-induced macrophage activation. Immunity 2002; 17:583-591.
102. Nakagawa R, Naka T, Tsutsui H et al. SOCS-1 participates in negative regulation of LPS responses. Immunity 2002; 17:677-687.
103. Wathelet MG, Lin CH, Parekh BS et al. Virus infection induces the assembly of coordinately activated transcription factors on the IFN-beta enhancer in vivo. Mol Cell 1998; 1:507-518.
104. Yie J, Senger K, Thanos D. Mechanism by which the IFN-beta enhanceosome activates transcription. Proc Natl Acad Sci USA 1999; 96:13108-13113.
105. Kim T, Kim TY, Lee WG et al. Signaling pathways to the assembly of an interferon-beta enhanceosome. Chemical genetic studies with a small molecule. J Biol Chem 2000; 275:16910-16917.
106. Agalioti T, Lomvardas S, Parekh B et al. Ordered recruitment of chromatin modifying and general transcription factors to the IFN-beta promoter. Cell 2000; 103:667-678.
107. Thanos D, Maniatis T. The high mobility group protein HMG I(Y) is required for NF-kappa B-dependent virus induction of the human IFN-beta gene. Cell 1992; 71:777-789.
108. Yie J, Liang S, Merika M et al. Intra- and intermolecular cooperative binding of high-mobility-group protein I(Y) to the beta-interferon promoter. Mol Cell Biol 1997; 17:3649-3662.
109. Falvo JV, Thanos D, Maniatis T. Reversal of intrinsic DNA bends in the IFN beta gene enhancer by transcription factors and the architectural protein HMG I(Y). Cell 1995; 83:1101-1111.
110. Yeow WS, Au WC, Juang YT et al. Reconstitution of virus-mediated expression of interferon alpha genes in human fibroblast cells by ectopic interferon regulatory factor-7. J Biol Chem 2000; 275:6313-6320.
111. Au WC, Pitha PM. Recruitment of multiple interferon regulatory factors and histone acetyltransferase to the transcriptionally active interferon a promoters. J Biol Chem 2001; 276:41629-41637.
112. Barnes BJ, Moore PA, Pitha PM. Virus-specific activation of a novel interferon regulatory factor, IRF- 5, results in the induction of distinct interferon alpha genes. J Biol Chem 2001; 276:23382-23390.

113. Bowie A, Kiss-Toth E, Symons JA et al. A46R and A52R from vaccinia virus are antagonists of host IL-1 and toll- like receptor signaling. Proc Natl Acad Sci USA 2000; 97:10162-10167.
114. Harte MT, Haga IR, Maloney G et al. The poxvirus protein A52R targets Toll-like receptor signaling complexes to suppress host defense. J Exp Med 2003; 197:343-351.
115. Ronco LV, Karpova AY, Vidal M et al. Human papillomavirus 16 E6 oncoprotein binds to interferon regulatory factor-3 and inhibits its transcriptional activity. Genes Dev 1998; 12:2061-2072.
116. Foy E, K Li, C Wang et al. Regulation of interferon regulatory factor-3 by the hepatitis C virus serine protease. Science 2003; 300:1145-1148.
117. Basler CF, Mikulasova A, Martinez-Sobrido L et al. The Ebola virus VP35 protein inhibits activation of interferon regulatory factor 3. J Virol 2003; 77:7945-7956.
118. Barnard P, Payne E, McMillan NA. The human papillomavirus E7 protein is able to inhibit the antiviral and anti-growth functions of interferon-alpha. Virology 2000; 277:411-419.
119. Zimring JC, Goodbourn S, Offermann MK. Human herpesvirus 8 encodes an interferon regulatory factor (IRF) homolog that represses IRF-1-mediated transcription. J Virol 1998; 72:701-707.
120. Gao SJ, Boshoff C, Jayachandra S et al. KSHV ORF K9 (vIRF) is an oncogene which inhibits the interferon signaling pathway. Oncogene 1997; 15:1979-1985.
121. Lin R, Genin P, Mamane Y et al. HHV-8 encoded vIRF-1 represses the interferon antiviral response by blocking IRF-3 recruitment of the CBP/p300 coactivators. Oncogene 2001; 20:800-811.
122. Symons JA, Alcami A, Smith GL. Vaccinia virus encodes a soluble type I interferon receptor of novel structure and broad species specificity. Cell 1995; 81:551-560.
123. Colamonici OR, Domanski P, Sweitzer SM et al. Vaccinia virus B18R gene encodes a type I interferon-binding protein that blocks interferon alpha transmembrane signaling. J Biol Chem 1995; 270:15974-15978.
124. Alcami A, Smith GL. Vaccinia, cowpox, and camelpox viruses encode soluble gamma interferon receptors with novel broad species specificity. J Virol 1995; 69:4633-4639.
125. Weihua X, Ramanujam S, Lindner DJ et al. The polyoma virus T antigen interferes with interferon-inducible gene expression. Proc Natl Acad Sci USA 1998; 95:1085-1090.
126. Young DF, Didcock L, Goodbourn S et al. Paramyxoviridae use distinct virus-specific mechanisms to circumvent the interferon response. Virology 2000; 269:383-390.
127. Andrejeva J, Young DF, Goodbourn S et al. Degradation of STAT1 and STAT2 by the V proteins of simian virus 5 and human parainfluenza virus type 2, respectively: Consequences for virus replication in the presence of alpha/beta and gamma interferons. J Virol 2002; 76:2159-2167.
128. Tan SL, Katze MG. Biochemical and genetic evidence for complex formation between the influenza A virus NS1 protein and the interferon-induced PKR protein kinase. J Interferon Cytokine Res 1998; 18:757-766.
129. Sharp TV, Moonan F, Romashko A et al. The vaccinia virus E3L gene product interacts with both the regulatory and the substrate binding regions of PKR: Implications for PKR autoregulation. Virology 1998; 250:302-315.
130. Rivas C, Gil J, Melkova Z et al. Vaccinia virus E3L protein is an inhibitor of the interferon (i.f.n.)- induced 2-5A synthetase enzyme. Virology 1998; 243:406-414.
131. Gale MJ, Korth MJ, Tang NM et al. Evidence that hepatitis C virus resistance to interferon is mediated through repression of the PKR protein kinase by the nonstructural 5A protein. Virology 1997; 230:217-227.
132. Martinand C, Montavon C, Salehzada T et al. RNase L inhibitor is induced during human immunodeficiency virus type 1 infection and down regulates the 2-5A/RNase L pathway in human T cells. J Virol 1999; 73:290-296.
133. Fitzgerald KA, Rowe DC, Barnes BJ et al. LPS-TLR4 signaling to IRF-3/7 and NF-kappaB involves the Toll adapters TRAM and TRIF. J Exp Med 2003; 198:1043-1055.

Chapter 6

The Induction of Dendritic Cell Activation and Maturation by Toll-Like Receptor Signaling

Tsuneyasu Kaisho and Shizuo Akira

Abstract

The host defense system of higher animals comprises both innate and adaptive immunity. Dendritic cells (DCs), professional antigen presenting cells, play critical roles in linking these two types of immunity in response to microbial stimuli. DCs express a group of type I transmembrane proteins, Toll-like receptors (TLRs), which can recognize a number of microorganism-derived molecular structures, including membrane components and nucleic acids. This recognition is a prerequisite for DC activation and maturation. Each TLR family member can provoke overlapping, but distinct biological consequences. In this chapter, we discuss how TLRs, as adjuvant receptors, instruct DCs to drive host defense.

Introduction

Innate and adaptive immunity should coordinately function to establish effective host defense in higher animals. Innate immunity consists mainly of antigen presenting cells such as macrophages or dendritic cells (DCs). DCs can activate naïve T cells and are especially important for linking innate and adaptive immunity.[1] Toll-like receptors (TLRs), a group of type I transmembrane proteins, are crucial for DCs to accomplish this function. The TLR family in humans and mice consists of ten and nine members, respectively (Fig. 1).*[2,3] Each TLR has the ability to discriminate microorganism-related molecular structures, originally termed pathogen-associated molecular patterns (PAMPs).[4] PAMPs are not found in the host and can thus be regarded as nonself. Furthermore, most PAMPs possess immune enhancing effects. In other words, TLRs are adjuvant receptors that sense nonself.

* Since going to press, TLR11 was identified, which recognizes a ligand expressed by uropathogenic bacteria. (Zhang D et al. Science 2004; 303:1522-1526.)

Toll and Toll-Like Receptors: An Immunologic Perspective, edited by Tina Rich. ©2005 Eurekah.com and Kluwer Academic / Plenum Publishers.

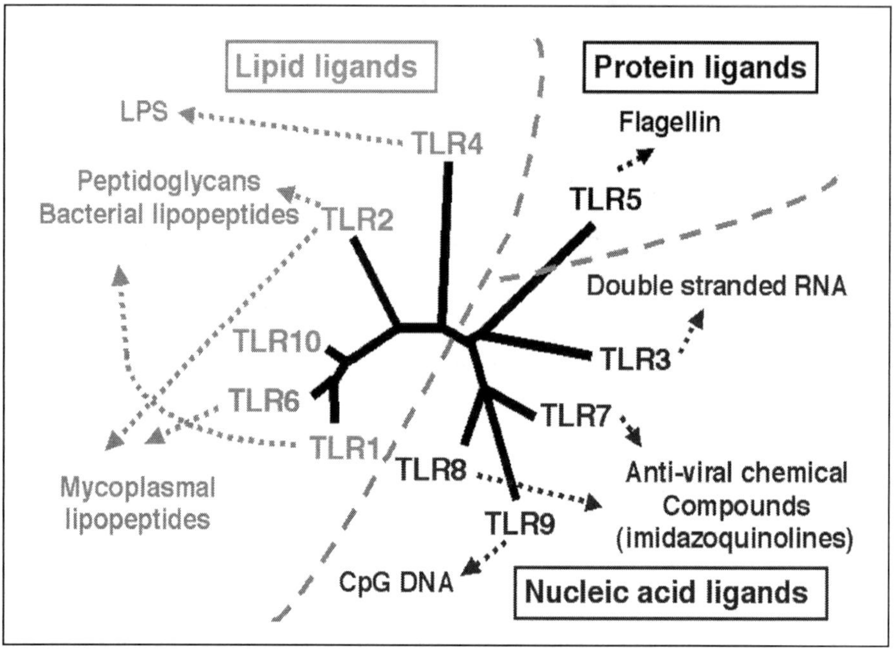

Figure 1. Human TLRs and their representative ligands. Phylogenetic tree depiction of human TLRs based on amino acid structures. TLRs and their representative ligands are connected by dotted arrows. TLR ligands are roughly categorized as lipid, protein, and nucleic acid as shown by broken lines.

Comparisons between TLRs and Toll

Similarities

TLRs were named after Toll as their immunoregulatory function and molecular structures were similar. Mutant flies with defective Toll succumb to fungal infection.[5] This susceptibility results from an inability of mutant Toll to induce the secretion of an antimicrobial peptide, Drosomycin. Janeway's group first described the immunoregulatory role of a human Toll-like protein, now known as TLR4.[6] TLR4 expression can induce the expression of cytokines and costimulatory molecules.

All TLRs and Toll are type I transmembrane proteins that possess leucine-rich repeats in their extracellular domains. In their intracytoplasmic regions, TLRs and Toll share a common motif, the Toll/IL-1R (TIR) domain, which is found in mammalian IL-1R family members. Furthermore, TLRs and Toll activate a closely related signal transduction machinery that leads to the expression of pro-inflammatory cytokines and antimicrobial peptides, respectively.[7]

Dissimilarity between TLRs and Toll

It is noteworthy to indicate how the TLR and Toll systems differ (Fig. 2). In *Drosophila*, fungal infection can lead to the activation of protease cascades that induce the cleavage of a secreted protein, Spaetzle.[8] Subsequently, cleaved Spaetzle is recognized by Toll. Consistent with this, mutant flies deficient for the serine protease inhibitor Spn43Ac manifest constitutive expression of cleaved Spaetzle and Drosomycin.[8] Thus, Toll does not recognize a microbial, but a host-derived endogenous product. This is in contrast to the TLR system, as TLRs are directly involved in microbial recognition (see below).

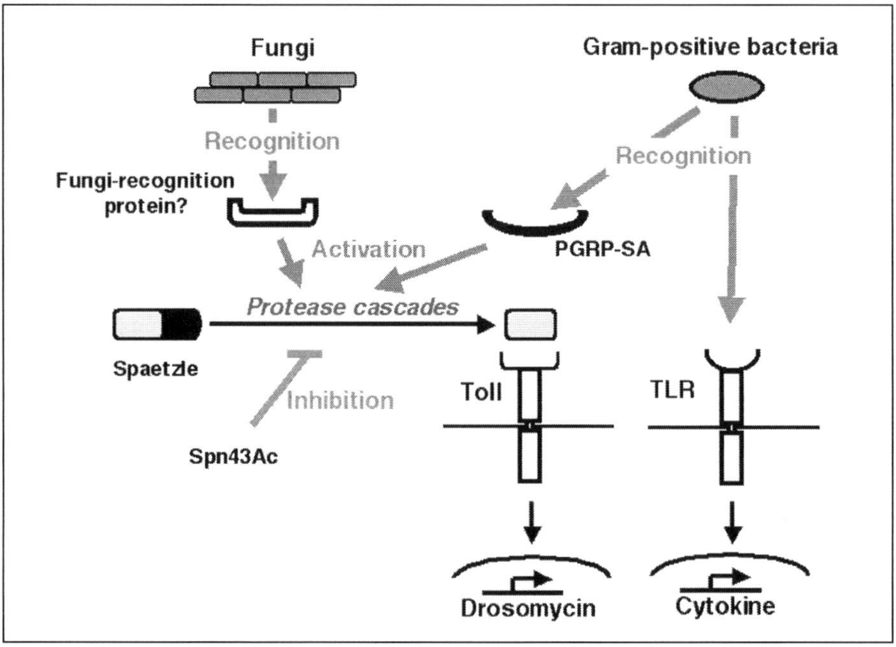

Figure 2. Comparison between Toll and TLR recognition system. In higher animals, membrane expressed TLRs directly recognize microorganism-derived molecular patterns. However, in *Drosophila*, microorganisms are discriminated in the hemolymph. PGRP-SA recognizes Gram-positive bacteria, whereas another unidentified soluble factor, is presumed to recognize fungi. Subsequently protease cascades are activated to cleave Spaetzle which is recognized by Toll. Thus, Toll on the membrane recognizes a host-derived product.

The Toll pathway is also activated by infection with Gram-positive bacteria (Fig. 2). A soluble peptidoglycan recognition protein, PGRP-SA, recognizes peptidoglycans from bacteria and activates protease cascades that lead to the cleavage of Spaetzle and subsequent Toll activation.[9] PGRP-SA-deficient flies still exhibit intact immune responses against fungi, implying the existence of a dedicated system to discriminate fungi. Both fungi and Gram-positive bacteria are detected in the body fluid, not at the membrane level. Similarly, the complement system in mammals recognizes invading microorganisms in serum and activates protease cascades. In this regard, it is conceivable that the complement system, rather than TLRs, is evolutionarily linked to *Drosophila* innate immunity.

Microbial Recognition by TLRs

Membrane Component Ligands

Bacterial membrane components are adjuvants of which lipopolysaccharide (LPS) has been the most extensively studied. LPS is a major component of the outer membranes of Gram-negative bacteria and its lipid portion, lipid A, is it's toxic constituent. LPS is bound to LPS binding protein, LBP, in the serum and transported to target cells. Another LPS-binding protein, CD14, expressed at the surface of target cells, receives LPS and presents it to a signal transducing receptor, TLR4 (Fig. 1). Critical roles of TLR4 were clarified by genetic analysis of two mutant mice, C3H/HeJ and C57BL10/ScCr.[10,11] C3H/HeJ mice have a point mutation in the TLR4 gene, which leads to an amino acid substitution in the intracytoplasmic region whereas C57BL10/ScCr mice have a deletion in the TLR4 gene locus. TLR4-deficient mice were found to lack responses to LPS.[12] Furthermore, some LPS-hyporesponsive patients also carry a TLR4 mutant which cannot bind LPS due to a missense mutation in the extracellular domain.[13]

TLR4 can associate with a host-derived small soluble factor, MD-2.[14] This association is critical for TLR4-mediated recognition of LPS, as was shown in genetic complementation experiments and the analysis of MD-2-deficient mice.[14,15] Gram-positive bacteria do not possess LPS, but instead are endowed with a thick layer of peptidoglycan in which a variety of lipoproteins and lipopeptides are buried. Mycoplasmas lack cell walls entirely and neither express LPS nor peptidoglycan. However, these pathogens still express lipoproteins and lipopeptides in their plasma membranes. These components can act as adjuvants, most of which require TLR2 as a signal transducer (Fig. 1).[2,16]

Heterodimerization is critical for TLR2 to recognize its diverse ligands. For example, TLR2 can interact with TLR6 and recognize Gram-positive bacteria or a yeast cell-wall particle, zymosan.[17] Furthermore, TLR2- and TLR6-deficient mice fail to respond to macrophage-activating lipopeptide-2 (MALP-2).[18,19] These results indicate a critical functional association of TLR2 with TLR6. Importantly, however, TLR2-, but not TLR6-, deficient mice lack responses against bacterial lipopeptides (BLP). Instead, TLR1-deficient mice show an impaired response to BLP.[20] Both BLP and MALP-2 share distinctive triacylated (BLP) and diacylated (MALP-2) cysteine residues at their amino-termini. Apparently, TLR2 can fine-tune the recognition of PAMPs by switching TLR partner (Fig. 1). Bacterial proteins can also act as TLR ligands. Some bacteria move by using a particular structure, the flagella. A protein component, flagellin, can stimulate intestinal epithelial cells to produce chemokines such as IL-8. TLR5 is critical for recognizing flagellin (Fig. 1).[21]

Nucleic Acid Ligands

DNA molecules derived from microorganisms can also act as immune adjuvants. Tokunaga et al first reported this effect when they demonstrated that an immunostimulatory activity in bacillus Calmette-Guerin (BCG) extracts could be attributed to DNA.[22] Bacterial DNA abundantly express unmethylated CpG motifs, which are rarely found in mammalian DNA.[23-25] Synthetic DNA containing an unmethylated CpG motif (CpG DNA), 20-30 base pairs in length, can also exhibit strong adjuvant activity. Thus, CpG DNA can be regarded as a nonself adjuvant. The typical responses to CpG DNA of cytokine production, B cell proliferation, DC maturation, and shock induction, were absent in TLR9-deficient mice.[26] It is known that human and murine immune cells respond to a different repertoire of CpG DNA. This species-specific response is reconstituted by expressing human or murine TLR9 in a human kidney-derived cell line.[27] Thus, TLR9 is essential and sufficient to recognize CpG DNA (Fig. 1). Unlike TLR2 and TLR4, TLR9 is not expressed on the cell surface. DNAs are first incorporated into the cell in a sequence-independent manner. Subsequently, only CpG DNAs interact with and stimulate TLR9 signaling in endosomal compartments. Inhibition of endosomal maturation can block the effects of CpG DNA, but not LPS.[28]

On screening a panel of synthetic CpG DNAs, immunostimulatory CpG DNAs were found to be divided into two types.[29-31] One is conventional CpG DNA, also termed K-type CpG DNA or CpG-B. Conventional CpG DNA consists of phosphorothioate-modified oligodeoxynucleotides. The second, termed D-type CpG DNA or CpG-A (A/D-type CpG DNA), carries a phosphorothioate-modified polyguanosine (polyG) stretch at the 5' and 3' ends and a phosphodiester backbone CpG motif in the central portion. Conventional CpG DNA can activate B cells to proliferate and secrete IL-6 more vigorously than A/D-type CpG

DNA. However, A/D-type CpG DNA exhibits greater ability to induce IFN-α from a DC subset, plasmacytoid DC (PDC), as well as IFN-γ from NK cells.[30,31] Thus, the TLR9 system is unique in the sense that its biological readout is ligand specific. However, the biological significance of these CpG motifs remains unclear and it is not known how the differential effects are provoked.

RNA virus infection can lead to the production of double-stranded RNAs (dsRNA). These are not present in the host and exhibit immune adjuvant activity. Synthetic dsRNAs, polyinosinic-polycytidylic acid (poly(I:C)), display similar activities. TLR3-deficient mice showed impaired responses to such dsRNAs, indicating that these RNAs are recognized by TLR3 (Fig. 1).[32] However, the response was not abolished in the absence of TLR3, indicating the involvement of other molecules such as RNA-dependent protein kinase (PKR).[33,34]

Synthetic Compound Ligands

In addition to microorganism components, synthetic immunostimulatory molecules also activate immune responses through TLRs. Imidazoquinoline derivatives, such as imiquimod or resiquimod (R848), can induce the production of a variety of cytokines including type I interferons (IFNs). Imiquimods are clinically used to treat genital warts induced by human papilloma virus. All activity of imidazoquinolines is abolished in TLR7-deficient mice (Fig. 1).[35] Certain guanosine derivatives, such as 7-allyl-8-oxo-guanosine (loxoribine) can also activate murine and human immune cells. In fact, Loxoribine is now promoted as an anticancer drug. Furthermore, 2-amin-5-bromo-6-phenyl-4(3)-pyrimidinone (bropirimine) is also an immunomodulator with an anti-tumor effect on superficial transitional cell carcinoma of the bladder. These compounds were also found to be ligands for TLR7.[36] Thus far, all the TLR7 ligands that have been identified are synthetic. They have similar molecular structures to nucleic acids. The possibility that the natural ligand for TLR7 may be viral in origin is provocative but remains to be proven.

TLR8 is structurally similar to TLR7 (Fig. 1). Indeed, human TLR8 also recognizes R848.[37] However, the lack of R848 responsiveness in TLR7-deficient mice would imply that murine TLR8 is not involved in recognizing R848. It remains unclear as to whether murine TLR8 is nonfunctional or whether it can recognize other, as yet unidentified ligands.*

Host-Derived Ligands

Infection can result in inflammation or tissue damage, which can subsequently lead to the release of endogeneous molecules. For example, heat shock proteins released from necrotic cells can provoke DC maturation. Heat shock proteins were shown to be recognized by TLR2 or TLR4.[38-40] Inflammation also triggers the activation of proteolytic cascades that degrade extracellular matrix components. Such degraded products, including fibrinogen or fragmented products of hyaluronic acids, can act through TLR4.[41,42] Furthermore, a small antimicrobial peptide, β-defensin-2, was also found to be a ligand of TLR4.[43] These findings raise the interesting possibility that, under 'alert' conditions, TLRs may be involved in the recognition of endogeneous host-derived products. However, this hypothesis still needs to be tested and it should always be kept in mind that contamination with even tiny amounts of endotoxin or bacterial products can cause significant adjuvant activity.[44]

* Since going to press murine TLR7 and human TLR8 have been shown to recognize ssRNA, both of viral and endogenous origin (Heil F et al. Science 2004; 303:1526-1529. Diebold SS et al. Science 2004; 303:1529-1531).

Signal Transduction of TLRs

Common Components of TLR Signaling

TLRs can activate signal transduction pathways through their intracytoplasmic TIR domains. Both TLRs and IL-1Rs associate with a cytoplasmic adapter, MyD88, through homophilic TIR domain interaction. MyD88 has a death domain at its N-terminal portion and can recruit IL-1R-associated kinase (IRAK) family members. Another adapter, tumor necrosis factor receptor-associated factor (TRAF6), is also recruited and, subsequently, two major signaling pathways, mitogen-activated protein kinase (MAPK) cascades and NF-κB, are activated.[2,16] The analysis of MyD88-deficient mice has clarified the significance of this particular signaling cascade.[45] IL-1 can stimulate T cell proliferation as well as the induction of acute phase proteins and cytokines. IL-18 can induce IFN-γ production and augment natural killer cell activity. All of these biological effects were abolished in the absence of MyD88.

Most TLRs also utilize MyD88 as a critical adapter. TLR2, TLR7, and TLR9 signaling fails to induce activation of NF-κB and MAPKs in MyD88-deficient cells. In accordance with this abolished biochemical response, MyD88-deficient cells are completely refractory to TLR2, TLR7, or TLR9 agonists.[18,35,46] Furthermore, a TLR5 ligand, flagellin, cannot induce cytokine production in MyD88-deficient mice.[21] Thus, the MyD88-mediated pathway is a common and critical pathway for IL-1R and TLR signaling.

Four IRAK family members, IRAK-1, IRAK-2, IRAK-M, and IRAK-4, have been identified thus far.[47] Amongst these, only IRAK-1 and IRAK-4 possess serine/threonine kinase activity. IRAK-4 is the most similar to *Drosophila* Pelle, a serine/threonine kinase required for Toll signaling. According to in vitro findings, IRAK-4 can activate NF-κB in a kinase-dependent manner and also phosphorylates IRAK-1.[48] Furthermore, IRAK-4 expression cannot rescue the response of IRAK-1 deficient cells, suggesting that IRAK-4 functions upstream of IRAK-1.[48] IRAK-1-deficient mice showed only partial impairment of IL-1 and TLR signaling.[49,50] In contrast, IRAK-4-deficient mice exhibited severely impaired responses to signaling not only through the IL-1 family, IL-1 and IL-18, but also through TLRs including TLR2, TLR4, and TLR9.[51] This phenotype is similar to that of MyD88-deficient mice. Thus, IRAK-4 is an essential IRAK for the MyD88-dependent pathway.

An alternatively spliced form of MyD88 called MyD88 short (MyD88s) lacks the short intermediate domain between the death and TIR domains. This domain is essential to recruit IRAK-4 into a MyD88-containing complex. Without this domain IRAK-1 is not phosphorylated and subsequent NF-κB activation is not induced.[52] The expression of MyD88s is augmented by TLR signaling,[53] suggesting that MyD88s is involved in a negative feedback loop. IRAK-M, which has no kinase activity, prevents the dissociation of IRAKs from MyD88 and the subsequent formation of IRAK-TRAF6 complexes.[54] IRAK-M deficient mice showed enhanced inflammatory responses upon bacterial infection as well as increased cytokine production through TLR signaling. Importantly, endotoxin tolerance is sub-optimal in the absence of IRAK-M. Thus, IRAK-M also negatively regulates TLR signaling.

Heterogeneity of TLR Signaling

Cytoplasmic adapters play critical roles in the heterogeneity of TLR signaling (Fig. 3). This was first revealed by the analysis of MyD88-deficient mice. Mutants retained the activation of NF-κB and MAPK in response to LPS.[55] Although the mutants lacked LPS-induced effects, such as cytokine production, B cell proliferation, and endotoxin shock, they still responded to LPS. First, MyD88-deficient DCs can mature in response to LPS (see below).[46] Second, expression of IFN-inducible genes such as IP-10 or GARG16 is retained in MyD88-deficient macrophages. This is due to the retained induction of IFN-β in LPS-stimulated MyD88-deficient cells.[56-58] Furthermore, LPS can activate caspase-1 and lead to the secretion of IL-18 in MyD88-deficient liver macrophages.[59]

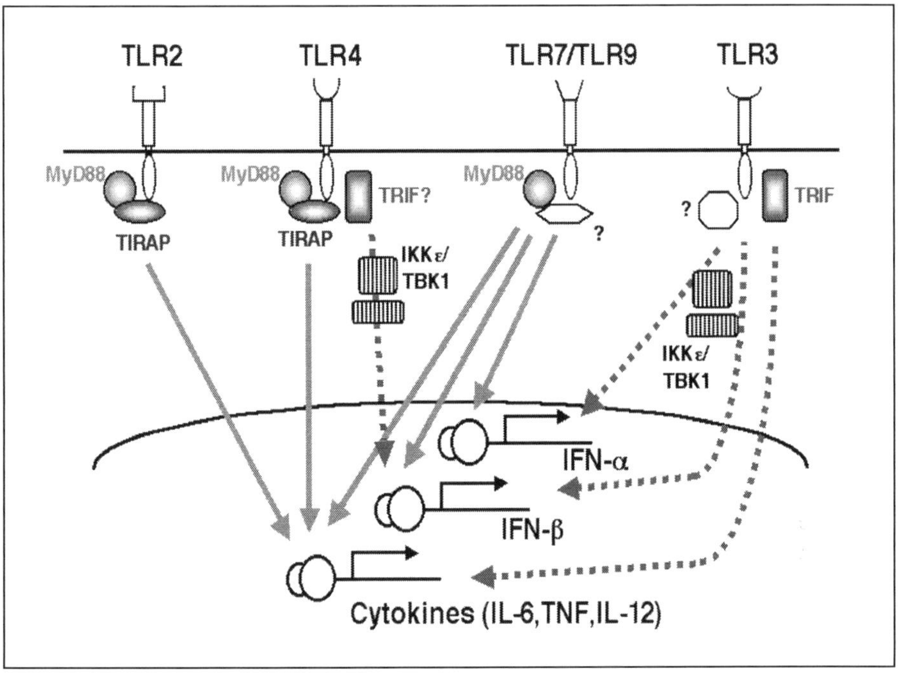

Figure 3. Cytoplasmic adapters of TLRs. Solid lines represent the MyD88-dependent pathway leading to cytokine production. TIRAP is cooperatively involved in this pathway through TLR2 and TLR4 signaling. In TLR7 and TLR9 signaling, all activities including type I IFN induction are dependent on MyD88. Dotted lines represent the MyD88-independent pathway, in which TRIF is most likely involved together with IKKε and TBK1. This pathway can lead to type I IFN induction through TLR3 and TLR4 signaling. Cytokine production by TLR3 seems to be induced in a MyD88-independent fashion.

Another adapter, TIR domain-containing adapter protein/MyD88-adapter-like (TIRAP/MAL), was found to associate with TLR4, but not with TLR9.[60,61] Both IFN-inducible gene expression and DC maturation by LPS were intact in TIRAP/MAL-deficient mice.[62,63] Furthermore, double knockout mice for MyD88 and TIRAP/MAL still retained the LPS-induced responses.[63] These results clearly demonstrate that TIRAP/MAL is involved in the TLR4-stimulated MyD88-dependent, rather than the MyD88-independent pathway (Fig. 3). TIRAP/MAL-deficient mice manifest defective TLR2 signaling, which does not activate the MyD88-independent pathway. In contrast, TLR3, TLR7, and TLR9 signaling is intact in the absence of TIRAP/MAL. Thus, TIRAP/MAL is involved specifically in TLR2 and TLR4 signaling which leads to cytokine production.[62,63] (Fig. 3). These results suggest that other adapter molecules involved in the MyD88-independent pathway may still exist. In addition to MyD88 and TIRAP/MAL, three more TIR domain-containing adapters have been identified.[64] One candidate molecule involved in the MyD88-independent pathway is TIR domain-containing adapter inducing IFN-β (TRIF), otherwise called TIR domain-containing adaptor molecule-1 (TICAM-1). TRIF/TICAM-1 can activate the IFN-β promoter and IRF-3.[65,66] Furthermore, TRIF/TICAM-1 can interact with a cytoplasmic domain of TLR3, which can activate IRF-3 in a MyD88-independent manner. Various lines of evidence were found to link two noncanonical IκB kinase homologs, IκB kinase-ε (IKKε) and TANK-binding kinase-1 (TBK1)[67-69] to TRIF/TICAM-1 and IRF-3. First, TRIF/TICAM-1 can associate with IKKε and TBK1.[70] Second, TRIF/TICAM-1-induced activation of IFN-β and NF-κB reporter activity was inhibited by dominant negative forms of IKKε and TBK1.[70] Furthermore, phosphorylation and nuclear localization of IRF-3 was induced by expression of IKKε and TBK1.[71] Thus, TRIF/TICAM-1 is most likely involved in the MyD88-independent pathway leading to IRF-3 activation through IKKε and TBK1 (Fig. 3). Although IRF-3 mediates antiviral gene expression by TLR3 and TLR4 signaling,[72] it is also notable that IRF-3 is differentially phosphorylated by TLR3 and TLR4 signaling.[73] This may indicate diverse molecular mechanisms in TLR-induced IRF-3 activation. TRIF-deficient mice were found to lack both TLR3- and TLR4-induced IFN-β expression and IRF-3 activation, indicating critical involvement of TRIF in the MyD88-independent pathway.[74,75] Furthermore, LPS-induced NF-κB activation was completely abolished in TRIF- and MyD88-double deficient cells. Analysis of mutant mice lacking IKKε or TBK1 will further clarify the TLR3 and TLR4 signaling mechanism.

TLR2, TLR7 and TLR9 exert all of their effects in a MyD88-dependent manner. However, although TLR2 is incapable of inducing the secretion of type I IFNs, TLR7 and TLR9 can induce both IFN-α and IFN-β. Therefore, we cannot exclude the possibility that other adapter molecule(s) is / are involved in TLR7 or TLR9 signaling (Fig. 3).

Regulation of Adaptive Immunity by DCs

TLRs and the Th1/Th2 Balance

DCs reside in peripheral tissues in an immature state. These immature DCs can take up pathogenic microorganisms and 'sense' infection through TLRs. TLR signaling subsequently leads to DC activation and maturation. Mature DCs alter their chemokine receptor expression and migrate to the secondary lymphoid organs. In lymphoid organs, mature DCs can present antigen to T cells. Ag presentation alone is not sufficient to induce T cell activation, but induces T cell anergy. However, TLR signaling can induce the up-regulation of costimulatory molecules necessary to induce clonal T cell expansion (Fig. 4). The subsequent T cell differentiation is crucial for host defense. $CD4^+$ T cells can differentiate into two distinct helper T (Th) subsets.[76] Th1 cells secrete mainly IFN-γ and are involved in cellular immunity against bacteria or viruses. Th2 cells produce mainly IL-4 or IL-13 and contribute to humoral immunity against helminth or allergic responses. DCs are critically involved in directing Th cell differentiation. Although multiple factors affect the ability of DCs to regulate the Th1/Th2 equilibrium,[77] the nature of the activating stimulus is critical.[78,79] Most TLR ligands, including LPS and CpG DNA, principally activate DCs to produce Th1-inducing cytokines such as IL-12 or IL-18. As a result, Th cell differentiation is biased towards Th1 (Fig. 4).

TLR-induced DC activation mechanisms are complex and ill understood. For example, MyD88-deficient DCs can mature in response to LPS.[46] However, in contrast to their wild type counterparts, mature MyD88-deficient DCs support Th2 rather than Th1 cell differentiation.[80] While their inability to produce IL-12 accounts for the failure to support Th1 differentiation, this defect is not, in itself, sufficient to explain why MyD88-deficient mature DCs induce Th2 differentiation. This altered DC function is believed to contribute to Th2-biased immune responses of MyD88-deficient mice immunized with adjuvants that are otherwise Th1-inducing.[80-83] Furthermore, DCs activated through the MyD88-independent pathway may account for the finding that TLR4 is involved in allergic Th2 responses.[84] In an allergic sensitization model, high and low doses of LPS can lead to Th1 and Th2 responses, respectively.[85] Thus, TLR4 signaling can variably activate adaptive immunity.

LPS derived from *Porphyromonas gingivalis* can trigger Th2 immune responses.[86] While *E. coli* LPS can activate $CD8\alpha^+$ DCs to produce IL-12, this is not the case for *P. gingivalis* LPS. Another TLR2 ligand, MALP-2, can also function as a Th2-inducing adjuvant.[87] Thus, TLR2 signaling can lead to Th2 immune responses. TLR2-induced Th2 cell differentiation likely occurs in a manner distinct from TLR4-induced MyD88-independent signaling, because TLR2

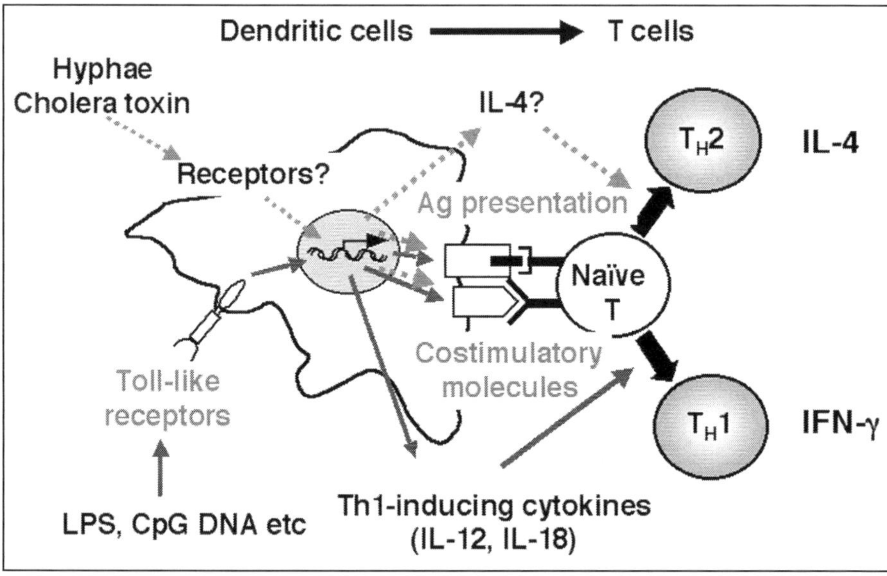

Figure 4. Interaction between DCs and T cells. DCs, stimulated by TLR ligands, mainly direct Th cell differentiation towards Th1 (solid arrows). Th2-inducing innate stimuli, although not yet fully characterized, instruct DCs to support Th2 cell differentiation (dotted arrows).

cannot activate the MyD88-independent pathway. In conclusion, TLR signaling can modulate the Th1/Th2 balance through multiple pathways.

Other microbial stimuli can also provoke Th2 immune responses. For example, nematode-derived secretion products and cholera toxin can activate DCs to support Th2 cell differentiation.[88,89] Interestingly, yeast and hyphal forms of *Candida albicans* can stimulate the Th1- and Th2-supporting ability of DCs, respectively.[90] Thus, this fungus qualitatively changes its adjuvanticity over its life cycle. Furthermore, in vivo injection of *Schistosoma mansoni* egg antigen licences DCs for Th2 priming.[91] However, in contrast to Th1-inducing innate stimuli, little is known as to how DCs are activated by Th2-inducing innate stimuli (Fig. 4).

The Control of Regulatory T Cell Function by DCs

DCs also regulate adaptive immunity through regulatory T cell function. TLR signaling in DCs can inhibit suppressive effects of regulatory T cells.[92] This is an alternative mechanism to activate immunity. IL-6 is suggested to be involved in this inhibition of regulatory T cell function. Stimulated by pathogens, DCs can also provoke regulatory T cell development. For example, filamentous hemagglutinin from *Bordetella pertussis* can stimulate DCs to induce IL-10 and inhibit IL-12 production, thereby augmenting regulatory T cell function.[93] Furthermore, *Shistosome*-derived phosphatidylserine stimulates DCs through TLR2 and subsequently induces regulatory T cell development.[94]

Type I IFN Induction by TLRs

TLR signaling can also stimulate DCs to produce type I IFNs. Type I IFNs can activate anti-viral mechanisms, induce DC maturation and function as immune adjuvants.[95-98] Thus, type I IFNs are critical cytokines linking innate and adaptive immunity. Type I IFNs consist of multiple members of the IFN-α and IFN-β family, although they act through a common receptor complex. IFN-αs and IFN-β require distinct transcription factors for their gene expression.[97] TLR3, TLR7, and TLR9, all of which can recognize nucleic acid ligands, can induce both IFN-α and IFN-β (Fig. 3). TLR2 cannot induce either (Fig. 3) and TLR4 is unique in that it cannot induce IFN-α but does induce IFN-β (Fig. 3). Thus, TLRs can be divided into subgroups according to their ability to induce type I IFN. Not only TLR expression, but also DC subsets are critical factors for TLR-induced type I IFN production. Two DC subsets, PDC and myeloid DC (MDC), have been identified.[99] PDC can produce high amounts of IFN-α upon viral infection.[100-102] In mice, both subsets express TLR9 and also increase expression levels of costimulatory molecules in response to the two types of CpG DNA, conventional and A/D-type CpG DNA. Furthermore, both DC subsets can secrete IL-12 in response to conventional CpG DNA. However, in response to A/D-type CpG DNA, PDCs produce predominantly IFN-α. This switches, in the main, to IL-12 on stimulation with conventional CpG DNA. All these effects are dependent on TLR9 and MyD88.[103] What exactly determines the differential responses to distinct TLR9 ligands remains an interesting question. Notably, nonPDC can also respond to intracellular dsRNA and secrete IFN-α.[104] This induction is independent of TLR3 and MyD88, but dependent on PKR. Thus, the type I IFN-inducing mechanism in DCs is complex and requires further study.

Crosstalk between Toll-Like and Other Receptors

It is important to clarify how the TLR system is regulated by other receptor signaling. A C-type lectin, dectin-1, which recognizes β-glucans, collaborates with TLR2 to enhance the innate immune response.[105,106] Meanwhile, other C-type lectins, such as the mannose receptor or blood DC antigen-2 (BDCA-2), can inhibit TLR-induced cytokine production.[107] Furthermore, a chromatin-IgG complex, which can engage the B cell receptor and TLR9 simultaneously, can activate B cells to produce autoantibodies.[108] This B cell activating mechanism can account for the predominance of autoantibodies against nucleic acids. Thus, other receptor stimuli critically affect TLR signaling, not only in host defense but also in pathologic situations.

Perspectives

Gene targeting experiments in mice have significantly contributed to the understanding of the function and signaling mechanisms of TLRs. Furthermore, human genetic analysis has revealed critical roles for TLRs. For example, IRAK-4 deficient mice and patients show no response to IL-1 and TLR ligands. The patients are in fact susceptible to infection with pyogenic bacteria.[109] However, much remains to be discovered. For example, TLR10 is still an orphan receptor. It is also far from clear as to the precise molecular mechanisms that underlie the function of each TLR-specific function. DCs are quite heterogeneous, expressing a variety of TLRs and playing multiple roles in regulating adaptive immunity. Therefore, DCs are valuable for further elucidation of the diverse function of TLRs. Knowledge about TLR function and signaling in DCs should enable us to acquire effective immunomodulatory tools to regulate infection, cancer and allergy.

Acknowledgements

We thank lab members for useful discussions and suggestions. This work was supported by grants, including the Funds of Advanced and Innovational Research program in Life Sciences, from the Ministry of Education, Culture, Sports, Science and Technology in Japan, Japan Science and Technology Corporation, and Uehara Memorial Foundation.

References

1. Banchereau J, Steinman RM. Dendritic cells and the control of immunity. Nature 1998; 392:245-252.
2. Medzhitov R. Toll-like receptors and innate immunity. Nat Rev Immunol 2001; 1:135-145.
3. Takeda K, Kaisho T, Akira S. Toll-like receptors. Annu Rev Immunol 2003; 21:335-376.
4. Medzhitov R, Janeway Jr CA. Innate immunity: The virtues of a nonclonal system of recognition. Cell 1997; 91:295-298.
5. Lemaitre B, Nicolas E, Michaut L et al. The dorsoventral regulatory gene cassette spatzle/Toll/cactus controls the potent antifungal response in Drosophila adults. Cell 1996; 86:973-983.
6. Medzhitov R, Preston-Hurlburt P, Janeway Jr CA. A human homologue of the Drosophila Toll protein signals activation of adaptive immunity. Nature 1997; 388:394-397.
7. Hoffmann JA, Reichhart JM. Drosophila innate immunity: An evolutionary perspective. Nat Immunol 2002; 3:121-126.
8. Levashina EA, Langley E, Green C et al. Constitutive activation of toll-mediated antifungal defense in serpin- deficient Drosophila. Science 1999; 285:1917-1919.
9. Michel T, Reichhart JM, Hoffmann JA et al. Drosophila Toll is activated by Gram-positive bacteria through a circulating peptidoglycan recognition protein. Nature 2001; 414:756-759.
10. Poltorak A, He X, Smirnova I et al. Defective LPS signaling in C3H/HeJ and C57BL/10ScCr mice: Mutations in Tlr4 gene. Science 1998; 282:2085-2088.
11. Qureshi ST, Gros P, Malo D. Host resistance to infection: Genetic control of Lipopolysaccharide responsiveness by Toll-like receptor genes. Trends Genet 1999; 15:291-294.
12. Hoshino K, Takeuchi O, Kawai T et al. Cutting edge: Toll-like receptor 4 (TLR4)-deficient mice are hyporesponsive to Lipopolysaccharide: Evidence for TLR4 as the Lps gene product. J Immunol 1999; 162:3749-3752.
13. Arbour NC, Lorenz E, Schutte BC et al. TLR4 mutations are associated with endotoxin hyporesponsiveness in humans. Nat Genet 2000; 25:187-191.
14. Shimazu R, Akashi S, Ogata H et al. MD-2, a molecule that confers Lipopolysaccharide responsiveness on Toll-like receptor 4. J Exp Med 1999; 189:1777-1782.
15. Nagai Y, Akashi S, Nagafuku M et al. Essential role of MD-2 in LPS responsiveness and TLR4 distribution. Nat Immunol 2002; 3:667-672.
16. Akira S, Takeda K, Kaisho T. Toll-like receptors: Critical proteins linking innate and acquired immunity. Nat Immunol 2001; 2:675-680.
17. Ozinsky A, Underhill DM, Fontenot JD et al. The repertoire for pattern recognition of pathogens by the innate immune system is defined by cooperation between toll-like receptors. Proc Natl Acad Sci USA 2000; 97:13766-13771.
18. Takeuchi O, Kaufmann A, Grote K et al. Cutting edge: Preferentially the R-stereoisomer of the mycoplasmal lipopeptide macrophage-activating lipopeptide-2 activates immune cells through a toll-like receptor 2- and MyD88-dependent signaling pathway. J Immunol 2000; 164:554-557.
19. Takeuchi O, Kawai T, Muhlradt PF et al. Discrimination of bacterial lipoproteins by Toll-like receptor 6. Int Immunol 2001; 13:933-940.
20. Takeuchi O, Sato S, Horiuchi T et al. Cutting edge: Role of Toll-like receptor 1 in mediating immune response to microbial lipoproteins. J Immunol 2002; 169:10-14.
21. Hayashi F, Smith KD, Ozinsky A et al. The innate immune response to bacterial flagellin is mediated by Toll- like receptor 5. Nature 2001; 410:1099-1103.
22. Tokunaga T, Yamamoto H, Shimada S et al. Antitumor activity of deoxyribonucleic acid fraction from Mycobacterium bovis BCG. I. Isolation, physicochemical characterization, and antitumor activity. J Natl Cancer Inst 1984; 72:955-962.
23. Wagner H. Bacterial CpG DNA activates immune cells to signal infectious danger. Adv Immunol 1999; 73:329-368.
24. Krieg AM. The role of CpG motifs in innate immunity. Curr Opin Immunol 2000; 12:35-43.
25. Yamamoto S, Yamamoto T, Tokunaga T. The discovery of immunostimulatory DNA sequence. Springer Semin Immunopathol 2000; 22:11-19.
26. Hemmi H, Takeuchi O, Kawai T et al. A Toll-like receptor recognizes bacterial DNA. Nature 2000; 408:740-745.

27. Bauer S, Kirschning CJ, Hacker H et al. Human TLR9 confers responsiveness to bacterial DNA via species-specific CpG motif recognition. Proc Natl Acad Sci USA 2001; 98:9237-9242.
28. Ahmad-Nejad P, Hacker H, Rutz M et al. Bacterial CpG-DNA and Lipopolysaccharides activate Toll-like receptors at distinct cellular compartments. Eur J Immunol 2002; 32:1958-1968.
29. Verthelyi D, Ishii KJ, Gursel M et al. Human peripheral blood cells differentially recognize and respond to two distinct CPG motifs. J Immunol 2001; 166:2372-2377.
30. Krug A, Rothenfusser S, Hornung V et al. Identification of CpG oligonucleotide sequences with high induction of IFN-alpha/beta in plasmacytoid dendritic cells. Eur J Immunol 2001; 31:2154-2163.
31. Ballas ZK, Krieg AM, Warren T et al. Divergent therapeutic and immunologic effects of oligodeoxynucleotides with distinct CpG motifs. J Immunol 2001; 167:4878-4886.
32. Alexopoulou L, Holt AC, Medzhitov R et al. Recognition of double-stranded RNA and activation of NF-kappaB by Toll- like receptor 3. Nature 2001; 413:732-738.
33. Yang YL, Reis LF, Pavlovic J et al. Deficient signaling in mice devoid of double-stranded RNA-dependent protein kinase. Embo J 1995; 14:6095-6106.
34. Chu WM, Ostertag D, Li ZW et al. JNK2 and IKKbeta are required for activating the innate response to viral infection. Immunity 1999; 11:721-731.
35. Hemmi H, Kaisho T, Takeuchi O et al. Small anti-viral compounds activate immune cells via the TLR7 MyD88- dependent signaling pathway. Nat Immunol 2002; 3:196-200.
36. Lee J, Chuang TH, Redecke V et al. Molecular basis for the immunostimulatory activity of guanine nucleoside analogs: Activation of Toll-like receptor 7. Proc Natl Acad Sci USA 2003; 100:6646-6651.
37. Jurk M, Heil F, Vollmer J et al. Human TLR7 or TLR8 independently confer responsiveness to the antiviral compound R-848. Nat Immunol 2002; 3:499.
38. Ohashi K, Burkart V, Flohe S et al. Cutting edge: Heat shock protein 60 is a putative endogenous ligand of the toll-like receptor-4 complex. J Immunol 2000; 164:558-561.
39. Basu S, Binder RJ, Suto R et al. Necrotic but not apoptotic cell death releases heat shock proteins, which deliver a partial maturation signal to dendritic cells and activate the NF-kappa B pathway. Int Immunol 2000; 12:1539-1546.
40. Vabulas RM, Wagner H, Schild H. Heat shock proteins as ligands of toll-like receptors. Curr Top Microbiol Immunol 2002; 270:169-184.
41. Smiley ST, King JA, Hancock WW. Fibrinogen stimulates macrophage chemokine secretion through toll-like receptor 4. J Immunol 2001; 167:2887-2894.
42. Termeer C, Benedix F, Sleeman J et al. Oligosaccharides of Hyaluronan activate dendritic cells via toll-like receptor 4. J Exp Med 2002; 195:99-111.
43. Biragyn A, Ruffini PA, Leifer CA et al. Toll-like receptor 4-dependent activation of dendritic cells by beta-defensin 2. Science 2002; 298:1025-1029.
44. Wallin RP, Lundqvist A, More SH et al. Heat-shock proteins as activators of the innate immune system. Trends Immunol 2002; 23:130-135.
45. Adachi O, Kawai T, Takeda K et al. Targeted disruption of the MyD88 gene results in loss of IL-1- and IL- 18-mediated function. Immunity 1998; 9:143-150.
46. Kaisho T, Takeuchi O, Kawai T et al. Endotoxin-induced maturation of MyD88-deficient dendritic cells. J Immunol 2001; 166:5688-5694.
47. Suzuki N, Suzuki S, Yeh WC. IRAK-4 as the central TIR signaling mediator in innate immunity. Trends Immunol 2002; 23:503-506.
48. Li S, Strelow A, Fontana EJ et al. IRAK-4: A novel member of the IRAK family with the properties of an IRAK-kinase. Proc Natl Acad Sci USA 2002; 99:5567-5572.
49. Thomas JA, Allen JL, Tsen M et al. Impaired cytokine signaling in mice lacking the IL-1 receptor-associated kinase. J Immunol 1999; 163:978-984.
50. Swantek JL, Tsen MF, Cobb MH et al. IL 1 receptor associated kinase modulates host responsiveness to endotoxin. J Immunol 2000; 164:4301-4306.
51. Suzuki N, Suzuki S, Duncan GS et al. Severe impairment of interleukin-1 and Toll-like receptor signalling in mice lacking IRAK-4. Nature 2002; 416:750-756.
52. Burns K, Janssens S, Brissoni B et al. Inhibition of interleukin 1 receptor/Toll-like receptor signaling through the alternatively spliced, short form of MyD88 is due to its failure to recruit IRAK-4. J Exp Med 2003; 197:263-268.

53. Janssens S, Burns K, Tschopp J et al. Regulation of interleukin-1- and Lipopolysaccharide-induced NF-kappaB activation by alternative splicing of MyD88. Curr Biol 2002; 12:467-471.
54. Kobayashi K, Hernandez LD, Galan JE et al. IRAK-M is a negative regulator of Toll-like receptor signaling. Cell 2002; 110:191-202.
55. Kawai T, Adachi O, Ogawa T et al. Unresponsiveness of MyD88-deficient mice to endotoxin. Immunity 1999; 11:115-122.
56. Toshchakov V, Jones BW, Perera PY et al. TLR4, but not TLR2, mediates IFN-beta-induced STAT1alpha/beta-dependent gene expression in macrophages. Nat Immunol 2002; 3:392-398.
57. Kawai T, Takeuchi O, Fujita T et al. Lipopolysaccharide stimulates the MyD88-independent pathway and results in activation of IFN-regulatory factor 3 and the expression of a subset of Lipopolysaccharide-inducible genes. J Immunol 2001; 167:5887-5894.
58. Hoshino K, Kaisho T, Iwabe T et al. Differential involvement of interferon-b in Toll-like receptor stimulated dendritic cell activation. Int Immunol 2002; 14:1225-1231.
59. Seki E, Tsutsui H, Nakano H et al. Lipopolysaccharide-induced IL-18 secretion from murine kupffer cells independently of myeloid differentiation factor 88 that is critically involved in induction of production of IL-12 and IL-1beta. J Immunol 2001; 166:2651-2657.
60. Fitzgerald KA, Palsson-McDermott EM, Bowie AG et al. Mal (MyD88-adapter-like) is required for Toll-like receptor-4 signal transduction. Nature 2001; 413:78-83.
61. Horng T, Barton GM, Medzhitov R. TIRAP: An adapter molecule in the Toll signaling pathway. Nat Immunol 2001; 2:835-841.
62. Horng T, Barton GM, Flavell RA et al. The adaptor molecule TIRAP provides signalling specificity for Toll-like receptors. Nature 2002; 420:329-333.
63. Yamamoto M, Sato S, Hemmi H et al. Essential role for TIRAP in activation of the signalling cascade shared by TLR2 and TLR4. Nature 2002; 420:324-329.
64. O'Neill LAJ, Fitzgerald KA, Bowie AG. The Toll-IL-1 receptor adaptor family grows to five members. Trends Immunol 2003; 24:286-290.
65. Yamamoto M, Sato S, Mori K et al. Cutting edge: A novel Toll/IL-1 receptor domain-containing adapter that preferentially activates the IFN-beta promoter in the Toll-like receptor signaling. J Immunol 2002; 169:6668-6672.
66. Oshiumi H, Matsumoto M, Funami K et al. TICAM-1, an adaptor molecule that participates in Toll-like receptor 3-mediated interferon-beta induction. Nat Immunol 2003; 4:161-167.
67. Peters RT, Liao SM, Maniatis T. IKKepsilon is part of a novel PMA-inducible IkappaB kinase complex. Mol Cell 2000; 5:513-522.
68. Shimada T, Kawai T, Takeda K et al. IKK-i, a novel Lipopolysaccharide-inducible kinase that is related to IkappaB kinases. Int Immunol 1999; 11:1357-1362.
69. Tojima Y, Fujimoto A, Delhase M et al. NAK is an IkappaB kinase-activating kinase. Nature 2000; 404:778-782.
70. Fitzgerald KA, McWhirter SM, Faia KL et al. IKKe and TBK1 are essential components of the IRF3 signaling pathway. Nat Immunol 2003; 4:491-496.
71. Sharma S, tenOever BR, Grandvaux N et al. Triggering the interferon antiviral response through an IKK-related pathway. Science 2003; 300:1148-1151.
72. Doyle S, Vaidya S, O'Connell R et al. IRF3 mediates a TLR3/TLR4-specific antiviral gene program. Immunity 2002; 17:251-263.
73. Servant MJ, Grandvaux N, TenOever BR et al. Identification of the Minimal Phosphoacceptor Site Required for in Vivo Activation of Interferon Regulatory Factor 3 in Response to Virus and Double-stranded RNA. J Biol Chem 2003; 278:9441-9447.
74. Yamamoto M, Sato S, Hemmi H et al. Role of adaptor TRIF in the MyD88-independent toll-like receptor signaling pathway. Science 2003; 301:640-643.
75. Hoebe K, Du X, Georgel P et al. Identification of Lps2 as a key transducer of MyD88-independent TIR signalling. Nature 2003; 424:743-748.
76. Abbas AK, Murphy KM, Sher A. Functional diversity of helper T lymphocytes. Nature 1996; 383:787-793.
77. Moser M, Murphy KM. Dendritic cell regulation of Th1-Th2 development. Nature Immunol 2000; 1:199-205.
78. Reis e Sousa C. Dendritic cells as sensors of infection. Immunity 2001; 14:495-498.

79. Rescigno M. Dendritic cells and the complexity of microbial infection. Trends Microbiol 2002; 10:425-461.
80. Kaisho T, Hoshino K, Iwabe T et al. Endotoxin can induce MyD88-deficient dendritic cells to support T(h)2 cell differentiation. Int Immunol 2002; 14:695-700.
81. Schnare M, Barton GM, Holt AC et al. Toll-like receptors control activation of adaptive immune responses. Nat Immunol 2001; 2:947-950.
82. Jankovic D, Kullberg MC, Hieny S et al. In the absence of IL-12, CD4(+) T cell responses to intracellular pathogens fail to default to a Th2 pattern and are host protective in an IL-10(-/-) setting. Immunity 2002; 16:429-439.
83. Muraille E, De Trez C, Brait M et al. Genetically resistant mice lacking MyD88-adapter protein display a high susceptibility to Leishmania major infection associated with a polarized Th2 response. J Immunol 2003; 170:4237-4241.
84. Dabbagh K, Dahl ME, Stepick-Biek P et al. Toll-like receptor 4 is required for optimal development of Th2 immune responses: Role of dendritic cells. J Immunol 2002; 168:4524-4530.
85. Eisenbarth SC, Piggott DA, Huleatt JW et al. Lipopolysaccharide-enhanced, toll-like receptor 4-dependent T helper cell type 2 responses to inhaled antigen. J Exp Med 2002; 196:1645-1651.
86. Pulendran B, Kumar P, Cutler CW et al. Lipopolysaccharides from distinct pathogens induce different classes of immune responses in vivo. J Immunol 2001; 167:5067-5076.
87. Rharbaoui F, Drabner B, Borsutzky S et al. The Mycoplasma-derived lipopeptide MALP-2 is a potent mucosal adjuvant. Eur J Immunol 2002; 32:2857-2865.
88. Whelan M, Harnett MM, Houston KM et al. A filarial nematode-secreted product signals dendritic cells to acquire a phenotype that drives development of Th2 cells. J Immunol 2000; 164:6453-6460.
89. Gagliardi MC, Sallusto F, Marinaro M et al. Cholera toxin induces maturation of human dendritic cells and licences them for Th2 priming. Eur J Immunol 2000; 30:2394-2403.
90. d'Ostiani CF, Del Sero G, Bacci A et al. Dendritic cells discriminate between yeasts and hyphae of the fungus Candida albicans. Implications for initiation of T helper cell immunity in vitro and in vivo. J Exp Med 2000; 191:1661-1674.
91. MacDonald AS, Straw AD, Bauman B et al. CD8- dendritic cell activation status plays an integral role in influencing Th2 response development. J Immunol 2001; 167:1982-1988.
92. Pasare C, Medzhitov R. Toll pathway-dependent blockade of CD4+CD25+ T cell-mediated suppression by dendritic cells. Science 2003; 299:1033-1036.
93. McGuirk P, McCann C, Mills KH. Pathogen-specific T regulatory 1 cells induced in the respiratory tract by a bacterial molecule that stimulates interleukin 10 production by dendritic cells: A novel strategy for evasion of protective T helper type 1 responses by Bordetella pertussis. J Exp Med 2002; 195:221-231.
94. van der Kleij D, Latz E, Brouwers JF et al. A novel host-parasite lipid cross-talk. Schistosomal lyso-phosphatidylserine activates toll-like receptor 2 and affects immune polarization. J Biol Chem 2002; 277:48122-48129.
95. Le Bon A, Schiavoni G, D'Agostino G et al. Type i interferons potently enhance humoral immunity and can promote isotype switching by stimulating dendritic cells in vivo. Immunity 2001; 14:461-470.
96. Biron CA. Interferons alpha and beta as immune regulators—a new look. Immunity 2001; 14:661-664.
97. Taniguchi T, Takaoka A. The interferon-alpha/beta system in antiviral responses: A multimodal machinery of gene regulation by the IRF family of transcription factors. Curr Opin Immunol 2002; 14:111-116.
98. Levy DE, Marie I, Prakash A. Ringing the interferon alarm: Differential regulation of gene expression at the interface between innate and adaptive immunity. Curr Opin Immunol 2003; 15:52-58.
99. Shortman K, Liu YJ. Mouse and human denedritic cell subtypes. Nature Rev Immunol 2002; 2:151-161.
100. Asselin-Paturel C, Boonstra A, Dalod M et al. Mouse type I IFN-producing cells are immature APCs with plasmacytoid morphology. Nat Immunol 2001; 2:1144-1150.
101. Nakano H, Yanagita M, Gunn MD. CD11c(+)B220(+)Gr-1(+) cells in mouse lymph nodes and spleen display characteristics of plasmacytoid dendritic cells. J Exp Med 2001; 194:1171-1178.

102. Colonna M, Krug A, Cella M. Interferon-producing cells: On the front line in immune responses against pathogens. Curr Opin Immunol 2002; 14:373-379.
103. Hemmi H, Kaisho T, Takeda K et al. The roles of Toll-like receptor 9, MyD88, and DNA-dependent protein kinase catalytic subunit in the effects of two distinct CpG DNAs on dendritic cell subsets. J Immunol 2003; 170:3059-3064.
104. Diebold SS, Montoya M, Unger H et al. Viral infection switches nonplasmacytoid dendritic cells into high interferon producers. Nature 2003; 424:324-328.
105. Gantner BN, Simmons RM, Canavera SJ et al. Collaborative Induction of Inflammatory Responses by Dectin-1 and Toll-like Receptor 2. J Exp Med 2003; 197:1107-1117.
106. Brown GD, Herre J, Williams DL et al. Dectin-1 mediates the biological effects of beta-glucans. J Exp Med 2003; 197:1119-1124.
107. Engering A, Geijtenbeek TB, van Kooyk Y. Immune escape through C-type lectins on dendritic cells. Trends Immunol 2002; 23:480-485.
108. Leadbetter EA, Rifkin IR, Hohlbaum AM et al. Chromatin-IgG complexes activate B cells by dual engagement of IgM and Toll-like receptors. Nature 2002; 416:603-607.
109. Picard C, Puel A, Bonnet M et al. Pyogenic bacterial infections in humans with IRAK-4 deficiency. Science 2003; 299:2076-2079.

Chapter 7

Pathogen Avoidance Using Toll Signaling in *C. elegans*

Nathalie Pujol and Jonathan J. Ewbank

Abstract

In the fruit fly *Drosophila melanogaster*, the Toll receptor plays two major roles, one is in early development and the other concerns resistance to infection. In mammals, the multiple Toll-Like Receptors (TLRs) are known to be key mediators of innate immunity. *Caenorhabditis elegans* possesses only one TLR (TOL-1) which is required during development, in a TIR-domain independent fashion. It also functions indirectly in defence, via a behavioral mechanism that keeps worms away from the pathogenic bacterium *Serratia marcescens*. We describe here that the *tol-1*-dependent avoidance behavior involves the recognition of a signal that includes contributions from both the bacteria and the worms, themselves. We outline the current understanding of how *C. elegans* detects bacteria and other worms and speculate on a possible link between chemosensation and immune recognition.

The evolutionary origins and ancestral function of the Toll-like receptors (TLRs) are still far from clear (for a review see refs. 1-2). TLRs are cell-surface receptors characterised by two types of protein domains, leucine rich repeats (LRRs) and the Toll-Interleukin 1 Receptor (TIR) domain. These domains are shared by multiple protein families with diverse functions (see Pfam entries 560 and 1582). Of the two, TIR domains are often associated with proteins that function in innate immunity either at the membrane or intracellularly, both in plants[3] and animals.[4] In vertebrates including mice and humans, multiple TLRs are thought to function directly in pathogen recognition and participate in branched signalling pathways that lead to the activation of transcription factors. One well-characterised branch that is conserved from *Drosophila* to vertebrates, and which can be considered as the canonical Toll/TLR pathway, leads to members of the nuclear factor κB (NF-κB) family. In animals, this pathway seems to have evolved after the first emergence of multicellularity. For example, the nearly complete genome sequence of the social amoeba *Dictyostelium*,[5] one of the simplest multicellular organisms with differentiated cell types, contains neither TLR nor NF-κB homologues. The genome of the simple nematode worm *Caenorhabditis elegans* does not contain an obvious NF-κB homologue either, but does include a single TLR gene, named *tol-1*. The corresponding protein, TOL-1, has an extracellular domain that contains 22 LRRs and an intracellular TIR domain. When compared to the family of *Drosophila* TLR proteins, TOL-1 is most similar to Toll-8, closely followed by Toll-6. TOL-1, however, lacks the C-terminal extension following the TIR domain, that is found in certain *Drosophila* Toll-family proteins.[6]

Toll and Toll-Like Receptors: An Immunologic Perspective, edited by Tina Rich.

Further phylogenetic analyses of TLR genes from diverse invertebrate species will be required to resolve the question of whether the presence of a single TLR in *C. elegans* reflects the fact that the common ancestor of flies and worms had a single TLR, or whether a lineage-specific loss of TLRs has occurred. This question cannot currently be resolved through analysis of the available nematode EST databases.

The Single *C. elegans* TLR TOL-1 Is Essential for Development

From a fundamental and comparative standpoint, it is interesting to study the role of TOL-1 in *C. elegans*. Two mutant alleles of *tol-1* have been generated, both recessive. The first, a complete loss of function allele, is associated with a temperature-sensitive lethality. At the restrictive temperature of 25°C, all mutant embryos die, while at 15°C, a small fraction survive and complete development to give marginally fertile adults. We have found that *tol-1* is necessary in early morphogenesis for the closure of the embryo (unpublished data) and are currently looking for genes that interact with *tol-1* during development. The second mutant allele (*nr2033*) corresponds to a truncated protein missing almost all of the intracellular TIR domain. In this case, mutants are viable and fertile both at 15°C and at 25°C, indicating that the TIR-domain is dispensable for the developmental role of TOL-1. The fact that TIR domain-independent functions for TLRs have not been seen in other organisms might be explained by a functional redundancy. In *Drosophila*, for example, Toll and Toll5 have overlapping spatial and temporal expression domains, as do 18-Wheeler and Toll8.[7] Detailed analyses of the developmental expression patterns of vertebrate TLRs to uncover potential developmental roles and possible functional redundancies would clearly be of great interest, particularly if the TLRs are to be used as targets for drug therapy in the future.

tol-1 Mutants Are Not Hypersusceptible to Fungal or Bacterial Infection

In *Drosophila* and vertebrates, activation of the canonical Toll/TLR pathway helps the host defend itself against pathogens. Thus the fly *Toll* mutants are hypersensitive to infection by fungi and Gram-positive bacteria,[8,9] while *Tlr4* mutant mice succumb to Gram-negative infections that wild type mice surmount.[10] In all cases, the antimicrobial role of these proteins is dependant upon an intact TIR domain. This raised the possibility that mutants homozygous for the *tol-1(nr2033)* allele, that encodes a protein lacking a TIR domain, might be less resistant to infection than wild-type worms. Apart from the famous 'nematode-trappers', only one fungal pathogen for *C. elegans* has been well characterised, the endoparasitic *Drechmeria coniospora* (reviewed in ref. 11). Removal of the TIR domain from TOL-1 was not associated with an altered susceptibility to infection with this fungus.[12] On the other hand, a number of bacterial pathogens of worms are known. In part, this is because they can be identified relatively easily. Indeed, when looking for potential bacterial pathogens of *C. elegans*, the worms' standard food, the *Escherichia coli* strain OP50 is simply replaced by another bacterial species. Some bacteria produce toxins that kill worms relatively rapidly.[13,14] Others colonize the worm's intestine and kill the host in a matter of days[11] and in one case, a nematode-specific bacterium that adheres to the worm's cuticle and provokes a pronounced swelling has been described.[15] For the time being, no significant change in susceptibility to a number of different types of bacterial infection has been recorded for the *tol-1(nr2033)* mutant,[16,12] arguing against a direct role for TOL-1 in resistance to infection.

TOL-1 Dependant Avoidance Behavior

In addition to their effects on the survival of worms, different bacteria can also influence worms' behavior, and can either repel or attract worms.[17,18] In the case of the Gram-negative enterobacterium *Serratia marcescens*, the situation is more complex. If given the choice between the *E. coli* strain OP50 and the pathogenic *S. marcescens* strain Db11, two thirds of worms will choose Db11. But contrary to worms on OP50, that remain in direct contact with the bacteria for extended periods of time, worms are progressively repelled by Db11, such that after 48 hours, less than 20% are found within the bacterial "lawn". This behavior is relatively specific, as it is not provoked by a mutant derivative of Db11, called Db1140. During tests of the resistance of *tol-1(nr2033)* mutants to *S. marcescens* infection, it became clear that the mutants exhibit a defect in this behavior. After 48 hours, two thirds of *tol-1(nr2033)* mutants are still found within the Db11 lawn. This phenotype can be rescued by reintroduction of the wild-type *tol-1* gene into the mutants demonstrating that it is caused by the loss of the TIR domain.[12]

Sensing a Bacterial Signal

What is the difference between Db11 and Db1140 that can account for their distinct effects on wild-type worms? At the molecular level, there are multiple differences between the two strains, and these include two interesting candidates. In contrast to Db11, Db1140 is non-motile, suggesting a defect in the production of flagellin, now known to be a ligand of mammalian TLR5. Secondly, Db1140 was known to be more sensitive to insect immune hemolymph, and we speculated that this might reflect a change in the structure of its lipopolysaccharide (LPS), the ligand of TLR4.[12] This possibility has now been confirmed, but our more recent work has reduced the likelihood that either LPS or flagellin acts as a TOL-1 ligand. We have generated a defined mutant of Db11, called 20C2, in which a homologue of the *wzm* gene that encodes the membrane component of an ABC-2 transporter specialized in the translocation LPS O-antigen is disrupted. Consistent with this, 20C2 has no O-antigen. It is also non-motile suggesting it lacks functional flagellae, presumably a secondary consequence of its LPS defect.[19] In contrast to Db1140, 20C2 still repels wild-type worms at 48 hours as strongly as Db11 (C. L. Kurz and JJE, unpublished results). In terms of their pathogenicity, Db1140 is less virulent than Db11. But as 20C2 is as attenuated in its virulence as Db1140, the intrinsic virulence of the bacteria does not appear to be a major determinant of this behavior either. There is one further way that Db1140 differs from Db11, as unlike the latter, it is incapable of secreting proteases. How this might bring about the observed (*tol-1*-dependent) difference of behavior between worms on Db11 and Db1140 is still obscure. One remark that can be made is that this difference has measurable effects on other aspects of the worms' defences against infection. While following infection of *C. elegans* by either Db11 or Db1140 a clear induction of the expression of the lysozyme gene *lys-1* is observed, only in the case of Db1140 do the levels of a corresponding reporter fusion protein increase. It would appear that Db11 is able to degrade this inducible antimicrobial protein.[20] It is likely that other *C. elegans* proteins undergo the same fate. Infection with Db11 could generate a signal, possibly derived from bacterial proteolysis of a host protein, which then acts as a stimulus to change worm behavior.

Communication Between Infected Worms

The idea that the worms themselves are in part the source of the repellent stimulus has been given support by recent work showing that the repulsion phenotype is dependent both on the age of the worms, and also on the density of the worm population (L. Pouyet and NP, unpublished results).

Relatively little is known about how worms communicate with each other. It has been known for a number of years that in the absence of food or when the worm population density is too high, young worms enter a diapause state, called the dauer larva, under the influence of a pheromone. This allows them to arrest development before resumption in more propitious times.[21] More recently, the existence of a hermaphrodite-specific pheromone that attracts males has been proposed.[22] But in neither case is the exact chemical nature of the pheromone known. These examples do, however, clearly set a precedent for inter-worm communication. On the other hand, quite a lot is already known about how *C. elegans* senses its chemical environment. Worms can detect a large range of inorganic and organic molecules, such as ions (e.g., Cl^-, Na^+, K^+), amino acids, cyclic nucleotides and volatile odorants such as alcohols, pyrazines and thiazoles. Of the 302 neurons that an adult worm possesses, some thirty are involved in chemosensation, grouped together in two chemosensory organs, the amphids in the head and the phasmids in the tail. The phasmid neurons act as negative modulators of repulsion, and it is possible that this allows worms to construct a head-to-tail spatial map of the chemical environment.[23]

Some of the amphid neurons, such as the pair of AWA neurons, are involved in sensing attractive compounds. Others, such as the two AWB neurons sense repulsive compounds. There are more than 700 putative olfactory receptors encoded by the *C. elegans* genome. In contrast to the situation in mammals, where each olfactive neuron expresses predominantly just one receptor, in the worm, as would be predicted on purely arithmetic grounds, several receptors are expressed in each chemosensory neuron.[24] Whether the worm reacts to a given compound by being attracted or repelled is not strictly linked to the compound or its receptor, but rather to the neuron in which the receptor is expressed. Thus, the receptor ODR-10 that is normally required for attraction of worms to diacetyl is expressed in the 'attraction' neurons AWA. But if ODR-10 is expressed ectopically in the 'repulsion' neurons AWB, diacetyl becomes a repulsive stimulus for the worms.[25] Moreover, behavioral tests have clearly shown that worms can discriminate between individual compounds. The two structurally similar AWC neurons can detect at least 5 different attractive odorants. Their capacity to discriminate between the two odorants pentanedione and butanone is linked to the asymmetric expression, in just one of the AWC pair, of the seven transmembrane receptor STR-2. If STR-2 is expressed in both AWC neurons, this capacity is lost.[26] The asymmetric expression of STR-2 is dependent upon a mitogen activated protein (MAP) kinase pathway involving SEK-1 and NSY-1. SEK-1 is a MAP kinase kinase (MAP2K), homologous to mammalian MKK3/MKK6, while NSY-1 is a MAP2K kinase (MAP3K), homologous to ASK1.[27,28] Interestingly, these two kinases have recently been shown to mediate resistance to infection by *Pseudomonas aeruginosa*,[29] and *Salmonella typhimurium*, in an LPS-dependent, but TOL-1-independent fashion.[16]

As well as being able to detect multiple chemical stimuli, allowing their detection of bacteria, worms can also sense bacteria physically, and move more slowly when they are in food, than out of it. This has been shown to be a dopamine-dependent mechanosensory process, modulated in a serotonin-dependent fashion by the worm's previous experience, such that starved and well-fed worms react differently.[30] Starvation also affects the expression of specific chemoreceptors, and consequently chemosensation.[31] One can imagine that these different behavioral and molecular mechanisms help the worm to eat the right type of bacteria in sufficient quantity to allow its survival and reproduction.

If all this seems relatively sophisticated for a nematode of less than 1000 cells, and just 302 neurons, it is important to note that these capacities are distinct from the TOL-1-dependent behavior described above, which therefore represents a further level of complexity. Thus *tol-1(nr2033)* mutants have normal amphid and phasmid chemosensory neurons, as judged by dye filling, and exhibit normal chemosensory behaviors in standard tests. They are also capable of distinguishing between the *E. coli* strain OP50 and *S. marcescens* strain Db11, and are preferentially attracted by the latter. Moreover, whether worms are wild type or *tol-1(nr2033)* mutants, if after 48 hours of exposure to Db11 (by which time the majority of wild-type worms are repelled by the bacteria), they are removed from the bacteria for an hour and presented anew with a choice between OP50 and Db11, in both cases, the "experienced" worms are preferentially attracted to Db11 (L. Pouyet and NP, unpublished results).

As a further distinction from standard chemosensation, in adults, *tol-1* is not expressed in the amphid or phasmid chemosensory neurons. Rather it exhibits a restricted pattern of expression, as judged by reporter gene activity, in the 6 mechanoreceptor cells, 6 interneurons, the head mesodermal cell (of unknown function) and the 4 URY neurons. Based on their morphology and connectivity, the URY neurons have previously been suggested to function as sensory receptors.[32] They may thus be involved in the altered behavior of the *tol-1(nr2033)* mutant. Such a hypothesis, however, awaits experimental confirmation. Further tests are also required to establish how general this *tol-1*-dependent aversion phenotype is, or whether it is really specific to *S. marcescens*, as is currently the case.

If the molecular and cellular basis of the *tol-1* repulsion phenotype and its integration with more general chemosensory mechanisms remain unclear, it is perhaps interesting to observe that there do exist certain parallels between pathogen recognition and chemosensation. For example, it is becoming clear that different pathogens have different 'signatures' that are recognised by different combinations of TLRs, leading to an appropriate response (reviewed in 33), much as different odorants are identified by a subset of chemoreceptors that they subsequently activate. Conversely, in their response to pheromones, rather than using a binary system of pheromone and corresponding receptor, insects and fish appear to respond to specific mixtures of chemical cues, that activate multiple receptors, thereby generating a response with the necessary diversity and specificity.[34] Such a system appears to apply equally to mammals, and in this case, very recent data have pointed to a tantalising link between the molecular mechanisms that underpin pheromone identification and immune recognition. In mice, proteins of the M10 family of major histocompatibility complex class Ib molecules have been shown to be specifically expressed in neurons of the vomeronasal organ, where they functionally interact with a particular family of pheromone receptors, the V2Rs. This association between M10 and V2R has been proposed to alter the mechanism and the specificity of pheromone recognition,[35] and perhaps supports the notion of an ancient link between the sensory and immune systems.

References

1. Beutler B, Rehli M. Evolution of the TIR, tolls and TLRs: functional inferences from computational biology. Curr Top Microbiol Immunol 2002; 270:1-21.
2. Ewbank JJ, Gorvel JP, Vivier E. A Widely Conserved Smell of Danger. Sci STKE 2003; PE9.
3. Jebanathirajah JA, Peri S, Pandey A. Toll and interleukin-1 receptor (TIR) domain-containing proteins in plants: a genomic perspective. Trends Plant Sci 2002; 7:388-91.
4. Imler JL, Hoffmann JA. Toll receptors in Drosophila: A family of molecules regulating development and immunity. Curr Top Microbiol Immunol 2002; 270:63-79.
5. Eichinger L, Noegel AA. Crawling into a new era-the Dictyostelium genome project. Embo J 2003; 22:1941-6.
6. Imler JL, Hoffmann JA. Toll and Toll-like proteins: an ancient family of receptors signalling infection. Rev Immunogenet 2002; 2:294-304.
7. Kambris Z, Hoffmann JA, Imler JL et al. Tissue and stage-specific expression of the Tolls in Drosophila embryos. Gene Expr Patterns 2002; 2:311-7.
8. Lemaitre B, Nicolas E, Michaut L et al. The dorsoventral regulatory gene cassette spatzle/Toll/cactus controls the potent antifungal response in Drosophila adults. Cell 1996; 86:973-83.
9. Rutschmann S, Kilinc A, Ferrandon D. The toll pathway is required for resistance to Gram-positive bacterial infections in Drosophila. J Immunol 2002; 168:1542-6.
10. Poltorak A, He X, Smirnova I et al. Defective LPS signaling in C3H/HeJ and C57BL/10ScCr mice: mutations in Tlr4 gene. Science 1998; 282:2085-2088.
11. Ewbank JJ. Tackling both sides of the host-pathogen equation with Caenorhabditis elegans. Microbes and Infection 2002; 4:247-56.

12. Pujol N, Link EM, Liu LX et al. A reverse genetic analysis of components of the Toll signalling pathway in Caenorhabditis elegans. Curr Biol 2001; 11:809-21.
13. Couillault C, Ewbank JJ. Diverse Bacteria Are Pathogens of Caenorhabditis elegans. Infect Immun 2002; 70:4705-7.
14. Darby C, Cosma CL, Thomas JH et al. Lethal paralysis of Caenorhabditis elegans by Pseudomonas aeruginosa. Proc Natl Acad Sci USA 1999; 96:15202-7.
15. Hodgkin J, Kuwabara PE, Corneliussen B. A novel bacterial pathogen, Microbacterium nematophilum, induces morphological change in the nematode C. elegans. Curr Biol 2000; 10:1615-1618.
16. Aballay A, Drenkard E, Hilbun LR et al. Caenorhabditis elegans innate immune response triggered by Salmonella enterica requires intact LPS and is mediated by a MAPK signaling pathway. Curr Biol 2003; 13:47-52.
17. Andrew PA, Nicholas WL. Effect of bacteria on dispersal of Caenorhabditis elegans (Rhabditidae). Nematologica 1976; 22:451-461.
18. Grewal PS, Wright DJ. Migration of Caenorhabditis elegans (Nematoda: Rhabditidae) larvae towards bacteria and the nature of the bacterial stimulus. Fundam Appl Nematol 1992; 15:159-166.
19. Kurz CL, Chauvet S, Andres E et al. Virulence factors of the human opportunistic pathogen Serratia marcescens identified by in vivo screening. Embo J 2003; 22:1451-1460.
20. Mallo GV, Kurz CL, Couillault C et al. Inducible antibacterial defence system in C. elegans. Curr Biol 2002; 12:1209-14.
21. Golden JW, Riddle DL. A pheromone-induced developmental switch in Caenorhabditis elegans: Temperature-sensitive mutants reveal a wild-type temperature-dependent process. Proc Natl Acad Sci USA 1984; 81:819-23.
22. Simon JM, Sternberg PW. Evidence of a mate-finding cue in the hermaphrodite nematode Caenorhabditis elegans. Proc Natl Acad Sci USA 2002; 99:1598-603.
23. Hilliard MA, Bargmann CI, Bazzicalupo P. C. elegans Responds to Chemical Repellents by Integrating Sensory Inputs from the Head and the Tail. Curr Biol 2002; 12:730-4.
24. Troemel ER, Chou JH, Dwyer ND et al. Divergent seven transmembrane receptors are candidate chemosensory receptors in C. elegans. Cell 1995; 83:207-18.
25. Troemel ER, Kimmel BE, Bargmann CI Reprogramming chemotaxis responses: sensory neurons define olfactory preferences in C. elegans. Cell 1997; 91:161-9.
26. Wes PD, Bargmann CI. C. elegans odour discrimination requires asymmetric diversity in olfactory neurons. Nature 2001; 410:698-701.
27. Sagasti A, Hisamoto N, Hyodo J et al. The CaMKII UNC-43 activates the MAPKKK NSY-1 to execute a lateral signaling decision required for asymmetric olfactory neuron fates. Cell 2001; 105:221-32.
28. Tanaka-Hino M, Sagasti A, Hisamoto N et al. SEK-1 MAPKK mediates Ca^{2+} signaling to determine neuronal asymmetric development in Caenorhabditis elegans. EMBO 2002; 3:56-62.
29. Kim DH, Feinbaum R, Alloing G et al. A conserved p38 MAP kinase pathway in Caenorhabditis elegans innate immunity. Science 2002; 297:623-6.
30. Sawin ER, Ranganathan R, Horvitz HR. C. elegans locomotory rate is modulated by the environment through a dopaminergic pathway and by experience through a serotonergic pathway. Neuron 2000; 26:619-31.
31. Peckol EL, Troemel ER, Bargmann CI. Sensory experience and sensory activity regulate chemosensory receptor gene expression in Caenorhabditis elegans. Proc Natl Acad Sci USA 2001; 98:11032-8.
32. White JG, Southgate E, Thomson JN et al. The structure of the nervous system of the nematode C. elegans. Philosophical Transactions of the Royal Society of London 1986; 314B:1-340.
33. Imler JL, Hoffmann JA. Toll signaling: the TIReless quest for specificity. Nat Immunol 2003; 4:105-6.
34. Sorensen PW, Christensen TA, Stacey NE. Discrimination of pheromonal cues in fish: Emerging parallels with insects. Curr Opin Neurobiol 1998; 8:458-67.
35. Loconto J, Papes F, Chang E et al. Functional expression of murine V2R pheromone receptors involves selective association with the M10 and M1 families of MHC class Ib molecules. Cell 2003; 112:607-18.

Chapter 8

Forward Genetic Analysis of TLR Pathways:

A Shared System for the Detection of Endotoxin and Viral Infection

Bruce Beutler, Kasper Hoebe, Philippe Georgel and Xin Du

The mammalian Toll-like receptors (TLRs) were first recognized as innate immune sensors when it was discovered that TLR4 is the key component of the mammalian endotoxin (lipopolysaccharide; LPS) receptor. This determination was made when a spontaneous mouse mutation, *Lps*, was positionally cloned and found to reside with the *Tlr4* locus.[1] In all, we now know of the existence of eleven mouse TLRs and ten human TLRs, which collectively serve as the principal sensors of the innate immune system. Without them, small inocula of microorganisms would pose a major threat to the host, growing unchecked for a long period of time before they are recognized. These TLRs are served by a collection of at least five adapter proteins, each with homology to the TLRs themselves, permitting homotypic interaction to occur. Since a pure forward genetic approach led to the identification of the LPS receptor, ENU mutagenesis has been applied to the identification of other critical components of TLR signaling pathways. This approach has revealed that one of the adapter proteins, Lps2 (also known as Trif or Ticam-1) is required for normal responses to double-stranded RNA and LPS. It now appears that two and only two branches of the LPS sensing pathway exist downstream of TLR4 in macrophages: one dependent upon the joint function of the adapter proteins MyD88 and MAL/Tirap; the other dependent upon Lps2. Poly I:C sensing, on the other hand, has but one recognizable branch, leading to type I interferon induction. Lps2 is an indispensable component of this branch. Destructive mutations affecting *Lps2* cause resistance to LPS toxicity, but also, heightened susceptibility to infection by mouse cytomegalovirus (mCMV). *Lps2* is therefore the most proximal component of a signal integration system required for innate immune responses to both viral and bacterial infections.

Introduction

Innate Immunity and the Endotoxin Mystery

All metazoan organisms have evolved complex immune defense systems whereby they repel invasive microbes that would parasitize or kill them. These immune systems are remarkably effective insofar as severe or sustained infections are quite rare. They are imperfect in that serious infections sometimes do occur, and also, in that immune responses may sometimes injure the host.

Toll and Toll-Like Receptors: An Immunologic Perspective, edited by Tina Rich. ©2005 Eurekah.com and Kluwer Academic / Plenum Publishers.

Innate immunity is the most universal, the most rapidly acting, and by some appraisals, the most important type of immunity. The elimination of even one subset of innate immune effector cells (for example, neutrophils) may be sufficient to cause a profound immunodeficiency state, more severe than that observed as a result of lymphoid aplasia. The innate immune system evolved long before the adaptive immune system, and in many respects, supports the function of the newer system. Without the vital antigen-presenting function of innate immune cells, and without the production of cytokines of innate immune origin (including IL-12, CD40L, IL-1, and TNF), adaptive immune responses are ineffectual. In this respect, while adaptive and innate immunity work hand in hand, adaptive immunity is subordinate to innate immunity.

So long as it has been known that adaptive immunity (but not innate immunity) is dependent upon clonal recognition, it has been obvious to all that the general strategy of innate immune detection must be one in which a limited number of receptors are dedicated to the recognition of microbial molecules that are conserved across broad taxa. The target molecules are indispensable components of the microbes, for which reason they are not readily altered by mutation and selection. The innate immune receptors must detect pathogen molecules within the microenvironment of the infectious inoculum, so as to permit interdiction of the infection before microbes proliferate, disseminate, and overwhelm the host. At least for the most part, the receptors must be indifferent to molecules of host origin (the basis of innate immune discrimination between self and non-self).

The search for the receptors that are required for innate immune detection began with the search for stimulatory ligands, and the first of these to be identified as a distinct molecular species was bacterial "endotoxin," so named by Pfeiffer more than 100 years ago (for a review, see ref. 2). In the wake of the identification of endotoxin as a component of Gram-negative bacteria, relevant to their pathogenic effect,[3] several decades were to elapse before endotoxin was chemically characterized as a lipopolysaccharide (LPS) component of Gram-negative bacteria. Ultimately, the lipid moiety was shown to be responsible for the toxic effects. LPS was found to have adjuvant properties, in that it would greatly enhance the ability of an animal to produce antibodies to a specific co-administered antigen. By the mid 1960s, much attention had focused on the nature of the LPS receptor. It was presumed that such a receptor must exist, since a single mutation identified in C3H/HeJ mice forbade all biological responses to LPS.

A number of other microbial molecules were found to share the activities of LPS. Among these were such bacterial and fungal components as the β-glucans of zymosan, lipopeptides, peptidoglycan, lipoteichoic acid, trehalose dimycolate, double-stranded RNA, and bacterial DNA. While generally less potent than LPS, all could cause fever, inflammation, and other changes characteristic of infection. Moreover, authentic infections seemed to cause rather similar effects even when structurally disparate pathogens were involved. Might it be that all pathogens stimulated similar receptors? The conclusion was virtually inescapable, but the receptors themselves remained elusive for a long period of time.

LPS occupied center stage in the study of innate immune signaling, because it was abundant, readily purified, and exceptionally potent. To a large extent, LPS mimicked all of the effects of an authentic Gram-negative infection; indeed, it mimicked the effects of most any infection. The key clue to the identification of the LPS receptor was a spontaneous mutation that occurred in the C3H/HeJ mouse, rendering this animal highly resistant to all biological effects of LPS. First noticed in 1965,[4] the so-called *Lpsd* mutation was shown to be allelic with a second defect, in C57BL/10ScCr mice[5] and mapped to chromosome 4 in 1978.[6] *Lpsd* became the object of intense study and speculation. The fact of its existence revealed that LPS sensing was entirely dependent upon the product of a single gene.

Before the nature of the LPS receptor was known, the C3H/HeJ mouse was used in a series of experiments that established a strong link between LPS and host immune defense, insofar as LPS insensitivity was associated with a worse prognosis following Gram-negative inoculation. Early sensing of LPS in a small inoculum of pathogenic organisms, it seemed, could prevent the development of an overwhelming infection.[7,8] Furthermore, endotoxin sensing and response appeared to be largely the duty of macrophages, which delivered the lethal effect of LPS if it was administered in a large dose.[9,10] Finally, TNF was taken as a relevant endpoint of LPS responses because it was synthesized in great abundance by macrophages induced with LPS,[11,12] and could cause many of the biological effects of LPS.[13]

A transfection-based approach was used to identify one of the components of the LPS receptor, though not the product of the *Lps* gene. In 1990, it was shown that CD14 expression would confer LPS sensitivity to some cells that ordinarily lacked it,[14] and several years later, it was demonstrated that mice lacking CD14 were highly (though not completely) resistant to LPS.[15] But because CD14 lacked a cytoplasmic domain, it was not clear how the protein could transduce LPS signals across the plasma membrane.

In 1998, Poltorak, et al determined the identity of the gene that was defective in C3H/HeJ mice.[1,16] As had been widely expected, the gene encoded a co-receptor capable of functioning in conjunction with CD14 to signal the presence of LPS. The gene encoded TLR4, a protein with no known function, and one of five paralogs that had, by that time, been identified in mammals (17-21) and suggested to play a role either in development[18] or in immunity,[19] as the *Drosophila* homolog Toll was known to do.[22] In C3H/HeJ mice, a point mutation (P712H) altered the cytoplasmic domain of the single-spanning TLR molecule, rendering it entirely inactive, although it is still expressed at the cell surface. In all, a total of ten TLR-encoding genes were soon identified in the human genome,[23-26] whereas eleven TLR proteins are synthesized by mice (see Figs. 1 and 2).

The modern mammalian TLRs are derived from ancient precursors, and are distinguished by their content of leucine-rich repeats (LRRs) in the ectodomain and TIR (Toll/IL-1 receptor) motifs in the cytoplasmic domain. The TIR domain is a particularly ancient motif, present even in bacteria. The TIR domains are the most conserved portion of the TLR molecules, and permit the construction of rather robust phylogenetic trees, whereby it may be seen that one of the *Drosophila* Tolls is closely related to the mammalian TLRs, where the others can be placed in a separate group. Moreover, the IL-1R/IL-18R group of receptors have TIR domains, but have substituted immunoglobulin-type folds for LRRs in the structure of their ectodomains. The transducer family members also seem to form a discrete group, contaminated, however, with one of the Ig-ectodomain family members (Fig. 2). In mammals, the TLRs have been adapted entirely for innate immune signaling, so far as is known. It is probable that this was the ancestral function for which the TIR evolved in eukaryotic systems, and therefore, that the developmental role subserved by the *Drosophila* Tolls is something of an aberration.

The Paralogous Status of the LPS Receptor Suggests that Each of the TLRs May Serve As a Discrete Microbial Sensor

A largely erroneous concept has beset the TLR field almost from its beginning. "Pattern recognition receptors" (PRRs) were declared to be those proteins that engage "pathogen-associated molecular patterns" (PAMPs) of microbial origin and activate the host innate immune response.[27] The terms never added anything to what was known of microbial sensing; rather, they encouraged a misunderstanding of how microbes are sensed. First, the notion that molecular "patterns" are required to activate cells is incorrect. Rather, at least for the most part, it is *molecules* that activate cells, and by the time "patterns" were suggested to be of key importance, the precise structure of the key molecules had been well established. Second, the signaling receptors may or may not have direct contact with the microbes or their components, and one must then extend the definition of "receptors" as such. It is possible, for

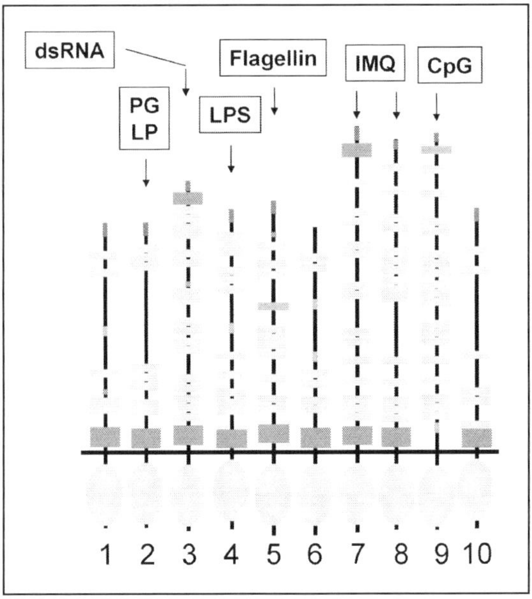

Figure 1. Schematic illustration of the ten human TLRs and their ligands, where known. Light gray rectangles indicate leucine-rich repeats (LRR) and ovals indicate TIR motifs. PG= peptidoglycan; LP= lipopeptide; dsRNA= double-stranded RNA; LPS= lipopolysaccharide; IMQ= imiquimod (similar to Resiquimod or R-848); CpG= unmethylated DNA bearing CpG dinucleotides repeats. TLR2 is known to form a heteromeric complex with either TLR1 or TLR6. The TLR1/2 complex senses tri-acylated lipopeptides; the TLR2/6 complex senses di-acylated lipopeptides, zymosan (a complex mixture of molecules), and probably several other discrete molecules of microbial origin. Note that mice do not have a TLR10 gene; however, they have and express two additional TLR genes (TLRs 11 and 12, not shown here).

example, that some host defenses are activated by proteolytic activities that originate in microbes, rather than as a result of any high-affinity interaction. Finally, not only pathogens, but microbes in general are recognized by the host.

In *Drosophila*, it had been shown that Toll was required for development,[28] and also for immune responses to fungi[22] and Gram-positive bacteria.[29] The latter were sensed through an elaborate apparatus consisting of microbial binding proteins, encoded by the *Semelweiss* gene[30] and the *Hades* gene (D. Ferrandon, personal communication) linked to proteolytic enzymes that generated a protein ligand, Spaetzle, which binds to Toll and activates it for signal transduction in time of infection. A proteolytic inhibitor of the serpin family, encoded by *necrotic*, blocks signaling from fungal activators, whatever they may be;[31] a serine protease encoded by *Persephone* is the target of this inhibitor.[32]

In mammals, the situation is quite different. LPS seems to have direct contact with TLR4, as judged by genetic complementation studies carried out independently by two groups,[33,34] and also, perhaps, with a small exteriorized protein known as MD-2, which binds to the TLR4 ectodomain.[35] CD14 appears to be involved in physical interaction with the receptor complex as well,[36] and as already noted, concentrates the LPS signal.[14,15] Similar studies have implicated DNA bearing unmethylated CpG motifs as a ligand for TLR9[37] and it has also been shown that a specific part of the flagellin protein enters into direct contact with TLR5.[38]

The tertiary and quaternary structure of the TLRs remains unsolved, although vigorous efforts to crystallize the proteins are underway in a number of laboratories. The cytoplasmic domain of TLRs 1 and 2 have been crystallized, and it is known that the equivalent mutation

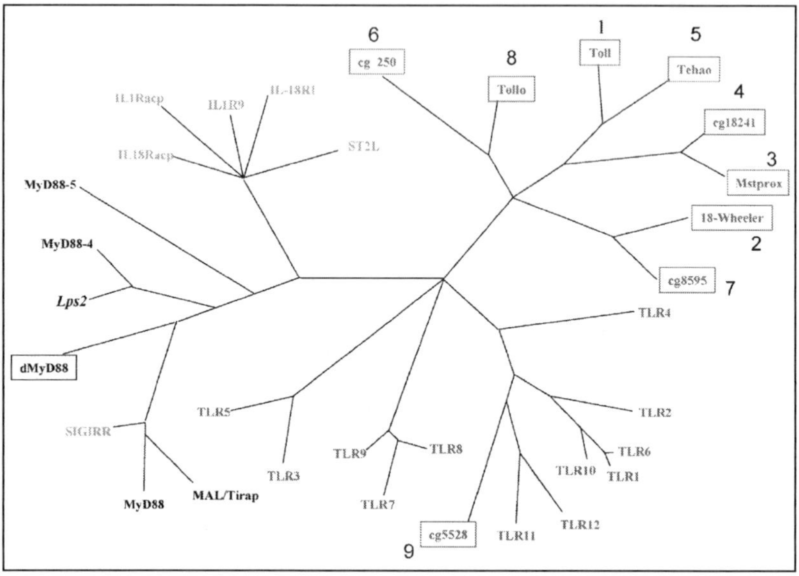

Figure 2. Unrooted tree showing TIR divergence in different proteins of *Drosophila* and human origin. *Drosophila* proteins are boxed, with their common names or Celera designations, together with more recently assigned numbers. Note that one member of the *Drosophila* family (Toll-9) falls among the mammalian TLRs, and that all TIR adapters, together with one member of the IL-1R/IL-18R clade (SIGIRR) form a separate clade of their own. Of central importance in this report, *Lps2* is also called Trif or Ticam-1; MyD88-4 is also called TIRP. Branch lengths are proportional to time. Generated based on TIR domain sequences only, using the technique of maximum parsimony.

to that representing the *Lpsd* allele of C3H/HeJ mice does not disrupt the structure of the protein, though it completely ablates signaling from the receptor. Because TLR4 seems to be constitutively active in an artificially enforced dimeric state,[19] and because Toll hypermorphs have been generated by altering membrane-proximal cysteine residues within the ectodomain,[39] it is likely that these two proteins are dimeric. The IL-1 and IL-18 receptors are known to be heterodimers, and TLR2 is believed to form heterodimers with either TLRs 1 or 6.[40] However, it may not be assumed that dimerization is brought about by ligand binding, and it would be reasonable to posit that the dimers are pre-formed, and that a conformational change is evoked by ligand binding. Du et al[41] showed that while overexpression of the ectodomain does not block signaling from the native TLR4 protein, overexpression of the cytoplasmic domain (in a membrane-tethered form) would do so. This may suggest that much of the binding affinity for maintenance of receptor quaternary structure is supplied by the cytoplasmic domain.

The Details of Signaling and the Role of Adapter Proteins

At least five TIR domain adapter proteins have been identified in the mammalian genome (Fig. 3). These proteins vary in size, and may display accessory motifs, such as SAM (Sterile Alpha Motif) in MyD88-5, or death domains in MyD88. These motifs likely foster interactions with distal components of the signaling cascade, and in the case of MyD88, are known to do so.

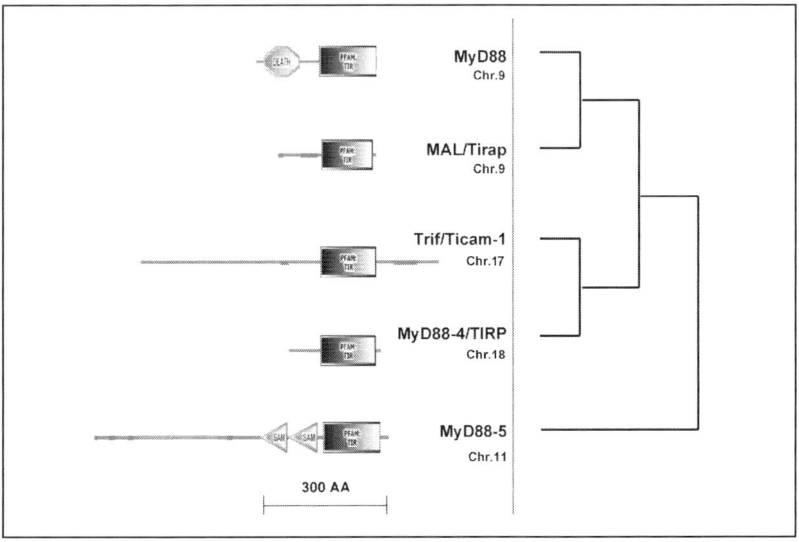

Figure 3. The TIR adapter family. A total of five TIR domain adapters can be identified by Hmmersearch of a translated EST database. These proteins have rather little similarity outside the confines of the TIR domain itself (gray rectangle). SAM= Sterile Alpha Motif, a common five-helix protein interaction motif, is found in MyD88-5 (depicted as a triangle). DEATH, a death domain, is found in MyD88 (octagon). Tree at right shows evolutionary relationship between the TIRs, but is not drawn to scale. Chromosome numbers refer to the mouse.

MyD88 was the first TIR motif protein to be identified.[42] Its role in signaling from the IL-1 and IL-18 receptors,[43] as well as several of the TLRs,[44-47] was established by gene knock-out. It became clear, however, that where TLR4 signaling was concerned, MyD88 was not the sole factor involved. Rather, the existence of an accessory "MyD88-independent" pathway was established.[44] Moreover, MyD88 is not required for signaling from the poly I:C receptor, TLR3 (Hoebe et al, in press).

The closest homolog of MyD88 is a second adapter, termed Mal[48] or Tirap.[49] While it was once proposed that Tirap served as the basis of MyD88-independent signaling from the LPS receptor,[48] it is now clear that this is not the case.[50,51] Rather, TLR activation causes the recruitment of both Mal (Tirap) and MyD88, and both serve identical roles in signaling, so far as is known. In other words, both contribute in an essential manner to the MyD88-dependent pathway.

On the basis of the pathway established in *Drosophila*, and as the result of additional knockout work in mammals, it became clear that LPS signals by sequentially causing the recruitment of MyD88[44] and Tirap[50,51] (though in an uncertain order); then IRAK4;[52] then Traf6;[53] then Tak1.[54] At this point, the signalosome (the complex of Nemo and Ikkα and Ikkβ) becomes activated, leading to degradation of IκB, and nuclear translocation of NF–κB. This results in the transcriptional activation of genes encoding such mediators as TNF, IL-1, IL-6, and other inflammatory cytokines that create the response to LPS. Also occurring, and perhaps contributing to the activation of cytokine gene expression, is the phosphorylation of proteins of the MAP kinase family and also PI3 kinase: very well known effects of macrophage activation by LPS. The kinase that links these enzymes to the core pathway remains to be established.

The MyD88-Dependent and Independent Signaling Pathways

As already mentioned, a MyD88-independent pathway of cell activation was defined by gene targeting studies.[44] In the absence of MyD88, LPS is still capable of evoking a response, albeit one in which phosphorylation of NF-κB and the MAP kinases was slightly delayed. Moreover, some endpoints of LPS signaling, such as the production of interferon-β, and downstream, the generation of chemokines such as GARG16 and IP10, were entirely unimpeded.[55,56] It was therefore suspected that an additional adapter molecule must be required for these events. While homology searches suggested three candidates, definitive identification of the relevant molecule came from a forward genetic approach.

The Forward Genetic Approach and the Identification of Lps2, Proximal Mediator of MyD88-Independent Signaling

Forward genetic analysis is that branch of genetics that begins with phenotype, and seeks to explain phenotype on the basis of mutation. Reverse genetic analysis begins with a sequence, and seeks to explain sequence function by such tools as gene knockout, or overexpression. The identification of the LPS receptor TLR4 was a purely forward genetic endeavor, and much opportunity for further discovery through forward genetic analysis exists, but must be facilitated by the creation of new monogenic phenotypes.

In 2002, a novel codominant germline mutation called *Lps2* (so named because it represented a partial phenocopy of the classical *Lps* locus) was created using the mutagen N-ethyl-N-nitrosourea (ENU).[57] This mutation was identified in a phenotypic screen in which mouse macrophages were harvested from germline mutant animals under anaesthesia and stimulated with TLR agonists. The *Lps2* homozygous founder was observed to have lost LPS-induced macrophage cytotoxicity, and drastically diminished TNF production in response to LPS and to poly I:C. Interferon-β production was abolished by the mutation almost entirely.

Signaling via other TLRs was unimpaired. Phenotypically consistent with a lesion of the MyD88-independent adapter predicted by Akira and his coworkers[44,55,56] the *Lps2* mutation proved early on that a shared adapter for TLR3 and TLR4 must exist.

The mutation was mapped on 1567 meioses to a 216 kb region of chromosome 17, and found to specify a distal frameshift error within the coding region of an adapter protein previously named Trif[58] or Ticam-1.[59] As it was the first germline mutation of this gene, *Lps2* was immediately informative with regard to the function of the protein, and its role in innate immune signaling, which to a large extent, contradicted the predictions made about the protein on the basis of the in vitro work just cited.

Lps2 is absolutely required for MyD88-independent signal transduction. It not only mediates a large part of the toxicity of LPS, but also, is required for type I interferon production in the course of an authentic viral infection. Therefore, if mice homozygous for the mutation are infected with mCMV, they fare poorly, usually succumbing to overwhelming growth of the virus in the liver and spleen, where titres are routinely found to be 1000-fold higher than those observed in normal mice. The effect is a broad one, and it may, as of now, be provisionally stated that many if not all viral infections are sensed through stimulation of the *Lps2* pathway.

Lps2 acts as a bridge between the TLR3 and TLR4 receptors and IRF-3, a transcription factor that is activated by either LPS or poly I:C, and acts to trigger interferon-β gene expression. It is, therefore, an adapter that responds not only to bacterial stimuli, but also to viral stimuli (in the latter case, probably to viral dsRNA, although a rigorous proof of this has not been ascertained). The similarity of clinical effects witnessed in viral and bacterial illnesses starts, in at least some cases, with this very protein.

The Nature of Signaling from the LPS Receptor: Two and Only Two Primary Rami

Mice homozygous for the *Lps2* mutation described show diminished TNF production in response to LPS (decreased by about 90%), and no TNF production in response to poly I:C. Mice lacking MyD88 show normal production of TNF in response to poly I:C, as well as normal interferon-β production, but very much diminished production of TNF in response to LPS (again, decreased by about 90%). These findings indicate that MyD88 serves TLR4 signaling but not TLR3 signaling. Lps2 serves both receptors.

As mentioned above, MyD88 and Mal/Tirap work in conjunction with one another to carry a part of the LPS signal from TLR4, and targeted deletion of either gene creates a phenocopy of the other deletion, so far as LPS signaling is concerned. Lps2 signaling would seem to be different, but to determine whether all LPS signaling is dependent upon two "branches" that emanate from the LPS receptor, it was necessary to create a double knockout mutation.

This was duly accomplished, and the LPS receptor was observed to be silent. Hence, a bifid transduction pathway would seem to exist. Since all signaling is silenced by the point mutation of TLR4 observed in C3H/HeJ mice, it may be imagined that both pathways arise not only from the same receptor, but from the same part of the receptor, and in all likelihood, the adapters serving each ramus of the LPS signaling pathway compete with one another for interaction with the receptor.

The Existence of Lps2-Dependent and Lps2-Independent Cell Populations

An added thread of complexity emerged when TNF production was examined by FACS analysis in cells derived from *Lps2* mutant and normal mice. Macrophages from Lps2 mutant mice show a biphasic distribution of TNF production, whereas wild-type cells show a monophasic distribution. A monophasic distribution of TNF production is also observed in MyD88-deficient cells stimulated with LPS, suggesting that all cells express MyD88 (although to be sure, they make far less TNF than wild-type cells).

Given these observations, it would seem most likely that Lps2-dependent and Lps2-independent macrophage populations exist. A separate adapter ("adapter X") is capable of substituting for Lps2 in some of the cells that lack it. This adapter is not likely to *be* MyD88, since MyD88 is evidently expressed in all cells. However, this adapter is dependent upon MyD88, since double deficiency (*Lps2* and MyD88 mutations) completely ablates LPS receptor signaling (Fig. 4).

While the identity of adapter X is not yet known, MyD88-4, the closest homolog of Lps2, would seem an excellent candidate. It cannot be excluded that this adapter might have many other functions as well. The functions of MyD88-5, the most distant adapter of the group, can only be guessed at. However, its SAM is intriguing in view of the fact that TLR12 also has a SAM, and by inference, might engage this adapter.

Two Populations of Macrophages Distinguished on the Basis of Responses to Poly I:C

Further evidence for afferent functional specialization is seen in the fact that only about half of peritoneal macrophages respond to poly I:C by producing TNF. It is not clear whether the bimodality of the response to poly I:C in normal cells is in any way tied to the bimodality of response to LPS in Lps2 mutant cells, and in all likelihood, it is not. Rather one might make

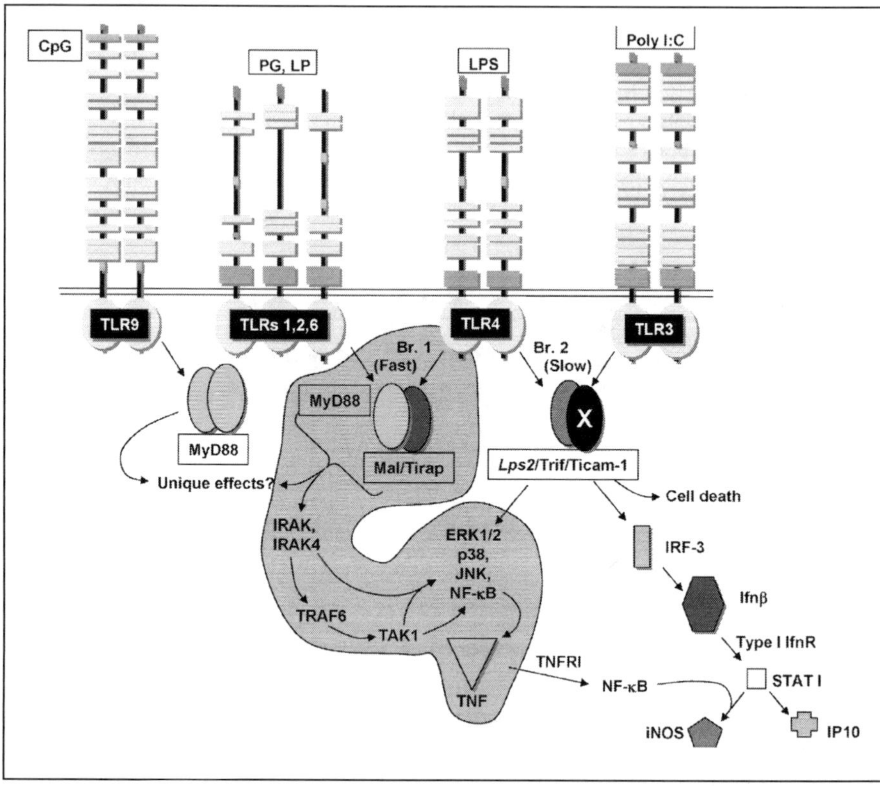

Figure 4. Current concept of TLR signal transduction leading to cytokine synthesis. TLRs are believed to be homodimers or heterodimers, and to initiate signaling following contact with microbial inducers (top). Ovals represent TIR adapters indicated; MyD88 (light gray), Mal/Tirap (dark gray) and Trif (dark gray). Two and only two signaling branches serve LPS signaling. One branch (which processes the signal quickly) depends on MyD88 and Mal/Tirap; the other depends either on Trif or on adapter X (black oval), which are expressed by different macrophage populations. The Trif pathway, which also serves TLR3, is quickly reunited with the Mal/Tirap pathway. Both pathways activate MAPK phosphorylation, IkB degradation, and cause TNF production. The Trif pathway alone, however, causes macrophage apoptosis, and is responsible for IRF-3 phosphorylation and all that follows this event. TLR2 signaling is MyD88 and Mal/Tirap dependent, but not Trif dependent. TLR9 (and also TLR7; not shown here) signals via MyD88, but neither Mal/Tirap nor Trif. It should be emphasized that other pathways, not yet clarified, are traversed to permit DC maturation.

the case that four types of macrophage exist. The basis of unresponsiveness to poly I:C might lie either at the level of the receptor, TLR3, or at a post-receptor level.

If similar differences in macrophage responses to other inducing stimuli can be observed, it might be offered that some cells are specialized for responses to one type of pathogen rather than another: an innate immune analog of clonality in the adaptive immune system. However, the degree of specialization remains to be established, and it is not at all clear that it approaches that of adaptive immune cells, which are exquisitely specific.

Limits of the Forward Genetic Approach and What May Be Expected of Innate Immune Signaling in the Future

Forward genetic analysis has the virtue of disclosing truly unexpected participants in the phenomenon that interests one most. Its value declines as more and more of the components of a system are identified. But "one only knows what one knows," and until true saturation of the genome is achieved, the participation of proteins yet unknown may still be sought. Genomic saturation is not readily achieved, however, and more than one million mice might be required to create hypomorphic alleles of all genes that are deemed important. Some genes might escape notice entirely, since it is not certain that ENU mutagenesis works a homogeneous effect. Moreover, some loci might prove so non-permissive for mutation, because of lethality, that it is difficult to create alleles that are viable, yet register in the screen that is applied.

Forward genetic screens are best applied in a very broad way: that is, to the analysis of a collection of pathways, or to the understanding of phenomena such as in vivo infections, containment of which may depend upon a large number of genes. And obviously, germline mutagenesis is, as a method, ideally suited to problems that defy in vitro analysis.

Because one need not be blind to the hints that are offered by the results of a forward genetic screen, it is wise to couple gene targeting-the premiere reverse genetic tool-with germline mutagenesis. In this manner, multiple paralogs of a family can be examined in a directed fashion, without waiting for mutations to be produced by a random process. Indeed, this has been the trend, and it is reasonable to think that many of the details of innate immune signaling will ultimately yield to these methods.

At present, we distinguish between afferent and efferent (or effector) systems in innate immunity. TLRs are clearly afferent components; NADPH oxidase is clearly an effector. When all of the genes that serve the innate immune response have been identified, the distinction may be seen as blurred. What, after all, are such intermediates as TNF? Do cytokines spread the word that infection has occurred? Or do they deal with the infection as effectors? It may also be seen that many proteins that are not "professional" components of the innate immune system discharge an innate immune function. Some examples may be seen already, in that mutant alleles of β-globin and G-6-PD can be very effective in combating malaria parasites.

What is the ultimate goal of understanding innate immunity? Practical approaches to both infection and autoimmunity might ultimately benefit. The innate immune system not only fights infection quite well by itself, but also sets the stage for an adaptive immune response. Adaptive immunity evolved atop innate immunity, developing in a world in which innate mechanisms were already refined and effective. We may look forward to a more precise understanding of how and why adaptive mechanisms succeed, and why they sometimes err.

Acknowledgement

This work was supported by funding from NIH grants GM067759, GM60031, U54A154523, 2P01-AI40682, and by a grant from the DARPA.

References

1. Poltorak A, He X, Smirnova I et al. Defective LPS signaling in C3H/HeJ and C57BL/10ScCr mice: mutations in Tlr4 gene. Science 1998; 282:2085-2088.
2. Beutler B, Rietschel ET. Timeline: Innate immune sensing and its roots: the story of endotoxin. Nat Rev Immunol 2003; 3(2):169-176.
3. Pfeiffer R. Untersuchungen über das Choleragift. Z Hygiene 1892; 11:393-412.
4. Heppner G, Weiss DW. High susceptibility of strain A mice to endotoxin and endotoxin red blood cell mixtures. J Bacteriol 1965; 90:696-703.
5. Coutinho A, Meo T. Genetic basis for unresponsiveness to lipopolysaccharide in C57BL/10Cr mice. Immunogenetics 1978; 7:17-24.

6. Watson J, Kelly K, Largen M et al. The genetic mapping of a defective LPS response gene in C3H/HeJ mice. J Immunol 1978; 120:422-424.
7. O'Brien AD, Rosenstreich DL, Scher I et al. Genetic control of susceptibility to Salmonella typhimurium in mice: role of the LPS gene. J Immunol 1980; 124:20-24.
8. Rosenstreich DL, Weinblatt AC, O'Brien AD. Genetic control of resistance to infection in mice. CRC Crit Rev Immunol 1982; 3:263-330.
9. Moore RN, Goodrum KJ, Berry LJ. Mediation of an endotoxic effect by macrophages. J Reticuloendothel Soc 1976; 19(3):187-197.
10. Michalek SM, Moore RN, McGhee JR et al. The primary role of lymphoreticular cells in the mediation of host responses to bacterial endotoxin. J Infec Dis 1980; 141:55-63.
11. Beutler B, Mahoney J, Le Trang N et al. Purification of cachectin, a lipoprotein lipase suppressing hormone secreted by endotoxin-induced RAW 264.7 cells. J Exp Med 1985; 161:984-995.
12. Beutler B, Greenwald D, Hulmes JD et al. Identity of tumour necrosis factor and the macrophage-secreted factor cachectin. Nature 1985; 316:552-554.
13. Beutler B, Milsark IW, Cerami A. Passive immunization against cachectin/tumor necrosis factor (TNF) protects mice from the lethal effect of endotoxin. Science 1985; 229:869-871.
14. Wright SD, Ramos RA, Tobias PS et al. CD14, a receptor for complexes of lipopolysaccharide (LPS) and LPS binding protein. Science 1990; 249:1431-1433.
15. Haziot A, Ferrero E, Kontgen F et al. Resistance to endotoxin shock and reduced dissemination of gram-negative bacteria in CD14-deficient mice. Immunity 1996; 4(4):407-414.
16. Poltorak A, Smirnova I, He XL et al. Genetic and physical mapping of the Lps locus- identification of the toll-4 receptor as a candidate gene in the critical region. Blood Cells Molecules & Diseases 1998; 24(17):340-355.
17. Nomura N, Miyajima N, Sazuka T et al. Prediction of the coding sequences of unidentified human genes. I. The coding sequences of 40 new genes (KIAA0001-KIAA0040) deduced by analysis of randomly sampled cDNA clones from human immature myeloid cell line KG-1. DNA Res 1994; 1(1):27-35.
18. Taguchi T, Mitcham JL, Dower SK et al. Chromosomal localization of TIL, a gene encoding a protein related to the Drosophila transmembrane receptor Toll, to human chromosome 4p14. Genomics 1996; 32(3):486-488.
19. Medzhitov R, Preston-Hurlburt P, Janeway CA, Jr. A human homologue of the Drosophila Toll protein signals activation of adaptive immunity. Nature 1997; 388(6640):394-397.
20. Chaudhary PM, Ferguson C, Nguyen V et al. Cloning and characterization of two Toll/Interleukin-1 receptor-like genes TIL3 and TIL4: evidence for a multi-gene receptor family in humans. Blood 1998; 91(11):4020-4027.
21. Rock FL, Hardiman G, Timans JC et al. A family of human receptors structurally related to Drosophila Toll. Proc Natl Acad Sci USA 1998; 95(2):588-593.
22. Lemaitre B, Nicolas E, Michaut L et al. The dorsoventral regulatory gene cassette spatzle/Toll/cactus controls the potent antifungal response in Drosophila adults. Cell 1996; 86(6):973-983.
23. Takeuchi O, Kawai T, Sanjo H et al. TLR6: A novel member of an expanding Toll-like receptor family. Gene 1999; 231(1-2):59-65.
24. Du X, Poltorak A, Wei Y et al. Three novel mammalian toll-like receptors: gene structure, expression, and evolution. Eur Cytokine Netw 2000; 11(3):362-371.
25. Chuang TH, Ulevitch RJ. Cloning and characterization of a sub-family of human toll-like receptors: hTLR7, hTLR8 and hTLR9. Eur Cytokine Netw 2000; 11(3):372-378.
26. Chuang T, Ulevitch RJ. Identification of hTLR10: a novel human Toll-like receptor preferentially expressed in immune cells. Biochim Biophys Acta 2001; 1518(1-2):157-161.
27. Janeway CA, Jr. The immune system evolved to discriminate infectious nonself from noninfectious self. Immunology Today 1992; 13(1):11-16.
28. Anderson KV, Bokla L, Nusslein Volhard C. Establishment of dorsal-ventral polarity in the Drosophila embryo: the induction of polarity by the Toll gene product. Cell 1985; 42(3):791-798.
29. Rutschmann S, Kilinc A, Ferrandon D. Cutting edge: the toll pathway is required for resistance to gram- positive bacterial infections in Drosophila. J Immunol 2002; 168(4):1542-1546.

30. Gottar M, Gobert V, Michel T et al. The Drosophila immune response against Gram negative bacteria is mediated by a peptidoglycan recognition protein. Nature 2002; 416(640):644.
31. Levashina EA, Langley E, Green C et al. Constitutive activation of toll-mediated antifungal defense in serpin- deficient Drosophila. Science 1999; 285(5435):1917-1919.
32. Ligoxygakis P, Pelte N, Hoffmann JA et al. Activation of Drosophila Toll during fungal infection by a blood serine protease. Science 2002; 297(5578):114-116.
33. Poltorak A, Ricciardi-Castagnoli P, Citterio A et al. Physical contact between LPS and Tlr4 revealed by genetic complementation. Proc Natl Acad Sci ,USA 2000; 97(5):2163-2167.
34. Lien E, Means TK, Heine H et al. Toll-like receptor 4 imparts ligand-specific recognition of bacterial lipopolysaccharide. J Clin Invest 2000; 105(4):497-504.
35. Akashi S, Nagai Y, Ogata H et al. Human MD-2 confers on mouse Toll-like receptor 4 species-specific lipopolysaccharide recognition. Int Immunol 2001; 13(12):1595-1599.
36. da Silva CJ, Soldau K, Christen U et al. Lipopolysaccharide is in Close Proximity to Each of the Proteins in Its Membrane Receptor Complex: Transfer from CD14 to TLR4 and MD-2. J Biol Chem 2001; 276(24):21129-21135.
37. Bauer S, Kirschning CJ, Hacker H et al. Human TLR9 confers responsiveness to bacterial DNA via species-specific CpG motif recognition. Proc Natl Acad Sci USA 2001; 98(16):9237-9242.
38. Mizel SB, West AP, Hantgan RR. Identification of a sequence in human toll-like receptor 5 required for the binding of Gram-negative flagellin. J Biol Chem 2003; 278: 23624 -23629.
39. Schneider DS, Hudson KL, Lin TY et al. Dominant and recessive mutations define functional domains of Toll, a transmembrane protein required for dorsal-ventral polarity in the Drosophila embryo. Genes Dev 1991; 5(5):797-807.
40. Ozinsky A, Underhill DM, Fontenot JD et al. The repertoire for pattern recognition of pathogens by the innate immune system is defined by cooperation between toll-like receptors. Proc Natl Acad Sci USA 2000; 97(25):13766-13771.
41. Du X, Poltorak A, Silva M et al. Analysis of Tlr4-mediated LPS signal transduction in macrophages by mutational modification of the receptor. Blood Cells Molecules & Diseases 1999; 25(21):328-338.
42. Hultmark D. Macrophage differentiation marker MyD88 is a member of the Toll/IL-1 receptor family. Biochem Biophys Res Commun 1994; 199(1):144-146.
43. Adachi O, Kawai T, Takeda K et al. Targeted disruption of the MyD88 gene results in loss of IL-1- and IL-18-mediated function. Immunity 1998; 9(1):143-150.
44. Kawai T, Adachi O, Ogawa T, et al. Unresponsiveness of MyD88-deficient mice to endotoxin. Immunity 1998; 11:115-122.
45. Takeuchi O, Hoshino K, Kawai T et al. Differential roles of TLR2 and TLR4 in recognition of Gram-negative and Gram-positive bacterial cell wall components. Immunity 1999; 11:443-451.
46. Hacker H, Vabulas RM, Takeuchi O et al. Immune Cell Activation by Bacterial CpG-DNA through Myeloid Differentiation Marker 88 and Tumor Necrosis Factor Receptor- Associated Factor (TRAF)6. J Exp Med 2000; 192(4):595-600.
47. Akira S. Toll-like receptors: Lessons from knockout mice. Biochem Soc Transact 2000; 28(5):551-556.
48. Fitzgerald KA, Palsson-McDermott EM, Bowie AG et al. Mal (MyD88-adapter-like) is required for Toll-like receptor-4 signal transduction. Nature 2001; 413(6851):78-83.
49. Horng T, Barton GM, Medzhitov R. TIRAP: an adapter molecule in the Toll signaling pathway. Nat Immunol 2001; 2(9):835-841.
50. Horng T, Barton GM, Flavell RA et al. The adaptor molecule TIRAP provides signalling specificity for Toll-like receptors. Nature 2002; 420(6913):329-333.
51. Yamamoto M, Sato S, Hemmi H et al. Essential role for TIRAP in activation of the signalling cascade shared by TLR2 and TLR4. Nature 2002; 420(6913):324-329.
52. Suzuki N, Suzuki S, Duncan GS et al. Severe impairment of interleukin-1 and Toll-like receptor signalling in mice lacking IRAK-4. Nature 2002; 416(6882):750-756.
53. Lomaga MA, Yeh WC, Sarosi I et al. TRAF6 deficiency results in osteopetrosis and defective interleukin-1, CD40, and LPS signaling. Genes Dev 1999; 13(8):1015-1024.

54. Lee J, Mira-Arbibe L, Ulevitch RJ. TAK1 regulates multiple protein kinase cascades activated by bacterial lipopolysaccharide. J Leukoc Biol 2000; 68(6):909-915.
55. Kawai T, Takeuchi O, Fujita T et al. Lipopolysaccharide stimulates the MyD88- independent pathway and results in activation of IFN-regulatory factor 3 and the expression of a subset of lipopolysaccharide-inducible genes. J Immunol 2001; 167(10):5887-5894.
56. Hoshino K, Kaisho T, Iwabe T et al. Differential involvement of IFN-beta in Toll-like receptor-stimulated dendritic cell activation. Int Immunol 2002; 14(10):1225-1231.
57. Hoebe K, Du X, Goode J et al. Lps2: A new locus required for responses to lipopolysaccharide, revealed by germline mutagenesis and phenotypic screening. J Endotoxin Res 2003; 9(4):250-255.
58. Yamamoto M, Sato S, Mori K et al. Cutting edge: a novel Toll/IL-1 receptor domain containing adapter that preferentially activates the IFN-beta promoter in the Toll-like receptor signaling. J Immunol 2002; 169(12):6668-6672.
59. Oshiumi H, Matsumoto M, Funami K et al. TICAM-1, an adaptor molecule that participates in Toll-like receptor 3-mediated interferon-beta induction. Nat Immunol 2003; 4:161-171.

CHAPTER 9

Agonists of Toll-Like Receptor 9:

Modulation of Host Immune Responses with Synthetic Oligodeoxynucleotides

Ekambar R. Kandimalla and Sudhir Agrawal

Innate immunity is the body's first line of defense against invading microbes. This component of our immune system relies on highly conserved pattern-recognition receptors (PRRs) to distinguish different pathogens.[1] Toll-like receptors (TLRs) are one class of PRR[1,2] of which ten (TLR1-10) have now been identified in mammals.[1-3] TLRs recognize pathogen-associated molecular patterns (PAMPs) and initiate the appropriate immune responses. These entail the activation of signal cascades leading to the secretion of cytokines, the activation of cell-surface molecules and the production of pathogen-specific immunoglobulins (Ig).[4] As TLRs are constitutively expressed on immune cells and are responsive to synthetic ligands, they provide us with a rational way with which to modulate the immune system; a strategy which can be seen as quite distinct from conventional vaccination.

At least three TLRs, TLR3, 7, and 9 recognize and respond to nucleosides, nucleotides, and oligo- and polynucleotides of natural and/or synthetic origin (Fig. 1). TLRs 3 and 9 recognize nucleic acid molecular patterns that are present in microbes but not mammals.[5-8] TLR3 recognizes viral RNA, synthetic polyI.polyC, and synthetic double-stranded (ds) RNA.[9] Short interfering (si) RNAs can also activate the interferon (IFN) system.[10,11] TLR7 recognizes small synthetic nucleoside-like fused-ring heterocyclic molecules[7] whilst TLR9 is specific for d(CpG) dinucleotides in specific sequence contexts (CpG motifs), whether in bacterial, plasmid or synthetic DNA.[5] Bacterial and synthetic CpG DNAs have direct mitogenic effects on B cells.[12,13] These DNAs can stimulate NK-cell activity in vitro and activate macrophages, dendritic cells (DCs), and monocytes, causing them to secrete cytokines, chemokines and to express cell surface molecules.[14-16] In vivo, CpG DNAs induce splenomegaly in mice (enlargement of the spleen)[17,18] accompanied by splenic B cell proliferation, the up-regulation of class-II major histocompatibility complex (MHC class II) antigens, the increased synthesis of RNA and DNA and the elevated production of cytokines and chemokines.[18]

CpG DNA does not directly activate T and NK cells as these cells do not express TLR9. However, CpG DNA can indirectly augment the DC mediated stimulation and cytokine production of other immune cells.[19] The expression of TLR9 in immune cells is required for these activities, but a direct interaction between CpG DNA and TLR9 has not yet been demonstrated. The ability of CpG DNA to induce strong innate then acquired immune responses is a clear indication of its potential as a pharmacophore and natural adjuvant.[20-23]

Toll and Toll-Like Receptors: An Immunologic Perspective, edited by Tina Rich. ©2005 Eurekah.com and Kluwer Academic / Plenum Publishers.

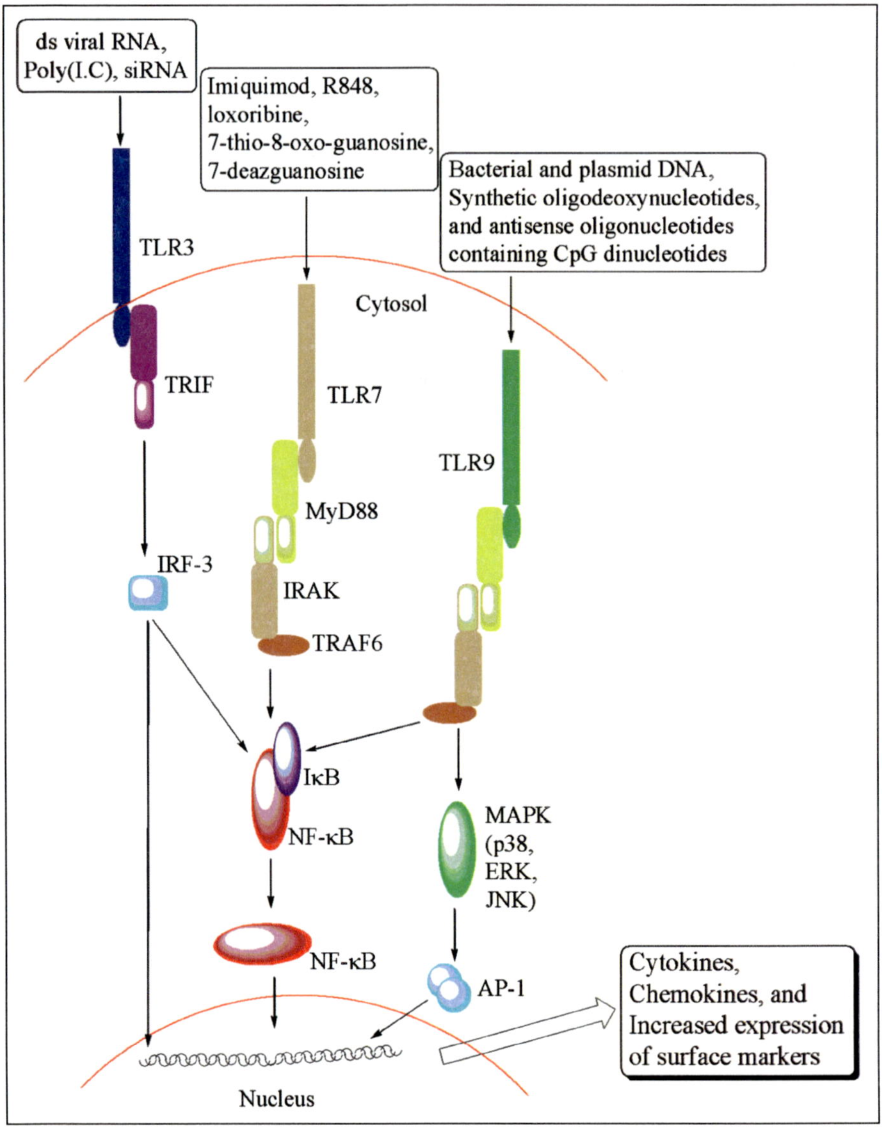

Figure 1. Three of the ten TLR family members that are known to recognize pathogen-associated nucleic acid or small nucleoside-like heterocyclic molecules. TLR 3 recognizes double-stranded viral and synthetic polyI.polyC RNAs. TLR 7 recognizes a number of synthetic nucleosides and small nucleoside-like molecules. No naturally occurring ligand for TLR 7 has yet been identified.* TLR 9 recognizes d(CpG) dinucleotides in specific sequential contexts (CpG motifs) in bacterial DNA, plasmid DNA, and synthetic oligodeoxynucleotides. TLR 9 also recognizes the synthetic YpG, CpR, YpR, and R'pG motifs discussed in the text. Key signaling components are shown. The activated transcription factors up-regulate the expression of a number of cytokines, chemokines, and costimulatory molecules.

* Since going to press murine TLR7 and human TLR8 have been shown to recognize ssRNA, both of viral and endogenous origin (Heil F et al. Science 2004; 303:1526-1529. Diebold SS et al. Science 2004; 303:1529-1531.)

TLRs are structurally and functionally similar to the IL-1 receptor. In general, TLRs contain an extracellular leucine-rich repeat and a cytoplasmic Toll/IL1-R (TIR) domain, connected by a transmembrane domain.[24] Although most TLRs are membrane receptors, evidence suggests that TLR9 is cytoplasmic and localises to lysosome-like vesicles.[25] Consequently, the cellular uptake and endosomal localization of CpG DNA is required for immune stimulation. In general, following ligand recognition and binding, TLR9 recruits MyD88 to its TIR domain, followed by the engagement of IRAK and TRAF6. Subsequent to the formation of the TLR9-adaptor protein complex, stress kinase c-Jun N-terminal kinase (JNK), p38, and IκB kinases are activated. These, in turn, activate the transcription factors AP-1 and NF-κB.[26] Several TLRs signal through MyD88-dependent and/or -independent pathways. However, the available evidence suggests that TLR9 signals exclusively by a MyD88-dependent pathway (Fig. 1). Broadly speaking, the immune responses to different TLR ligands are similar, consisting of cytokine secretion and surface-molecule expression leading to strong adaptive immune responses. However, the cytokine profiles that they induce can be quite different, as is the expression of TLRs in different immune cells and mammalian species.[27,28]

Tokunaga and colleagues were the first to show that bacterial DNA could induce NK-cell activity and type-I interferon secretion.[14,29,30] Subsequently, the same group showed that synthetic palindromic phosphodiester DNA containing CG dinucleotides could activate immune cells.[30-32] A number of other groups also showed that bacterial DNA (not mammalian) and specific synthetic phosphorothioate oligodeoxynucleotides (containing unmethylated deoxycytidine) could activate immune cells.[12,17,18,33-37] The latter studies emphasized that it was unmethylated CpG dinucleotides in specific sequence contexts (CpG motifs) that were responsible for immune activity.[13] The immunostimulatory effects of CpG DNAs are multifactorial and reflect their sequence, the nature of their DNA backbone and the presence of specific structural motifs. Based on the type of immune-cell populations that are activated and the cytokine profiles produced, synthetic CpG DNAs can be split into three classes, A/D, B/K, and C (Fig. 2). As each class of CpG DNA forms a different secondary structure, it seems likely that specificity in cell activation and the generation of distinct immune profiles arise, in part, because of the specific secondary structures involved, as well as the CpG motif itself.

Class A/D CpG DNAs comprise CpG dinucleotides within palindromic sequences that contain a phosphodiester backbone[30-32] or flanking poly(dG) sequences.[33,34] This class of CpG DNA principally activates plasmacytoid DCs (pDCs) and NK cells, leading to the secretion of type-I interferons (IFNs) and IFN-γ. The downstream effects of class A/D CpG DNAs are similar to those achieved by bacterial or plasmid DNA. Indeed this class of CpG DNAs are close mimics of the palindromic immunostimulatory sequences described by Tokunaga and colleagues,[30-32] but with poly(dG) nucleotide stretches at their termini to provide additional nuclease protection and to enhance scavenger receptor-mediated cellular uptake.[36] However, poly(dG) sequences can, by themselves, induce immune responses and bind to a number of proteins, so inhibiting their action.[37] Moreover, problems associated with poly(dG) synthesis, purification, quality control, and the in-vivo pharmacokinetics of poly(dG) sequence-containing oligonucleotides have decreased their potential as therapeutic molecules.[35,38]

Phosphorothioate oligonucleotides containing CpG dinucleotides in specific sequence contexts fall into the category denoted as class B/K. These CpGs principally activate B cells, macrophages, monocytes and DCs, inducing cytokine secretion.[13] In general, class B/K CpG DNAs do not contain secondary structure and it is the two nucleotides that flank the CpG that influence its activity.[13] The lack of secondary structure prevents them from activating plasmacytoid DCs (pDCs), the principle source of CpG DNA-induced IFN-α. Although not classified as CpG DNA, a number of first-generation antisense phosphorothioate oligonucleotides that contained CpG dinucleotides were found to induce immune responses in vitro and in vivo.[39-46]

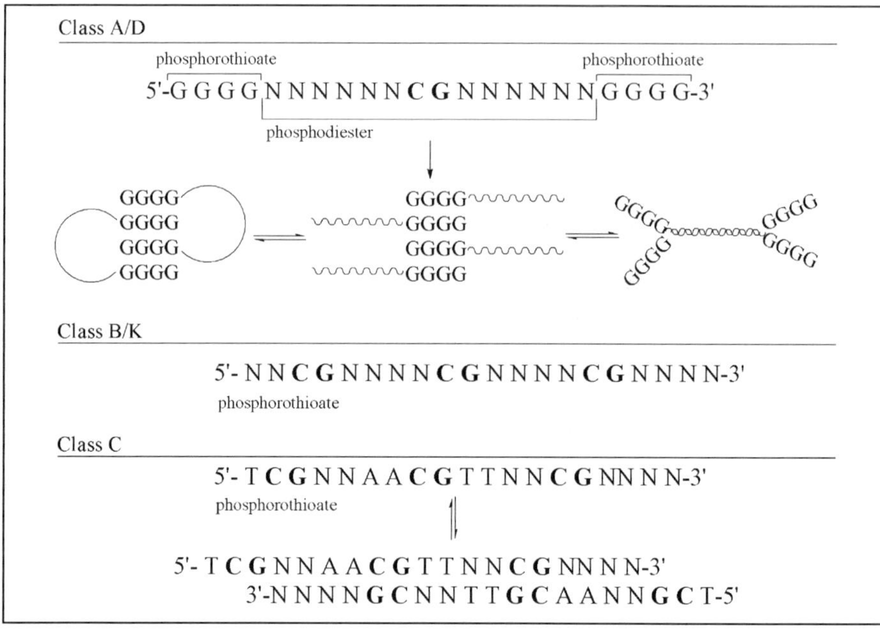

Figure 2. Different classes/types of CpG DNA and their possible secondary structures. N stands for 2'-deoxy-A, -C, -G, or -T. Class A/D CpG DNAs activate and induce the maturation of plasmacytoid dendritic cells as well as stimulating IFN-α, and TNF-α production. IFN-α activates NK cells, γδT cells, CD+ T cells and induces monocyte-derived DC maturation. Class A/D CpG DNAs are poor activators of B cells. Class B/K CpG DNAs activate B cells, induce cytokine secretion and the production of immunoglobulins, induce pDC maturation and TNF-α and IL-6 (not IFN-α) production. Class C CpG DNAs exhibit properties of both class A/D and B/K CpG DNAs.

Several first- and second-generation antisense oligonucleotides that are currently being tested, or have been subjected to clinical trial,[45,46] (Table 1) also contain CpG motifs that could be classified as class B/K CpG DNAs on the basis of their immunostimulatory activities.[47,48]

A new class of CpG DNA, class C, which mimics the effects of both class A/D and B/K CpG DNAs has recently been described.[49,50] Class C CpG DNA contains uniform phosphorothioate backbone modifications, as for class B/K CpG DNA, with both palindromic and non-palindromic segments for secondary structure formation (as for class A/D CpG DNA). Class C CpG DNAs induce IFN-α secretion and activate B cells in-vitro in human cell-based assays.

Though TLR9 appears to be the only receptor that is involved in the CpG DNA immunostimulatory pathway for all classes of CpG DNA, it is not yet clear how the different classes of CpG DNA can induce their specific immune profiles. One hypothesis would be that other receptors or co-receptors are involved.[51-53] In the following sections we describe the recent progress made in our laboratory in the design and development of second-generation immunomodulatory oligonucleotides and synthetic stimulatory motifs to modulate the TLR9-mediated immune response.

CpG DNA Structure

Though a CpG motif is requisite for TLR9 recognition and activation, the activity of CpG DNA also depends on the nucleotides adjacent to the CpG dinucleotide, the nature of the nucleotide backbone, and secondary structure. A number of other chemical modifications

Table 1. Antisense oligonucleotides that were/are in clinical trials

Target[a]	Sequence/Number
AA1R	5′GATGGAGGGCGGCATGGCGGG-3′
Bcl-2	5′-TCTCCCAGCGTGCGCCAT-3′
Bcr-abl	5′-CGCTGAAGGGCTTCTTCCTTATTGAT-3′
CMV	5′-GCGTTTGCTCTTCTTCTTGCG-3′
CMV	5′-TGGGGCTTACCTTGCGAACA-3′[b]
c-myb	5′-TATGCTGTGCCGGGGTCTTCGGGC-3′
c-myc	5′-GCTAACGTTGAGGGGGCAT-3′
DNA metase	MG 98[b,c]
HIV-I	5′-CTCTCGCACCCATCTCTCTCCTTCT-3′
HPV	5′-TTGCTTCCATCTTCCTCGTC-3′[b]
HCV	5′-GTGCTCATGGTGCACGGTCT-3′[d]
H-ras	5′-TCCGTCATCGCTCCTCAGGG-3′
ICAM-1	5′-GCCCAAGCTGGCATCCGTCA-3′
PKA	5′-GCGTGCCTCCTCACTGGC-3′[c]
PKC-α	5′-GTTCTCGCTGGTGAGTTTCA-3′
RNR	GTI 2040[c]
RNR	GTI 2501[c]
TGF-β2	AP 12009[c]
TNF-α	5′-GCTGATTAGAGAGAGGTCCC-3′[e]

[a] AA1R= Adenosine A1 receptor; CMV= cytomegalovirus; DNA metase= DNA methyltransferase; HIV-I= human immunodeficiency virus-I; HPV= human papilloma virus; HCV= Hepatitis C virus; ICAM-1= intracellular adhesion molecule-1; PKA= protein kinase A; PKC= protein kinase C; RNR= ribonucleotide reductase; TGF= transforming growth factor; TNF= tumor necrosis factor; [b] 2′-O-methyl-RNA/DNA mixed backbone; [c] Sequence not disclosed; [d] all Cs are methylated; [e] 2′-methoxyethyl-RNA/DNA mixed backbone.

can also influence activity, depending on their nature and the site of modification (Fig. 3).[54-57] Chemical modification of the CpG dinucleotide impairs activity,[58] suggesting that specific recognition of the dinucleotide by TLR9 and/or other (co)receptors is requisite for activation.

Role of Flanking Sequences

The Effect of Phosphate Charge in Flanking Sequences

The negative charge on internucleotide phosphates plays a significant role in the recognition and interaction of DNA with receptors. Investigating this further, we incorporated non-ionic methylphosphonate internucleotide linkages (Fig. 3a) at specific positions in sequences that flank a CpG dinucleotide. Charge neutralization at the fifth or sixth linkage, 5' of the CpG dinucleotide significantly enhanced immunostimulatory activity, possibly reflecting tighter receptor binding.[59] In contrast, the insertion of non-ionic internucleoside linkages 3'of the CpG dinucleotide had no significant effect,[59] whilst their incorporation in the 5'-flanking sequence generated a cytokine secretion profile that was markedly different from that achieved using unmodified CpG DNA (Fig. 4).[46]

The Effect of 2'-O-Alkyl Substitutions

Deoxyribonucleotides adopt a 2'-exo conformation whilst ribo or 2'-substituted ribonucleotides assume a 2'-endo conformation, which distinguishes the two types of natural

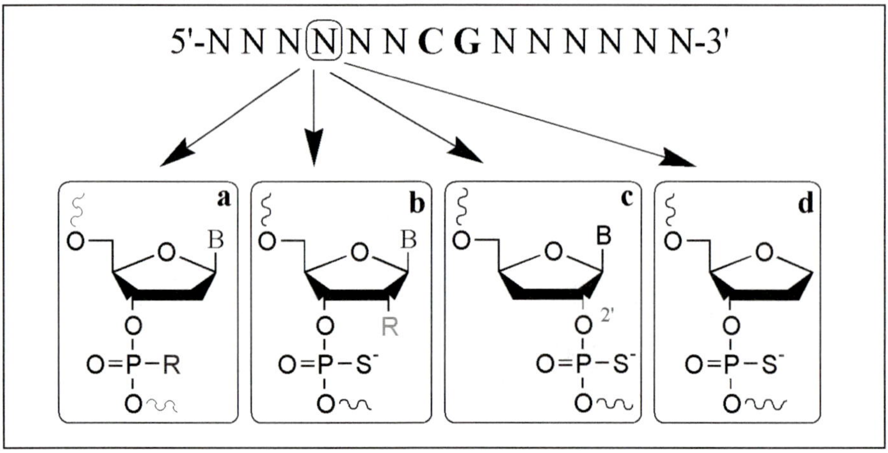

Figure 3. Some of the site-specific modifications studied for their immunostimulatory effects. a) CpG DNA structures showing site-specific modifications. Phosphate backbone modifications. R= O^- (natural phosphodiester), S^- (phosphorothioate), or CH_3 (methylphosphonate). Incorporation of a methylphosphonate linkage in CpG DNA neutralizes the anionic charge ($-S^-$) and modulates immunostimulatory activity. b) Sugar modifications at 2'-position. R = O-alkyl. Substitution of a 2'-deoxyribonucleoside with 2'-O-alkylribonucleoside in CpG DNA modulates immune responses. c) Substitution of 2'-deoxyribonucleosides with 3'-deoxyribonucleosides. The incorporation of 3'-deoxyribonucleosides results in the formation of 2'-5'-internucleotide linkages in contrast to natural 3'-5'-linkages. N stands for A, C, G, or T 2'-deoxyribonucleotide. B stands for heterocyclic nucleobase. d) Abasic or 1'-2'-dideoxyribonucleotide.

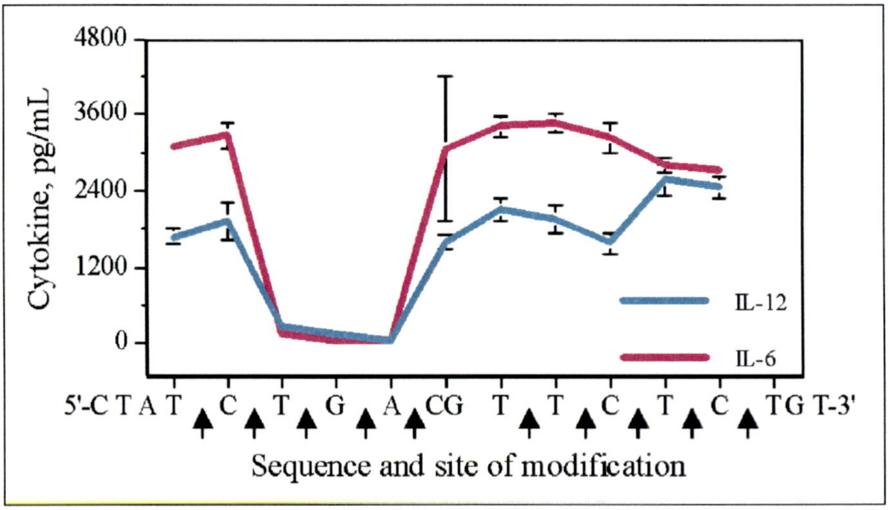

Figure 4. Immunostimulatory activity of CpG DNAs containing a methylphosphonate internucleotide linkage at the position indicated by arrow. Secretion of IL-12 and IL-6 by BALB/c mouse spleen cell cultures elicited by 1.0 μg/ml CpG DNA after a 24 hr incubation. Each value is an average of three or four replicates. CG indicates parent CpG DNA without modification.

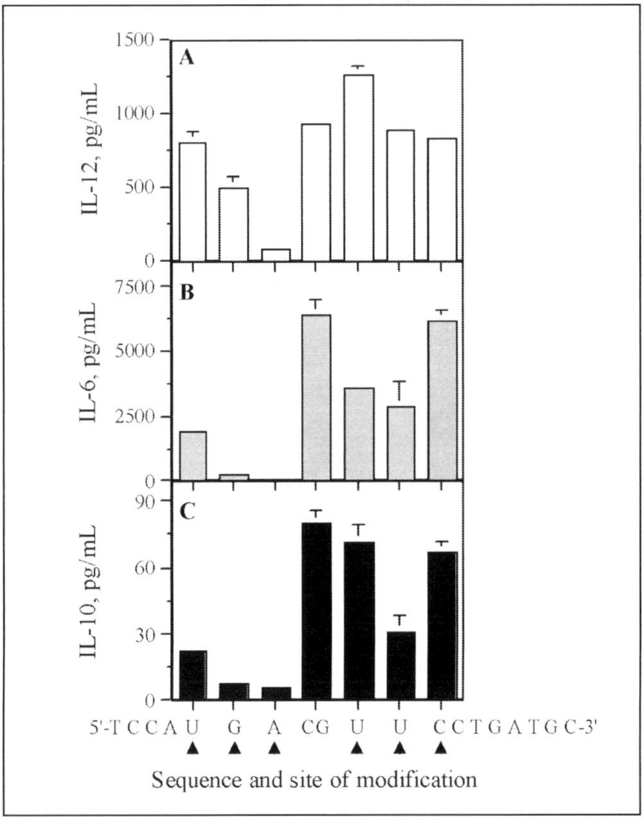

Figure 5. Immunostimulatory activity of CpG DNAs containing a single 2'-O-methylribonucleoside substitution at the position indicated with arrow. Secretion of A) IL-12, B) IL-6, and C) IL-10 in BALB/c mouse spleen cell cultures, treated with 1.0 μg/ml CpG DNA after a 24 hr incubation. Each value represents the average of three or four replicates. CG indicates parent CpG DNA without modification.

nucleotide. In general, TLR9 recognizes deoxyribonucleotides whilst TLR3 recognizes ribonucleotides. The incorporation of 2'-O-methyl or –methoxyethyl ribonucleosides (Fig. 3b) in the flanking sequences distal to the CpG dinucleotide has different effects depending on the position of the substitution.[60,61] In general, the substitution of 2'-O-alkylribonucleoside distal to the CpG dinucleotide, on either side, enhances spleen-cell proliferation and cytokine secretion in vitro, with the production of distinct cytokine secretion profiles (Fig. 5).[46]

The Effect of 2'-5' Internucleoside Linkages

The incorporation of unnatural 3'-deoxynucleosides results in the formation of 2'-5'-internucleotide linkages in an otherwise 3'-5'-linked DNA (Fig. 3c). The presence of 3'-deoxynucleosides distal to the CpG dinucleotide in the 5'-flanking sequence enhances activity; levels of secreted IL-6 and IL-10 are elevated whilst IL-12 production is comparable to that achieved with the parental CpG DNA. The incorporation of the same modification, distal to the CpG, in the 3'-flanking sequence results in the secretion of less IL-6 and IL-10, though again, both the modified and parental CpG generate similar amounts of IL-12.[62] Similar results were found with 3'-O-methyl-ribonucleotide substitutions (Fig. 6).[46]

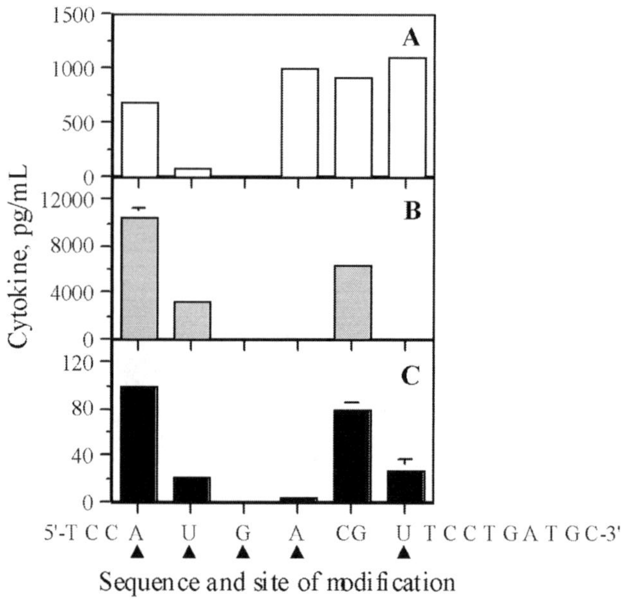

Figure 6. Immunostimulatory activity of CpG DNAs containing a single 3'-O-methylribonucleoside substitution at the position indicated with arrow. Secretion of A) IL-12, B) IL-6, and C) IL-10 in BALB/c mouse spleen cell cultures at 1.0 μg/ml concentration of CpG DNAs after a 24 hr incubation. Each value is an average of three or four replicates. CG indicates parent CpG DNA without modification.

The Role of Nucleobases in the Recognition of CpG DNA

Although a hexameric motif containing a central CpG dinucleotide is important for immune stimulation, a hexanucleotide by itself is not active, suggesting that the flanking sequence plays a significant role in immune stimulation. However, the extent of the involvement of these sequences in receptor recognition and immune stimulation is unclear. To address these issues we incorporated 1',2'-dideoxyribonucleotides (Fig. 3d) in a site-specific manner into CpG DNA to assess their effects on immune stimulation. Deletion of one or two nucleobases, lying three or more bases 5' of the CpG, increased activity and altered the cytokine secretion profile.[63] A similar deletion, in the 3'-flanking sequence, had no significant affect, suggesting that nucleobases downstream of the CpG are not involved in recognition (Fig. 7).[63]

Substitution of Nucleosides with Non-Nucleosidic Linkers

The entire nucleoside in a CpG dinucleotide can be replaced with a non-nucleosidic linker in certain positions (Fig. 8).[64] The nature and number of linkers substituted as well as their position influences immune stimulation.[64] Though a C3-linker optimally enhances immunostimulatory activity, longer ethylene glycol- and branched alkyl-linkers can also enhance immune stimulation. In general, a linker substitution in the 5' flanking sequence of the CpG dinucleotide enhances activity whilst the same substitution, in the 3'-flanking sequence, has no effect. Linker substitutions in the 5'-flanking sequence increase IL-6 secretion several-fold over that seen with unmodified CpG DNA. However, linker-substituted CpG DNAs induce IL-12 secretion at levels similar to those achieved with the parental CpG DNA which are, none-the-less, sufficient to produce a Th1 immune response. As high levels of IL-6 can induce

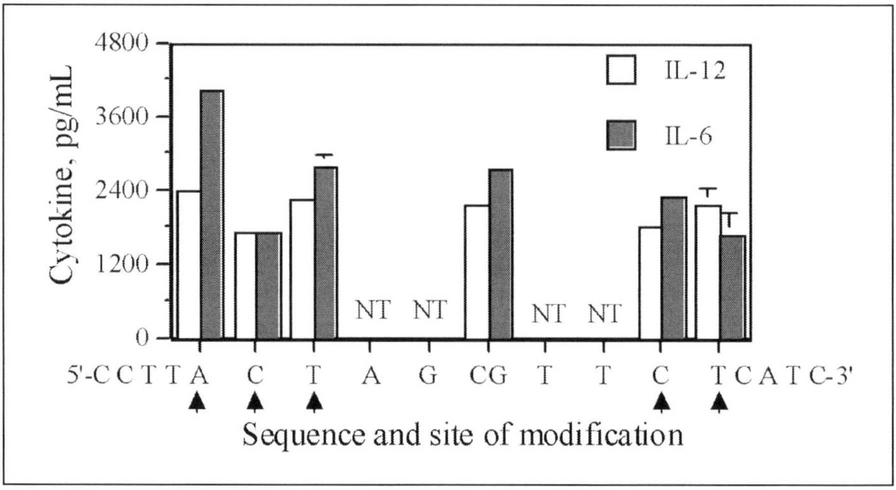

Figure 7. Immunostimulatory activity of CpG DNAs containing a single 1',2'-dideoxyribonucleoside substitution at the position indicated with arrow. Secretion of IL-12 and IL-6 in BALB/c mouse spleen cell cultures at 1.0 μg/mL concentration of CpG DNAs after 24 hr incubation. Each value is an average of three or four replicates. CG indicates parent CpG DNA without modification.

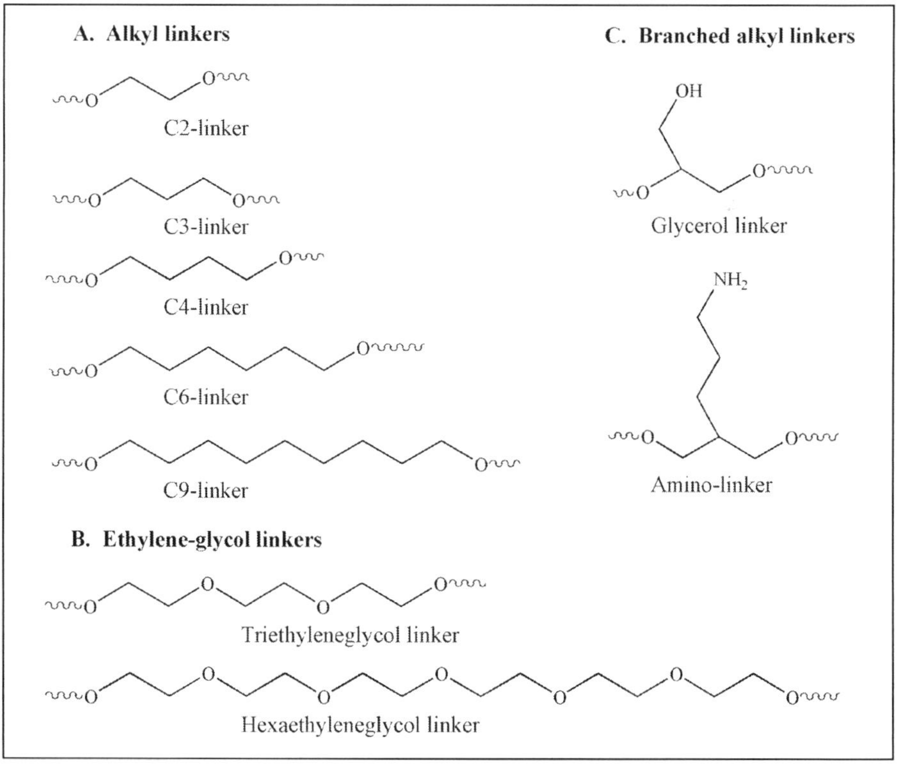

Figure 8. Some of the non-nucleosidic linkers studied as substitutes for nucleosides in CpG DNA.

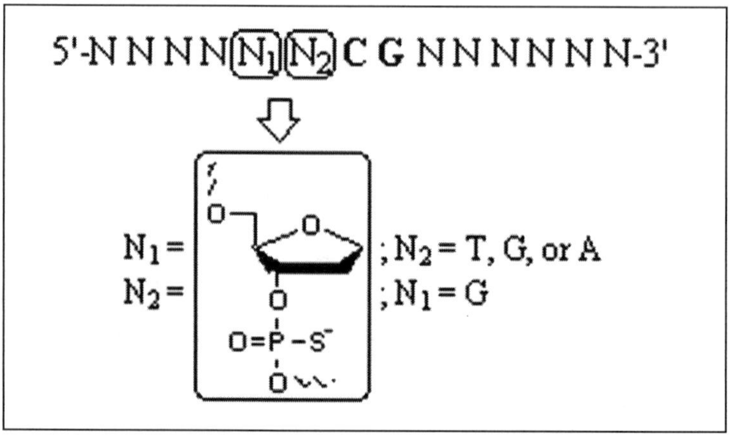

Figure 9. Significance of nucleobases adjacent to the CpG dinucleotide. An abasic nucleoside can be substituted at either N_1 or N_2 depending on the nucleoside in the adjacent position. N stands for A, C, G, or T 2'-deoxyribonucleotide.

the maturation of B and T-cell-specific humoral immune responses, these compounds could serve as useful adjuvants.

The Role of Nucleotides Adjacent to the CpG Dinucleotide in DNA

A hexameric $N_1N_2CGN_3N_4$ motif is the minimal motif required for immunostimulatory activity. TLR9 recognition of a wide range of CpG motifs is governed by their sequence, structure and the cell types and species involved. The molecular basis for this discrimination is however unclear. We examined the effects of charge, ribonucleotide and 3'-deoxyribonucleotide substitutions, and nucleobase (N) deletion adjacent to the CpG dinucleotide.[66] Our results suggested that a negative charge at these positions is required for activity[66] whilst 3'-deoxyribonucleotide substitutions at adjacent nucleotide positions severely impaired immunostimulatory activity.[66] 2'-O-alkyl-ribonucleotide substitutions at the same positions were tolerated to some extent.[46,60,61]

Both the murine and human TLR9 can discriminate the nucleotide at N2 of a typical CpG motif $N_1N_2CGN_3N_4$. Using an abasic linker (1',2'-dideoxyribonucleotide; X) at N_1-N_4 we have shown that a nucleobase is requisite at positions N_3 and N_4, whilst X is permitted in place of N_1 or N_2 depending on the neighboring base (Fig. 9). Additionally, we found that the 'GXCGTT' motif had an intermediate activity between 'GACGTT' and 'GTCGTT' in mouse-cell cultures.[66] The results presented in the preceding sections suggest that the activity of CpG DNA not only depends on the presence of the CpG dinucleotide but also the nucleotides 5'of the CpG, more so than the 3' sequence. Moreover, a number of other characteristics such as the structure of the CpG DNA and the chemical nature of the stimulatory motif plays a significant role in immunostimulation, as discussed below.

Immunomers

TLR9 Reads DNA Sequence from the 5'-End

CpG DNAs linked at their 5' termini either fail to activate, or only weakly activate immune cells, in spite of their dual CpG motifs.[67] The same DNAs, linked by their 3'-termini (immunomer) retain activity and can induce cytokine profiles that are distinct from those generated using parent CpG without 3'-3'-linkages.[67-69] These are the first studies to suggest that an accessible 5'-end is required for CpG DNA activity and that the receptor interprets DNA sequence from its 5'-end. We recently showed that the conjugation of fluorescein to the 5'-end of a CpG DNA impeded its immunostimulatory activity but not its cellular uptake,[68] an observation that has since been confirmed by other groups.[70-72] Whilst the nature and length of the linker plays a role in immune stimulation,[69] the length of the immunomer itself can also effect cytokine secretion (Fig. 10).[69] Further studies have shown that the kinetics of NF-κB activation in J774 macrophages exposed to immunomers is different from that achieved using CpG DNAs with a single 5'-end.[73] In fact immunomers accelerate NF-κB activation whilst leaving the kinetics of the MAP kinase pathway comparable with that seen with conventional CpG.[73] These studies suggest that the CpG motif is required for immune stimulation whilst receptor recognition and activation can be modulated by the appropriate immunostimulatory motif.[73] Based on these studies we hypothesized that the cellular recognition of CpG DNA occurs from its 5'-end (Fig. 11A). Further, CpG DNAs that contain multiple CpG dinucleotides are no more stimulatory than a single CpG, unless the CpGs are appropriately spaced with non-stimulatory sequences.[73] This is because the DNA molecule still only has a single accessible 5'-end. Consequently, the recognition of one CpG precludes recognition of a second, if the motifs are sufficiently close.[73] In contrast, the simultaneous binding of two receptors at both the accessible 5'-ends of an immunomer trigger the rapid activation of transcription factors and immune stimulation (Fig. 11B). CpG DNA that does not contain accessible 5'-ends may not be recognized by the receptor in spite of the appropriate CpG motifs (Fig. 11C). An additional consideration is that the dual 5'-termini of immunomers may facilitate receptor (Fig. 12) dimerization, possibly explaining their ability to accelerate NF-κB activation and induce higher levels of cytokine production.[73]

Short Immunomers

Previous studies have shown that the minimal sequence required for the recognition of a DNA sequence by immune cells is a PuPu(Py)CGPyPy hexameric sequence motif.[13] Immunomer design (Fig. 13A) permitted us to develop CpG DNAs comprising strands of only five or six nucleotides which did not require the 'PuPu(Py)CGPyPy' hexameric motif.[74] Surprisingly, these short immunomers induced high IL-12 and minimal IL-6 secretion in murine spleen-cell and peripheral blood mononuclear cell cultures.[74] Unlike conventional CpG DNAs, the novel short-hexa- (Fig. 13B) and pentanucleotide immunomers (Fig. 13C) optimally stimulate NF-κB and stress-activated pathways in murine cells and induce cytokine secretion from both murine and human cells.[74] These short immunomers are the first CpG DNAs to be recognized by both murine and human cells without sequence modification and will allow us complete preclinical and clinical testing with identical molecules. The economic viability of short immunomers as therapeutic agents is underscored by their short synthetic cycle, potent biological activity and favorable safety profile. Studies performed by other groups support our findings that immunomers permit the design of short oligonucleotides.[71]

Immunomer Design Enhances the Metabolic Stability of CpG DNA

Phosphodiester single-stranded CpG DNA can activate immune cells but only after repeated transfection at high concentration. This is due to nucleolytic degradation of the natural backbone of these molecules.[75] Phosphodiester CpG DNAs that contain palindromic structures[31,32] and poly(dG) termini[33,34] are more commonly used because of their superior stability against nucleases. As discussed above, these CpG DNAs (class A/D) can effectively activate NK cells and pDCs to produce IFNs.

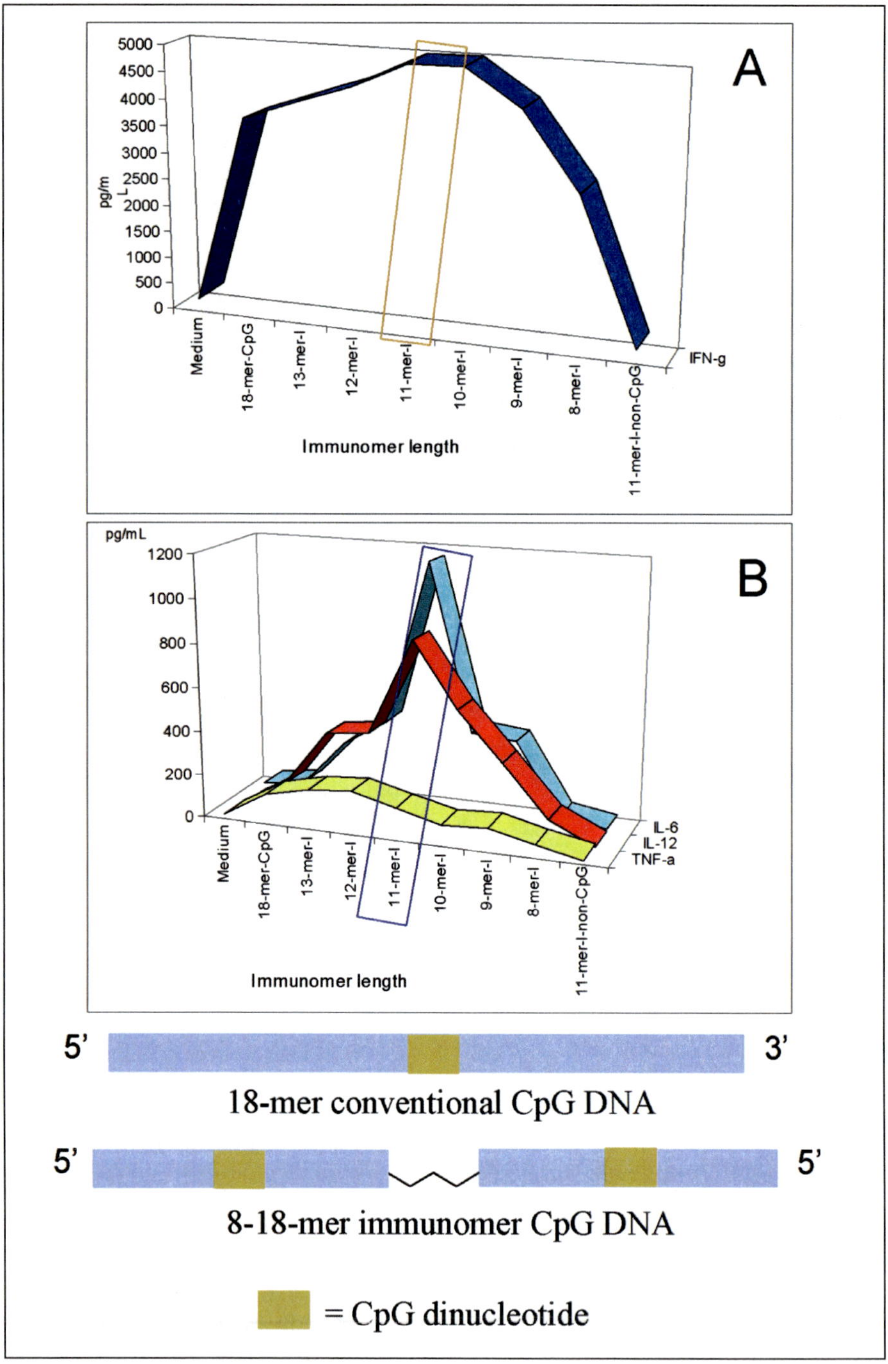

Figure 10. Effect of immunomer length on cytokine induction in BALB/c mouse spleen cell cultures at 1 μg/ml concentration. A) IFN-γ, and B) IL-12, IL-6 and TNF-α. Line drawings of conventional CpG DNA and immunomers are shown. In the case of immunomers (I), length noted is for each arm.

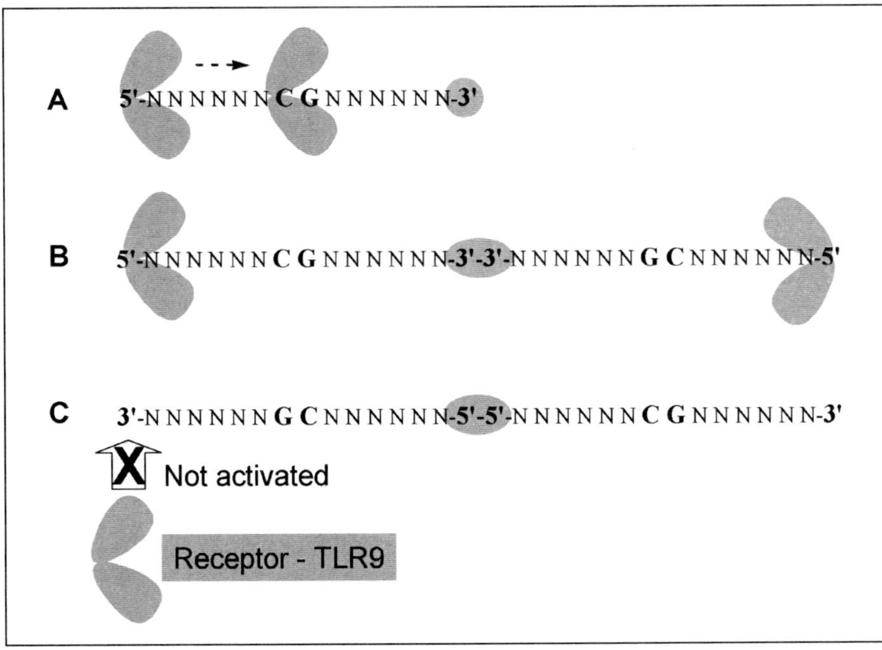

Figure 11. Hypothetical models showing the recognition of CpG DNA and immunomers by TLR9. A) TLR9 recognizes the CpG DNA sequence from its 5'-end. Dotted arrow indicates the reading direction of TLR9. B) Recognition of a 3'-3'-attached immunomer by two TLR9s independently from both the 5'-ends. C) Immunomers attached through 5'-5'-linkage do not activate the immune system. The receptor is shown in grey.

We recently showed that immunomers containing a phosphodiester backbone could induce potent immune responses in the absence of poly(dG) palindromic sequences or multiple transfection.[76] Phosphodiester immunomers induce higher levels of IL-12 than phosphorothioate CpG DNAs and minimal to no IL-6 secretion, which distinguishes them as a separate class of immune activator. Immunomers also display optimal TLR9 recognition characteristics as well as a greater stability against nucleases and, as a consequence, a greater in vivo antitumor activity.[76]

The Significance of d(CpG) Dinucleotides and the Role of Functional Groups of Cytosine and Guanine in Immune Stimulation

Although the flanking sequences of a CpG dinucleotide, its accessible 5'-end and backbone chemistry all contribute to TLR9 activation, it is the CpG dinucleotide itself that participates in receptor recognition whether of single or double-stranded phosphodiester or phosphorothioate CpG DNA. Chemical modification of the CpG dinucleotide to alter its structure or conformation will abolish receptor recognition and any subsequent immune response. Chemical modifications at the 2'-position of the sugar ring of either the C or G nucleotide also abrogates immunostimulatory activity, suggesting that the receptor recognizes not only the heterocyclic bases of C and G, but also their conformation.[58] The Rp diastereomer of a phosphorothioate CpG oligo is more active than the Sp diastereomer, suggesting that TLR9 can distinguish between these isomers.[77] In terms of charge, the negative charges on the phosphates (p) at the -pCpGp- positions are critical for activity, and their neutralization with

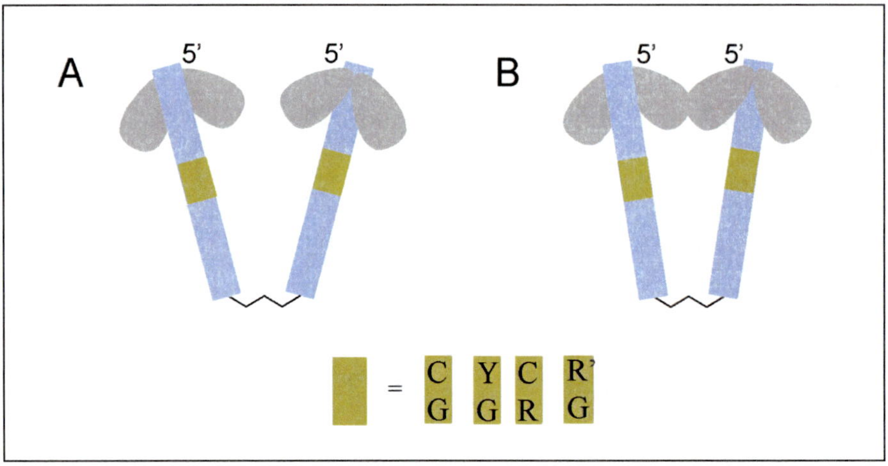

Figure 12. Models depicting dimerization-independent (A) and -dependent (B) (homo or hetero-dimeric) receptor recognition of immunomers. Brown rectangle indicates CpG, YpG, CpR, R'pG dinucleotides.

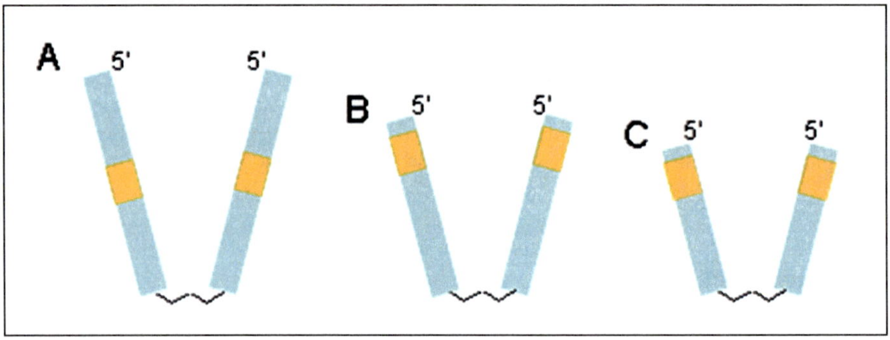

Figure 13. Comparison of (A) immunomers with short (B) hexa- and (C) penta-nucleotide immunomers that are active in both murine and human systems without a need to modify the sequences. Brown rectangle indicates the location of the CpG dinucleotide in the sequence. The immunostimulatory activity of immunomers depends on their length.

methylphosphonate linkages ablates activity.[58] Likewise, the incorporation of a non-ionic linkage (methylphosphonate)[59] or a 2'-5'-linkage between the C and G of a CpG dinucleotide also abrogates activity,[62] suggesting that both the negative charge and a natural 3'-5'-internucleotide linkage is required for immune stimulation. Both nucleobases C and G of the CpG dinucleotide are absolutely required for immune stimulation as evidenced by the loss of activity seen when either is substituted with a 1',2'-dideoxyribonucleotide.[63] A methyl substitution at the 5-position of C also interferes with recognition and/or the activity of CpG DNA, [58] a fact that vertebrates have exploited to distinguish self DNA from invading bacterial DNA, which contains a higher percentage of unmethylated CpG motifs.

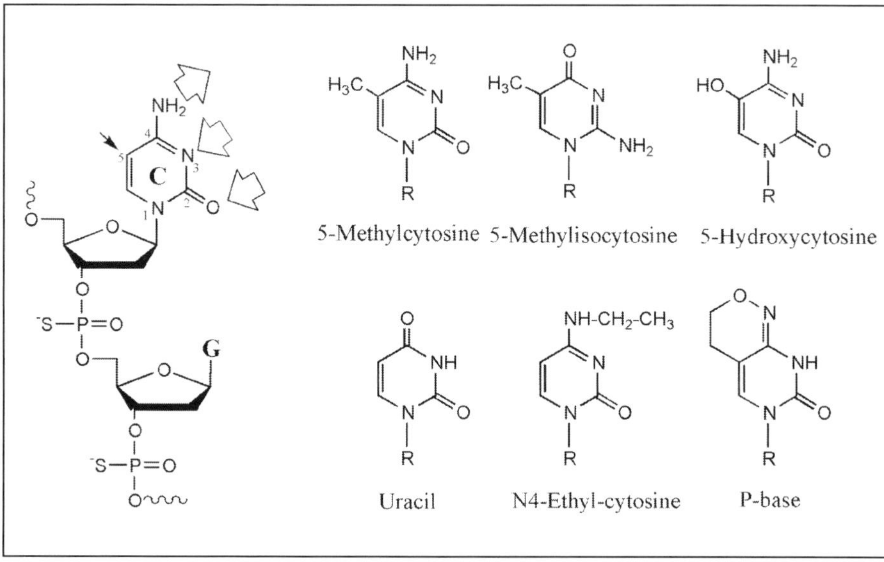

Figure 14. Structure of 2'-deoxycytidine showing the hydrogen bond acceptor (inward arrows) and donor (outward arrows) groups that could be involved in hydrogen bond interactions with TLR9. The 5-position of cytosine is indicated by the solid arrow. Methylation of the 5-positon of cytosine in the CpG motif leads to a loss of immunostimulatory activity. Some of the synthetic analogs (Y) of cytosine studied as substitutes for natural cytosine are shown.

Functional Groups of Cytosine

Cytosine contains 2-keto, 3-imino, and 4-amino functional groups that could serve as specific receptor recognition sites (Fig. 14). By using various pyrimidine analogs (Y) in place of C in the CpG dinucleotide, we studied the effects of deleting or substituting functional groups on immune stimulation.[78] Some of the analogs used included 5-methylcytosine, 5-methylisocytosine, 5-hydroxycytosine, uracil, N4-ethylcytosine, and P-base-nucleoside (Fig. 14). Deletion or substitution of any of the functional groups at positions 2 and/or 4 resulted in a loss of activity.[78] An alkyl substitution on the 4-amino group did not block recognition whilst a methyl substitution at the 5-position did. However, a hydroxy substitution at position-5 had no effect on activity. The results suggest that the receptor can recognize a variety of structural motifs and have led to the development of alternate synthetic nucleotide motifs (YpG) with immunostimulatory activity.[78]

Functional Groups of Guanine

In the case of G, two hydrogen-bond acceptor groups at positions 6 (keto-oxygen; O6) and 7 (nitrogen; N7) and two hydrogen-bond donor groups at positions 1 (imino nitrogen; N1) and 2 (amino group; 2-NH_2) could serve as potential sites for receptor recognition and interaction (Fig. 15). A number of modified purine nucleobases (R) were used to study the role of functional groups in CpG DNA-stimulated activity.[78] These 2'-deoxy-analogs included inosine, 2- aminopurine, 7-deazaguanine, nebularine, isoguanine, 2-aminoadenine, K-base, and 7- deazaxanthine (Fig. 15). The deletion or modification of hydrogen-bond acceptor and donor groups at the 1-, 2-, or 6-positions of G, but not the 7-position, resulted in a loss of activity.[78] These studies provide the first important clues as to which functional groups in a

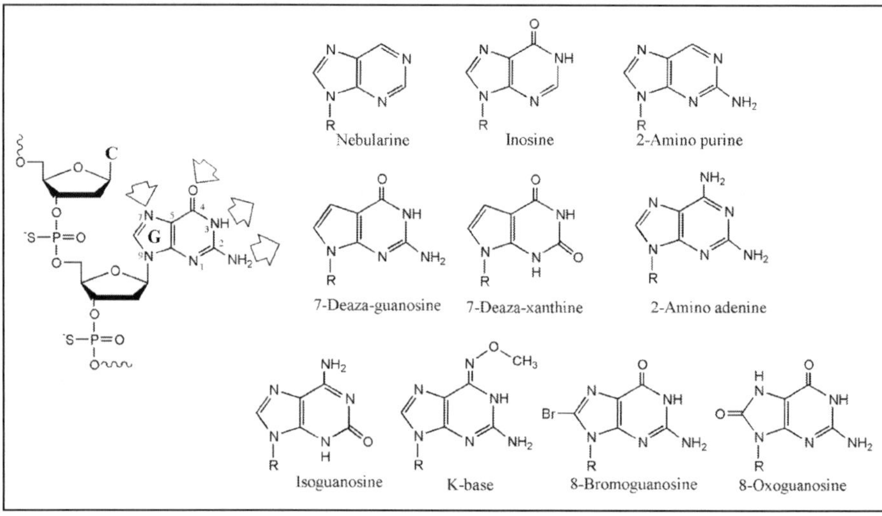

Figure 15. Structure of 2'-deoxyguanosine showing the hydrogen bond acceptor (inward arrows) and donor (outward arrows) groups that could be involved in hydrogen bond interactions with TLR9. Some of the synthetic analogs (R) of guanine that were used as substitutes for natural guanine are shown.

CpG dinucleotide are required for receptor recognition and allow us to develop YpG, CpR, and YpR immunostimulatory motifs with distinct cytokine induction profiles.

Recognition of a Bicyclic Heterobase at the C-Position and the Negation of Species Specificity

We recently reported the development of a synthetic nucleoside with a bicyclic heterobase [1-(2'-deoxy-β-D-ribofuranosyl)-2-oxo-7-deaza-8-methyl-purine; R'] to replace the C in CpG, giving an R'pG dinucleotide (Fig. 16).[79] Oligos containing the R'pG motif induce potent cytokine secretion in mouse spleen-cell cultures and induce human B cells to proliferate, suggesting that both species of TLR9 can recognize this motif. Immunomers containing R'pG dinucleotides also show activity in HEK293 cells that stably express mouse TLR9, which would suggest a direct involvement of TLR9 in R'pG recognition. In J774 macrophages, both the R'pG and CpG motifs activate NF-κB and MAP kinase pathways. In this respect, the immune signals elicited by R'pG motifs seem to be transduced via signaling pathways that are comparable to those triggered by natural CpG DNAs.[79] This was the first report of the recognition of a bicyclic heterobase in place of C in a CpG motif.

The Nucleotide Motif Recognition Pattern (NMRP)

While TLR7 recognizes a variety of structurally dissimilar synthetic ligands, TLR9 recognizes CpG dinucleotides in various sequence, structure and conformational contexts. Our medicinal chemistry shows that TLR9 can recognize a broad range of synthetic nucleotide motifs, YpG, CpR, R'pG, and YpR, as well as the natural CpG dinucleotide. We also recently showed that TLR9 recognizes CpR and R'pG motifs in immunomers, but that these motifs induce different cytokine secretion profiles compared to the natural CpG motif.[73,79] In both in vitro and in vivo studies, an immunomer containing the natural CpG induced higher levels of IL-12 and IL-6 than conventional CpG DNA. However, an immunomer containing a synthetic CpR motif induced significantly lower levels of IL-6 though a comparable amount of

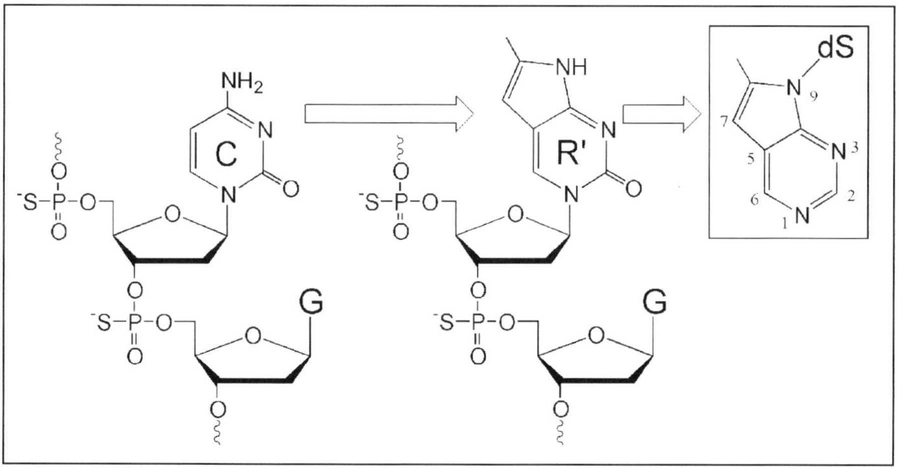

Figure 16. Structures of natural CpG and synthetic R'pG immunostimulatory dinucleotides. Structure of a 7-deaza-purine with 9-glycoside is shown in inset. dS stands for 2'-deoxyribosugar.

IL-12 to that induced by natural CpGs.[73] Whilst the kinetics of activation of NF-κB are similar for immunomers containing both natural CpG and synthetic CpR motifs, distinct downstream effects, including cytokine secretion and antigen-specific IgG2a/IgG1 antibody ratios are observed.[73] The recognition of natural CpG and synthetic CpR and R'pG motifs would suggest multiple NMRPs for TLR9.

Secondary Structure in CpG DNA Affects Immunostimulatory Activity

The 5'-end of a CpG oligonucleotide should be accessible for receptor recognition and subsequent immune stimulation. If blocked, for example by a 5'-5' linkage, immunostimulatory activity is abrogated. Our studies have shown that 5'-terminal secondary structures as well as conjugates affect activity to a greater extent than 3'-modifications.[80] CpG oligos containing 3'-hairpin structures also induce lower levels of IL-6 secretion than conventional CpG DNAs. The need for an accessible 5'-end suggests that the receptor responsible for immune stimulation interprets the DNA sequence from this end.

Modulation of the Immune Response through Cell Specific Activation

We recently proposed a rational combination of stimulatory and structural domains in CpG DNA to optimally activate TLR9-positive immune cell subsets (pDCs and B cells).[81] These studies allowed us to delineate the structures required to activate different immune cells and led to the development of a novel class of immunomodulatory oligonucleotides. These oligonucleotides, comprising a short stimulatory domain with CpG motif and a 3' hairpin-loop structure, are termed self-stabilized CpG DNAs (Fig. 17). This class of CpG DNAs can activate human B cells and induce pDCs to secrete high levels of IFN-α. We found that both the stimulatory and secondary structure domains in CpG DNAs are required for pDC activation,[81] whilst B-cell activation required only a stimulatory domain. In fact, the presence of structural motifs in CpG DNA may interfere with B-cell activation to some extent. Interestingly, CpG motifs are not required for activity in the hairpin duplex region. Further modification of the hairpin duplex with a mixture of oligodeoxynucleotides and oligo-2'-O-methylribonucleotides permits the activation of both human B cells and pDCs. It is not clear, however, as to how the same TLR9, in either a B cell or pDC, can use different

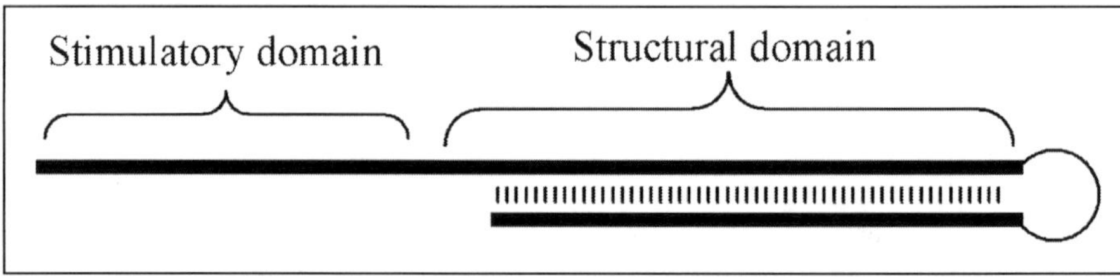

Figure 17. Schematic drawing of the novel CpG DNA design called self-stabilized CpG DNA, showing essential stimulatory and structural domains. The stimulatory domain contains an appropriate CpG motif but no structural domain. 2'-O-methylribonucleotides can be incorporated in the structural domain. Self-stabilized CpG DNAs stimulate human B cells and also activate human plasmacytoid dendritic cells to secrete IFN-α.

structural ligand cues to trigger the immune system. The possibility remains that different co-receptor and/or adaptor molecules for TLR9 are involved in signaling in pDCs and B cells.

Species Specific Recognition of CpG DNA

Although the presence of an unmethylated CpG dinucleotide in DNA is essential for its immunostimulatory activity, the sequences that flank the CpG dinucleotide also play a significant role. The initial studies by Tokunaga and colleagues using a number of short sequences represented in mycobacterial DNA suggested that palindromic sequences containing an unmethylated CpG dinucleotide in certain sequence contexts could activate NK cells and induce interferon secretion.[31,32] The latter studies suggested that a palindromic sequence in phosphorthioate modified DNA was not required for activity whereas the bases both preceding and following the CpG dinucleotide could influence activity.[13] In general, the CpG dinucleotides that are preceded by a C or followed by a G stimulate lower immune responses compared with other natural nucleosides in these positions.[13] The CpG dinucleotide flanked by two purine bases on the 5'-side and two pyrimidine bases on the 3'-side, such as 'GACGTT', efficiently activate the murine immune system.[13] In contrast human cells fail to respond well to this sequence and require different sequence motifs such as 'GTCGTT' or 'TTCGTT' for optimal stimulation.[82] Certain other sequences, such as the palindromic 'AACGTT' sequence, induce pronounced immunostimulation in both mouse and human systems.[65] Reports of the activation of the immune system of fish,[83-85] chicken,[79,86-90] dog,[90,91] cat,[90-92] cattle,[93,94] pig,[79,90,95,96] horse,[79,90] rabbit,[90] goat,[79,90] and sheep[79,90] by CpG DNAs have also recently appeared. Several of these species appear to recognize the human specific GACGTT motif, more so than other motifs. However, it is important to note that TLR9 of different species recognize CpG dinucleotides flanked by a variety of sequences, though to different extents. As a result it is difficult to compare results obtained with different species.

Immunomers containing 'GTR'GTT' (human-specific) and 'GAR'GTT' (mouse-specific) motifs are recognized to a similar extent by both murine and human immune systems.[79] Additionally, both mouse- and human-specific R'pG-immunomers potently stimulate the proliferation of peripheral blood mononuclear cells obtained from diverse vertebrate species, including cynomolgus monkey, pig, horse, sheep, goat, rat, and chicken. This suggests that both motifs are recognized without a bias for their flanking sequence (Fig. 18),[79] which would make them the first chemical modifications to be so recognized. This modification may also permit the rapid development of DNA-based immunomodulatory agents for veterinary and aquatic applications.

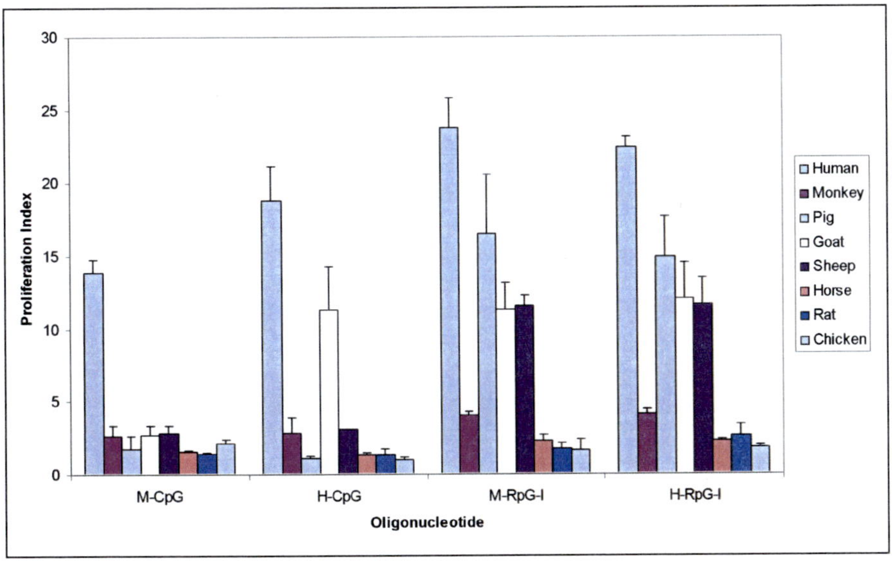

Figure 18. Proliferation of B cells (human) or PBMCs isolated from the blood obtained from different vertebrate species at a concentration of 1 μg/ml of each immunomer. Each value is an average of three replicates and the results are representative of one or two independent experiments, except in the case of human. Human data are representative of six donors. The sequences of M-CpG, H-CpG, M-R'pG-I, and H-R'pG-I are 5'-CTATCT<u>GACGTT</u>CTCTGT-3', 5'-CTATCT<u>GTCGTT</u>CTCTGT-3', 5'-TCT<u>GAR'GTT</u>CT-L-TC<u>TTGR'AG</u>TCT-5', and 5'-TCT<u>GTR'GTT</u>CT-L-TC<u>TTGR'TG</u>TCT-5', respectively. Mouse- and human-specific motifs are shown underlined. 'I' denotes immunomer and 'L' indicates a glycerol linker. The structure of R' is shown in Figure 16.

Therapeutic Applications of CpG DNA

CpG DNA has a variety of effects on the immune system. It induces B-cell proliferation, activates macrophages, monocytes, and DCs to produce cytokines and chemokines, and induces the expression of costimulatory molecules. CpG DNA also activates NK cells either directly or indirectly to secrete IFN-γ as well as promoting NK-cell lytic activity and T cell function. In addition, CpG DNA induces the maturation of immature DCs and the secretion of IFN-α/β, TNF-α, IL-6, and IL-12. CpG DNA induces a Th1-type immune response, reversing the allergen-induced Th2-biased response and promoting the production of strong total IgG and antigen-specific IgG2a antibodies in serum. These immunopharmacological properties permit the use of CpGs as therapeutic agents against infection, cancer and allergic asthma alone or in combination with vaccines, antigens, and monoclonal antibodies (Fig. 19).

Several studies have shown that the cytokines secreted (by activated innate immune cells) in response to CpG DNA can produce a strong immune response against bacterial, viral, and parasitic infections, including anthrax, listeria, herpes, HIV-1, cytomegalovirus, leishmania, and malaria (Table 2).[43,45,97-142] Our previous studies using an antisense oligo against human papillomavirus (HPV) showed that the presence of an intact CpG dinucleotide provided antiviral protection.[43] This protection is considerably lower in IL-12 p40 KO mice, IFN-γ depleted mice, and Beige (bg/bg) NK-lytic-deficient mice than in immunocompetent mice, suggesting the important immunoprotective role of these cytokines in vivo.[43] Our earlier studies also showed that DNA containing CpG dinucleotides could inhibit cytomegalovirus (CMV)

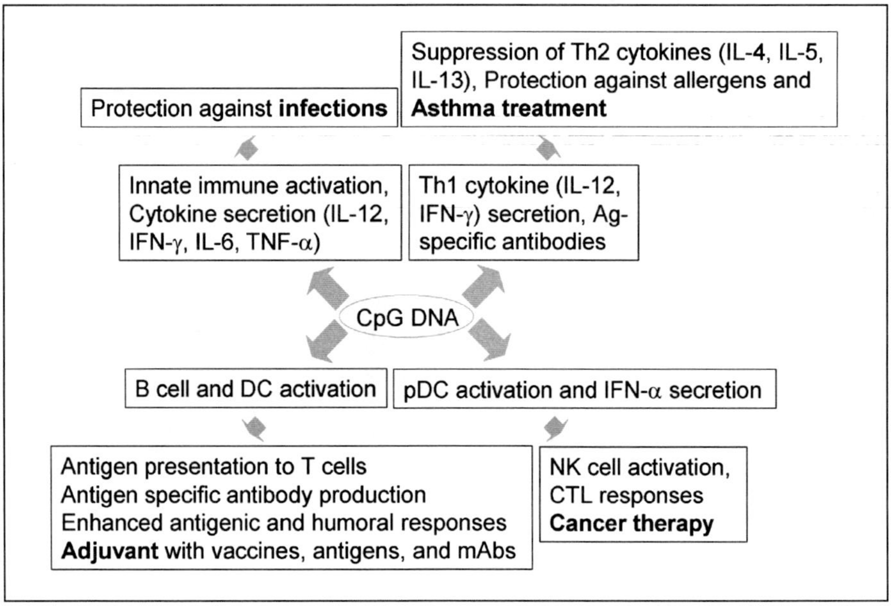

Figure 19. Possible applications of CpG, YpG, CpR, YpR, and R'pG DNAs based on their immunostimulatory profiles.

and prolong the survival of CMV-infected mice.[45] The coordinated secretion of cytokines as a result of CpG activated innate immune cells was proven with the use of neutralizing antibodies to a number of cytokines.[15] These cytokines provide non-specific protection against infections, which suggests their potential as therapeutic agents against a range of infectious diseases in preclinical (Table 2) and clinical (Table 3) studies.

On the contrary, co-stimulation of certain TLRs, including TLR2, and TLR9, with soluble *Mycobacterium tuberculosis* factor (STF) or CpG DNA, respectively, has been shown to induce HIV replication.[143,144] Ex-vivo stimulation of spleen cells from HIV-1 transgenic mice via TLRs 4, 2, and 9 using LPS, STF, and CpG DNA respectively, induced p24 Ag production in a dose-dependent manner, suggesting that the activation of TLR pathways by these ligands induces HIV-1 replication.[143,144] Consistent with these results, an increased viral load was observed in the plasma of HIV positive subjects following the continual intravenous infusion (8 days in a blinded dose-escalation phase II study) of a first-generation phosphorothioate antisense oligonucleotide (Table 1) complementary to the *gag* gene of HIV-1 which contained a CpG motif (GEM91 (Fig. 20)).[145] Previously, GEM91 had been shown to be a potent inhibitor of HIV-1 replication in cell cultures.[146] These results suggest that CpG DNAs could, in a virus specific fashion, promote viral replication though the molecular mechanisms of this effect is unknown.

CpG DNA induces strong Th1 responses, the rationale for its use as an anticancer agent.[147-149] Its ability to activate the NK cells and macrophages within a tumor mass and to stimulate antibody-dependent cell cytotoxicity (ADCC) is further evidence for its usefulness in cancer therapy.[150] Furthermore, the Th1-biased IL-12 and IFN-γ production induced by CpG DNA can promote tumor-specific cytotoxic T lymphocyte (CTL) activity by the production of tumor-specific IgG2a antibodies. CpG DNA-treated tumor-free mice effectively reject a challenge with the same tumor cell, suggesting that tumor-specific adaptive immune responses

Table 2. CpG DNA applications for infections

Infectious Agent[a]	Use	Combination Agent
Viruses		
CMV	Monotherapy[45]	-
	Adjuvant[99]	DNA vaccine
DEN-2	Adjuvant[100]	DNA vaccine
FIV	Adjuvant[101]	DNA vaccine
FMDV	Adjuvant[102]	DNA vaccine
Friend RV	Monotherapy[103]	-
HDV	Adjuvant[104]	HDAg vector
HBV	Adjuvant[105-107]	HB surface Antigen (HBsAg)
HCV	Adjuvant[108,109]	DNA vaccine/HCV-LP
HIV-1	Adjuvant[110]	rHIV-1gag/DNA vaccine
	Adjuvant[111]	HIV-1 immunogen
HPV	Monotherapy[43]	-
HSV-2	Monotherapy[112,113]	-
Influenza	Monotherapy[114]	-
Measles	Adjuvant[115]	Ag
Pneumonia	Adjuvant[116]	Vaccine
Pneumonia (SPn)	Adjuvant[117]	Type 2 Ag
RSV	Adjuvant[118]	Fusion protein
SHIV	Adjuvant[119]	Part of expression vector
Bacteria		
Anthrax	Adjuvant[120]	rPA
B. abortus	Adjuvant[121]	p39 and BFR Ag
H. pylori	Monotherapy[122]	-
Listeria	Monotherapy[123,124]	-
Listeria/Francisella	Monotherapy[125]	-
Meningitis	Adjuvant[126]	Ag
TB	Adjuvant[127,128]	BCG vaccine/Ag
Parasites		
F. hepatica	Adjuvant[129]	Ag
Leishmania	Monotherapy[130-132]	-
Malaria	Monotherapy[133]	-
	Adjuvant[133-137]	Ag/Vaccine
Trypnosoma	Adjuvant[138,139]	Ag
Fungi		
Aspergillosis	Adjuvant[140,141]	Asp-f 6
C. neoformans	Adjuvant[142]	P13 Ag

[a] Ag= antigen; B. abortus= *Brucell abortus*; CMV= cytomegalovirus; C. Neoformans= *Cryptococcus neoformans*; DEN-2= Dengue virus type-2; F. hepatica= *Faciola hepatica*; FIV= feline immunodeficiency virus; FMDV= Foot and mouth disease virus; Friend RV= Friend retro virus; H. pylori= *Helicobacter pylori*; HDV= hepatitis delta virus; HBV= Hepatitis B virus; HCV= Hepatitis C virus; HIV-1= human immunodeficiency virus-I; HPV= human papilloma virus; HSV-2= herpes simplex virus type-2; RSV= respiratory syncytial virus; SHIV= simian/human immunodeficiency virus; SPn= *Streptococcus pneumoniae*; TB= *Mycobacterium tuberculosis*.

Table 3. First- and second-generation CpG oligodeoxynucleotides that are in clinical trials

Agent	Sequence	Disease Indication[a]
Promune™	5′-TCGTCGTTTTGTCGTTTTGTCGTT-3′[b,c]	NSCLC (with chemo), Melanoma (mono and with chemo), CTCL, RCC, NHL (with Rituxan), and BCC.
Actilon™	Not disclosed[c]	Hepatitis C
VaxImmune™	Not disclosed[c]	Cancers (with cancer vaccines)
ISS 1018	5′-TGACTGTGAACGTTCGAGATGA-3′[c]	Allergy, Asthma, Infections (HBV; as adjuvant), Cancer (NHL; with mAb)
IMOxine™	Not disclosed[d]	Cancers (monotherapy)
Amplivax™	Not disclosed[d]	As adjuvant[e]

[a] BCC, basal cell carcinoma; CTCL= cutaneous-T cell lymphoma; HBV= Hepatitis B virus; NHL= non-Hodgkins lymphoma; NSCLC= non-small cell lung carcinoma; RCC= renal cell carcinoma; [b] From published patent application No. WO 2004/005476; [c]= First-generation CpG DNA; [d] Second-generation immunomodulatory oligonucleotide; [e]= Completed phase 1 study in healthy human volunteers.

develop in treated animals. CpG DNAs have been studied alone or in combination with chemotherapeutic agents, monoclonal antibodies and antigens in a number of preclinical tumor studies.[147-149] Clinical trials are currently underway to test the ability of CpG DNA alone, or in combination with monoclonal antibodies, to treat cancers (Table 3). Recently, we evaluated in vivo the immunopharmacological and antitumor properties of second-generation immunomodulatory oligonucleotide (IMO) immunomers containing CpG or CpR motifs alone or in combination with chemotherapeutic agents.[151] Repeated peritumoral administration of CpG or CpR immunomers inhibited CT26 colon tumor or B16.F0 melanoma growth at 1 mg/kg in mice. In these studies immunomers induced tumor-specific CTL responses compared with treatment with a control non-CpG DNA or PBS. These responses correlated with the secretion of theTh1 cytokine IFN-γ, but not theTh2 cytokine IL-4 in IMO treated mice.[151] A 5-fold increase in β-gal specific IgG2a antibodies was found in mice treated with immunomers, which significantly increased the IgG2a/IgG1 ratio. Immunomers showed similar anti-tumor activity in both wild-type and IL-6 knock-out C57BL/6 mice but failed to elicit activity in IL-12 null C57BL/6 mice. Again, tumor-free mice from the immunomer treated cohort rejected the same tumor cell when re-challenged, suggesting that these mice had established an adaptive immune response.[151] Moreover, naïve mice quickly developed specific antitumor responses (without immunomer treatment) following adoptive transfer of splenocytes from tumor free mice that had been subject to immunomer treatment. The co-administration of immunomers with the chemotherapeutic agents, docetaxel and doxorubicin, resulted in a synergistic antitumor effect in both the B16.F0 melanoma and 4T1 breast carcinoma models.[151]

CpG DNA also has great potential to treat allergy and asthma, both of which provoke Th2-type responses. CpG DNA induces strong Th1 responses that include the production of IL-12 and IFN-γ. IFN-γ is a potent inhibitor of IL-4 and IL-5, both of which are required in the Th2 type immune response.[152] CpG DNA can be used as a potent adjuvant to prevent airway inflammation and asthma as it can directly activate B cells, stimulating them to produce strong antigen-specific responses coupled with the Th1-type response. Indeed the immune

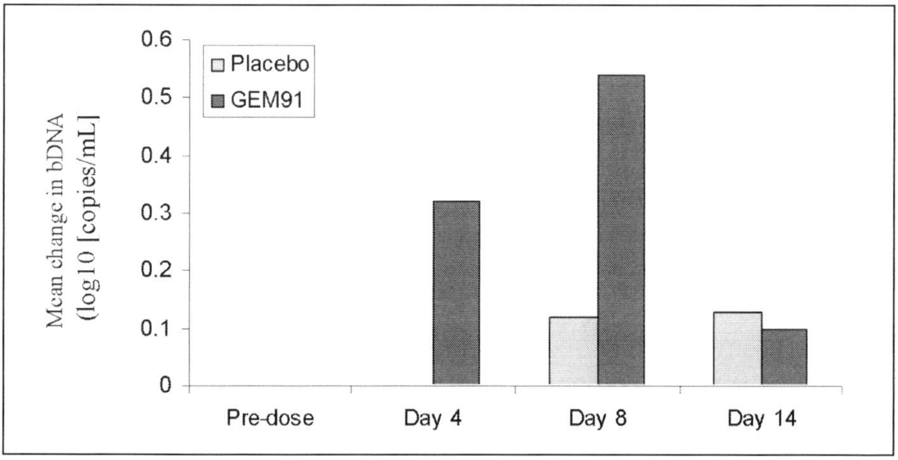

Figure 20. Plasma HIV-1 load, expressed as mean $\log_{10}$ copies/ml by the branched DNA (bDNA) method, are compared to pretreatment values for the six GEM91-treated subjects at 4.0 mg/kg/day dose. Results for the three placebo-treated subjects in each group are pooled. Plasma bDNA for placebo-treated subjects remained within 0.13 $\log_{10}$ of pretreatment values while mean plasma bDNA increased at 4 and 8 days for the GEM91-treated group. However, by day 14, six days after the end of GEM91 infusion, the bDNA was similar in GEM91 and placebo-treated subjects.

response to CpG DNA is associated with an increase in total and antigen-specific IgG2a, a decrease in IgE and IgG1, and the maturation of T cells in the serum. CpG DNA has been found to be effective in both preventative and therapeutic modes in a number of in vivo murine asthma models[153-156] and CpG DNAs are currently being evaluated in clinical trials for allergic asthma (Table 3). Recently, immunomers containing CpG or CpR motifs were shown to be potent inhibitors of OVA induced airway hyper-responsiveness and also reduced bronchial eosinophilia and decreased OVA-specific IgE in a mouse model.[157]

Several lines of evidence now suggest that CpG DNA can enhance the immune response against any type of antigen. Consequently, it is possible that CpG DNA could be used as a universal adjuvant to improve the efficacy of prophylactic and therapeutic vaccines, peptides, antibodies, allergens, antigens, and DNA vaccines in preclinical (Table 2) and clinical (Table 3) studies. The immunogenicity of antigens is increased when CpG DNA is directly conjugated to antigens or allergens, as shown by greater antigen uptake and DC activation by CpG conjugated antigen.[158,159] However, caution must be exercised regarding the site of conjugation as 5'-accessible CpG is an absolute requirement for optimal recognition and activation of TLR9.[68] CpG DNA also enhances the activity of other adjuvants such as alum, chitosan, Quil A, QS21, and MPL, most of which induce Th1-type immune responses.[160]

Plasmid vectors or DNA vaccines that express specific antigens may not induce strong immune responses unless the vector contains CpG motifs.[21,161-163] Bacterial products containing unmethylated DNA, especially from bacillus Calmette-Guerin (BCG) and *M. vaccae*, have been shown to up-regulate the cell mediated Th1 immune response[21,164] and a number of studies have shown the therapeutic efficacy of bacterial extracts for the treatment of allergen-induced asthma, both in preclinical and clinical studies.[165-169] Live bacteria and extracts of BCG have also been used as standard therapies against bladder cancer and other carcinomas and infections.[170-173]

CpG DNAs in Clinical Trials

In the last few years significant progress has been made in understanding the immunological and pharmacological effects of the first-generation CpG DNAs. Data from a number of CpG DNA clinical trials have now started to appear in the literature[105,174] and a number of other CpG DNAs are in early-stage clinical trials as monotherapies against lymphoma, melanoma, and basal cell carcinoma (Table 3). The safety and immunopharmacology of a second-generation CpR immunomer has recently been studied in a dose-escalation study in healthy human volunteers.[175] The results of this phase-I clinical study suggests that there are no safety concerns in the dose range studied (0.005 to 0.16 mg/kg). The safety of the same second-generation CpR immunomer is currently being tested in phase-I clinical trials in cancer patients.[176]

The Safety of CpG DNA

The primary safety concern surrounding the use of CpG DNA as a therapeutic agent is that the immunogenicity of self antigens could be enhanced, which could trigger autoimmune disorders and cause septic shock. The possible development of neutralizing antibodies to the therapeutic agent is another concern. Extensive data collected in clinical trials with antisense oligonucleotides (AOs) which were administered to humans at doses ranging from 2 to 20 mg/kg by continuous intravenous infusion for over 2 years, have shown no long-term drug-related toxicity.[177] Some of the AOs being tested in clinical trials also possess one or more CpG dinucleotides, providing us with valuable pharmacokinetic and safety profiles for the administration of CpG DNA to humans. None of these studies have shown any indication that neutralizing antibodies to DNA were generated. Further, as antisense DNA is used at mg/kg doses, whilst CpG DNA elicits effects at µg/kg doses, the likelihood of severe dose-related side effects would appear to be minimal.

Conclusions

CpG DNA is a valuable tool with which to modulate the host immune response. Researchers have made tremendous progress in their understanding of the molecular mechanisms of CpG DNA action and in developing potent second-generation agents. Other than the CpG motif, a number of other features of CpG DNA sequences appear to influence the immunostimulatory profile. First-generation CpG DNAs that target a number of diseases are currently being tested in human clinical trials and second-generation CpG DNAs are poised to enter clinical trials. This new generation of CpG DNAs are designed to induce selective immune responses and to target disease. The ability of CpG DNAs to induce rapid innate immune responses, leading to adaptive responses, make CpG DNAs an attractive partner with which to combat disease.[178]

References

1. Akira S, Hemmi H. Recognition of pathogen-associated molecular patterns by TLR family. Immunol Lett 2003; 85:85-95.
2. Lien E, Ingalls RR. 2002. Toll-like receptors. Crit Care Med 2002; 30 (1 Supp):S1-S11.
3. Akira S. Mammalian Toll-like receptors. Cur Opin Immunol 2003; 15:5-11.
4. Underhill DM. Toll-like receptors: networking for success. Eur J Immunol 2003; 33:1767-1775.
5. Hemmi H, Takeuchi O, Kawai T et al. A Toll-like receptor recognizes bacterial DNA. Nature 2000; 408:740-745.
6. Hemmi H, Kaisho T, Takeuchi O et al. Small anti-viral compounds activate immune cells via the TLR7 MyD88-dependent signaling pathway. Nat Immunol 2002; 3:196-200.

7. Lee J, Chuang TH, Redecke V et al. Molecular basis for the immunostimulatory activity of guanine nucleoside analogs: Activation of Toll-like receptor 7. Proc Natl Acad Sci USA 2003; 100:6646-6651.
8. Heil F, Ahmad-Nejad P, Hemmi H et al. The Toll-like receptor 7 (TLR7)-specific stimulus loxoribine uncovers a strong relationship within the TLR7, 8 and 9 subfamily. Eur J Immunol 2003; 33:2987-2997.
9. Alexopoulou L, Holt AC, Medzhitov R et al. Recognition of double-stranded RNA and activation of NF-κB by Toll-like receptor 3. Nature 2001; 413:732-738.
10. Bridge AJ, Pebernard S, Ducraux A et al. Induction of an interferon response by RNAi vectors in mammalian cells. Nature Genetics 2003; 34:263-264.
11. Sledz CA, Holko M, deVeer MJ et al. Activation of the interferon system by short-interfering RNAs. Nature Cell Biol 2003; 5:834-839.
12. Messina JP, Gilkeson GS, Pisetsky DS. Stimulation of in vitro murine lymphocyte proliferation by bacterial DNA. J Immunol 1991; 147:1759-1764.
13. Krieg AM, Yi AK, Matson S et al. CpG motifs in bacterial DNA trigger direct B-cell activation. Nature 1995; 374:546-549.
14. Yamamoto S, Kuramoto E, Shimada S et al. In vitro augmentation of natural killer cell activity and production of interferon-α/β and -gamma with deoxyribonucleic acid fraction from Mycobacterium bovis BCG. Jpn J Cancer Res 1988; 79:866-873.
15. Klinman DM, Yi AK, Beaucage SL et al. CpG motifs present in bacteria DNA rapidly induce lymphocytes to secrete interleukin 6, interleukin 12, and interferon γ. Proc Natl Acad Sci USA 1996; 93:2879-2883.
16. Yamamoto S, Yamamoto T, Iho S et al. Activation of NK cell (human and mouse) by immunostimulatory DNA sequence. Springer Semin Immunopathol 2000; 22:35-43.
17. Branda RF, Moore AL, Mathews L et al. Immune stimulation by an antisense oligomer complementary to the rev gene of HIV-1. Biochem Pharmacol 1993; 45:2037-2043.
18. Zhao Q, Temsamani J, Zhou RZ et al. Pattern and kinetics of cytokine production following administration of phosphorothioate oligonucleotides in mice. Antisense Nucleic Acid Drug Dev 1997; 7:495-502.
19. Sun S, Zhang X, Tough D et al. Multiple effects of immunostimulatory DNA on T cells and the role of type I interferons. Springer Semin Immunopathol 2000; 22:77-84.
20. Kandimalla ER, Agrawal S. Therpeutic potential of synthetic CpG DNA – Current status and future directions. IDrugs 2001; 4:963-966.
21. Yamamoto S, Yamamoto T, Nojima Y et al. Discovery of immunostimulatory CpG-DNA and its application to tuberculosis vaccine development. Jpn J Infect Dis 2002; 55:37-44.
22. Krieg AM. CpG motifs in bacterial DNA and their immune effects. Annu Rev Immunol 2002; 20:709-760.
23. Dalpke A, Zimmermann S, Heeg K. Immunopharmacology of CpG DNA. Biol Chem 2002; 383:1491-1500.
24. Bell JK, Mullen GE, Leifer CA et al. Leucine-rich repeats and pathogen recognition in Toll-like receptors. Trends Immunol 2003; 24:528-533.
25. Ahmad-Nejad P, Hacker H, Rutz M et al. Bacterial CpG-DNA and lipopolysaccharides activate Toll-like receptors at distinct cellular compartments. Eur J Immunol 2002; 32:1958-1968.
26. Hacker H. Signal transduction pathways activated by CpG-DNA. Curr Top Microbiol Immunol 2000; 247:77-92.
27. Hornung V, Rothenfusser S, Britsch S et al. Quantitative expression of toll-like receptor 1-10 mRNA in cellular subsets of human peripheral blood mononuclear cells and sensitivity to CpG oligodeoxynucleotides. J Immunol 2002; 168:4531-4537.
28. Kadowaki N, Ho S, Antonenko S et al. Subsets of human dendritic cell precursors express different toll-like receptors and respond to different microbial antigens. J Exp Med 2001; 194:863-869.
29. Shimada S, Yano O, Tokunaga T. In vivo augmentation of natural killer cell activity with a deoxyribonucleic acid fraction of BCG. Jpn J Cancer Res 1986; 7:808-816.
30. Yamamoto S, Yamamoto T, Tokunaga T. The discovery of immunostimulatory DNA sequence. Springer Semin Immunopathol 2000; 22:11-19.

31. Yamamoto S, Yamamoto T, Kataoka T et al. Unique palindromic sequences in synthetic oligonucleotides are required to induce INF and augment INF-mediated natural killer activity. J Immunol 1992; 148:4072-4076.
32. Tokunaga T, Yano O, Kuramoto E et al. Synthetic oligonucleotides with particular base sequences from the cDNA encoding proteins of Mycobacterium bovis BCG induce interferons and activate natural killer cells. Microbiol Immunol 1992; 36:55-66.
33. Verthelyi D, Ishii KJ, Gursel M et al. Human peripheral blood cells differentially recognize and respond to two distinct CPG motifs. J Immunol 2001; 166:2372-2377.
34. Krug A, Rothenfusser S, Hornung V et al. Identification of CpG oligonucleotide sequences with high induction of IFN-alpha/beta in plasmacytoid dendritic cells. Eur J Immunol 2001; 31:2154-2163.
35. Agrawal S, Iadarola PL, Temsamani J et al. Effect of G-rich sequences on the synthesis, purification, binding, cell uptake, and hemolytic activity of oligonucleotides. Bioorg Med Chem Lett 1996; 6:2219-2224.
36. Kimura Y, Sonehara K, Kuramoto E et al. Binding of oligoguanylate to scavenger receptors is required for oligonucleotides to augment NK cell activity and induce IFN. J Biochem 1994; 116:991-994.
37. Rando RF, Hogan ME. Biological activity of guanosine quartet-forming oligonucleotides. In Stein CA, Krieg AM, eds. Applied Antisense Oligonucleotide Technology. New York: Wiley-Liss, 1998:335-352.
38. Agrawal S, Tan W, Cai Q et al. In vivo pharmacokinetics of phosphorothioate oligonucleotides containing contiguous guanosines. Antisense Nucleic Acid Drug Dev 1997; 7:245-249.
39. Tanaka T, Chu CC, Paul WE. An antisense oligonucleotide complementary to a sequence in I gamma 2b increases gamma 2b germline transcripts, stimulates B cell DNA synthesis, and inhibits immunoglobulin secretion. J Exp Med 1992; 175:597-607.
40. McIntyre KW, Lombard-Gillooly K, Perez JR et al. A sense phosphorothioate oligonucleotide directed to the initiation codon of transcription factor NF-kappa B p65 causes sequence-specific immune stimulation. Antisense Res Dev 1993; 3:309-322.
41. Mojcik CF, Gourley MF, Klinman DM et al. Administration of a phosphorothioate oligonucleotide antisense to murine endogenous retroviral MCF env causes immune effects in vivo in a sequence-specific manner. Clin Immunol Immunopathol 1993; 67:130-136.
42. Pisetsky DS, Reich CF. Stimulation of murine lymphocyte proliferation by a phosphorothioate oligonucleotide with antisense activity for herpes simplex virus. Life Sci 1994; 54:101-107.
43. Lewis EJ, Agrawal S, Bishop J et al. Non-specific antiviral activity of antisense molecules targeted to the E1 region of human papillomavirus. Antiviral Res 2000; 48:187-196.
44. Krieg AM. Leukocyte stimulation by oligodeoxynucleotides. In Stein CA, Krieg AM, eds. Applied Antisense Oligonucleotide Technology. New York: Wiley-Liss, 1998:431-448.
45. Agrawal S, Kandimalla ER. Antisense therapeutics: Is it as simple as complementary base recognition? Mol Med Today 2000; 6:72-81.
46. Agrawal S, Kandimalla ER. Antisense and/or immunostimulatory oligonucleotide therapeutics. Curr Cancer Drug Targets 2001; 1:197-209.
47. Jahrsdorfer B, Jox R, Muhlenhoff L et al. Modulation of malignant B cell activation and apoptosis by bcl-2 antisense ODN and immunostimulatory CpG ODN. J Leukoc Biol 2002; 72:83-92.
48. Lai JC, Benimetskaya L, Santella RM et al. G3139 (oblimersen) may inhibit prostate cancer cell growth in a partially bis-CpG-dependent non-antisense manner. Mol Cancer Ther 2003; 2:1031-1043.
49. Hartmann G, Battiany J, Poeck H et al. Rational design of new CpG oligonucleotides that combine B cell activation with high IFN-α induction in plasmacytoic cells. Eur J Immunol 2003; 33:1633-1641.
50. Marshall JD, Fearon K, Abbate C et al. Identification of a novel CpG DNA class and motif, which optimally stimulate B cell and plasmacytoid dendritic cell functions. J Leukoc Biol 2003; 73:781-792.
51. Chu W, Gong X, Li Z et al. DNA-PKcs is required for activation of innate immunity by immunostimulatory DNA. Cell 2000; 103:909-918.

52. Ishii KJ, Takeshita F, Gursel M et al. Potential role of phosphatidylinositol 3 kinase, rather than DNA-dependent protein kinase, in CpG DNA-induced immune activation. J Exp Med 2002; 196:269-274.
53. Bandholtz L, Guo Y, Palmberg C et al. Hsp90 binds CpG oligonucleotides directly: implications for hsp90 as a missing link in CpG signaling and recognition. Cell Mol Life Sci 2003; 60:422-429.
54. Agrawal S, Kandimalla ER. Medicinal chemistry and therapeutic potential of CpG DNA. Trends Mol Med 2002; 8:114-121.
55. Kandimalla ER, Yu D, Agrawal S. Towards optimal design of second-generation immunomodulatory oligonucleotides. Curr Opin Mol Ther 2002; 4:122-129.
56. Kandimalla ER, Zhu FG, Bhagat L et al. Toll-like receptor 9: Modulation of recognition and cytokine induction by novel synthetic CpG DNAs. Biochem Soc Trans 2003; 31:653-657.
57. Agrawal S, Kandimalla ER. Modulation of Toll-like receptor 9 responses through synthetic immunostimulatory motifs of DNA. Ann NY Acad Sci 2003; 1002:30-42.
58. Zhao Q, Temsamani J, Iadarola PL et al. Effect of different chemically modified oligodeoxynucleotides on immune stimulation. Biochem Pharmacol 1996; 51:173-182.
59. Yu D, Kandimalla ER, Zhao Q et al. Immunostimulatory activity of CpG oligonucleotides containing non-ionic methylphosphonate linkages. Bioorg Med Chem 2001; 9:2803-2808.
60. Zhao Q, Yu D, Agrawal S. Site of chemical modifications in CpG containing phosphorothioate oligodeoxynucleotide modulates its immunostimulatory activity. Bioorg Med Chem Lett 1999; 9:3453-3458.
61. Zhao Q, Yu D, Agrawal S. Immunostimulatory activity of CpG containing phosphorothioate oligodeoxynucleotide is modulated by modification of a single deoxynucleoside. Bioorg Med Chem Lett 2000; 10:1051-1054.
62. Yu D, Kandimalla ER, Zhao Q et al. Immunostimulatory properties of phosphorothioate CpG DNA containing both 3'-5'- and 2'-5'-internucleotide linkages. Nucleic Acids Res 2002; 30:1613-1619.
63. Yu D, Kandimalla ER, Zhao Q et al. Modulation of immunostimulatory activity of CpG oligonucleotides by site-specific deletion of nucleobases. Bioorg Med Chem Lett 2001; 11:2263-2267.
64. Yu D, Kandimalla ER, Cong Y et al. Design, synthesis, and immunostimulatory properties of CpG DNAs containing alkyl-linker substitutions: role of nucleosides in the flanking sequences. J Med Chem 2002; 45:4540-4548.
65. Van Uden J, Raz E. Introduction to immunostimulatory DNA sequences. Springer Semin Immunopathol 2000; 22:1-9.
66. Yu D, Kandimalla ER, Zhao Q et al. Requirement of nucleobase proximal to CpG dinucleotide for immunostimulatory activity of synthetic CpG DNA. Bioorg Med Chem 2003; 11:459-464.
67. Yu D, Zhao Q, Kandimalla ER et al. Accessible 5'-end of CpG-containing phosphorothioate oligodeoxynucleotides is essential for immunostimulatory activity. Bioorg Med Chem Lett 2000; 10:2585-2588.
68. Kandimalla ER, Bhagat L, Yu D et al. Conjugation of ligands at the 5'-end of CpG DNA affects immunostimulatory activity. Bioconjug Chem 2002; 13:966-974.
69. Yu D, Kandimalla ER, Bhagat L et al. 'Immunomers'--novel 3'-3'-linked CpG oligodeoxyribonucleotides as potent immunomodulatory agents. Nucleic Acids Res 2002; 30:4460-4469.
70. Marshall JD, Hessel EM, Gregorio J et al. Novel chimeric immunomodulatory compounds containing short CpG oligodeoxyribonucleotides have differential activities in human cells. Nucleic Acids Res 2003; 31:5122-5133.
71. Fearon K, Marshall JD, Abbate C et al. A minimal human immunostimulatory CpG motif that potently induces IFN-gamma and IFN-alpha production. Eur J Immunol 2003; 33:2114-2122.
72. Narayanan S, Dalpke AH, Siegmund K et al. CpG oligonucleotides with modified termini and nicked dumbbell structure show enhanced immunostimulatory activity. J Med Chem 2003; 46:5031-5044.
73. Kandimalla ER, Bhagat L, Wang D et al. Divergent synthetic nucleotide motif recognition pattern: design and development of potent immunomodulatory oligodeoxyribonucleotide agents with distinct cytokine induction profiles. Nucleic Acids Res 2003; 31:2393-2400.
74. Bhagat L, Zhu FG, Yu D et al. CpG penta- and hexadeoxyribonucleotides as potent immunomodulatory agents. Biochem Biophys Res Commun 2003; 300:853-861.

75. Dalpke AH, Zimmermann S, Albrecht I et al. Phosphodiester CpG oligonucleotides as adjuvants: polyguanosine runs enhance cellular uptake and improve immunostimulative activity of phosphodiester CpG oligonucleotides in vitro and in vivo. Immunology 2002; 106:102-112.
76. Yu D, Zhu FG, Bhagat L et al. Potent CpG oligonucleotides containing phosphodiester linkages: in vitro and in vivo immunostimulatory properties. Biochem Biophys Res Commun 2002; 297:83-90.
77. Yu D, Kandimalla ER, Roskey A et al. Stereo-enriched phosphorothioate oligodeoxynucleotides: synthesis, biophysical and biological properties. Bioorg Med Chem 2000; 8:275-284.
78. Kandimalla ER, Yu D, Zhao Q et al. Effect of chemical modifications of cytosine and guanine in a CpG-motif of oligonucleotides: structure-immunostimulatory activity relationships. Bioorg Med Chem 2001; 9:807-813.
79. Kandimalla ER, Bhagat L, Zhu FG et al. A dinucleotide motif in oligonucleotides shows potent immunomodulatory activity and overrides species-specific recognition observed with CpG motif. Proc Natl Acad Sci USA 2003; 100:14303-14308.
80. Kandimalla ER, Bhagat L, Cong YP et al. Secondary structures in CpG oligonucleotides affect immunostimulatory activity. Biochem Biophys Res Commun 2003; 306:948-953.
81. Cong YP, Song SS, Bhagat L et al. Self-stabilized CpG DNAs optimally activate human B cells and plasmacytoid dendritic cells. Biochem Biophys Res Commun 2003; 310:1133-1139.
82. Hartmann G, Weeratna RD, Ballas ZK et al. Delineation of a CpG phosphorothioate oligodeoxynucleotide for activating primate immune responses in vitro and in vivo. J Immunol 2000; 164:1617-1624.
83. Jorgensen JB, Johansen LH, Steiro K et al. CpG DNA induces protective antiviral immune responses in Atlantic salmon (Salmo salar L.). J Virol 2003; 77:11471-11479.
84. Lee CH, Jeong HD, Chung JK et al. CpG motif in synthetic ODN primes respiratory burst of olive flounder Paralichthys olivaceus phagocytes and enhances protection against Edwardsiella tarda. Dis Aquat Organ 2003; 56:43-48.
85. Bridle AR, Butler R, Nowak BF. Immunostimulatory CpG oligodeoxynucleotides increase resistance against amoebic gill disease in Atlantic salmon, Salmo salar L. J Fish Dis 2003; 26:367-371.
86. Gomis S, Babiuk L, Godson DL et al. Protection of chickens against Escherichia coli infections by DNA containing CpG motifs. Infect Immun 2003; 71:857-863.
87. Xie H, Raybourne RB, Babu US et al. CpG-induced immunomodulation and intracellular bacterial killing in a chicken macrophage cell line. Dev Comp Immunol 2003; 27:823-834.
88. Babiuk IA, Gomis S, Hecker R. Molecular approaches to disease control. Poult Sci 2003; 82:870-875.
89. Vleugels B, Ververken C, Goddeeris BM. Stimulatory effect of CpG sequences on humoral response in chickens. Poult Sci 2002; 81:1317-1321.
90. Rankin R, Pontarollo R, Ioannou X et al. CpG motif identification for veterinary and laboratory species demonstrates that sequence recognition is highly conserved. Antisense Nucleic Acid Drug Dev 2001; 11:333-340.
91. Leutenegger CM, Boretti FS, Mislin CN et al. Immunization of cats against feline immunodeficiency virus (FIV) infection by using minimalistic immunogenic defined gene expression vector vaccines expressing FIV gp140 alone or with feline interleukin-12 (IL-12), IL-16, or a CpG motif. J Virol 2000; 74:10447-10457.
92. Wernette CM, Smith BF, Barksdale ZL et al. CpG oligodeoxynucleotides stimulate canine and feline immune cell proliferation. Vet Immunol Immunopathol 2002; 84:223-236.
93. Zhang Y, Palmer GH, Abbott JR et al. CpG ODN 2006 and IL-12 are comparable for priming Th1 lymphocyte and IgG responses in cattle immunized with a rickettsial outer membrane protein in alum. Vaccine 2003; 21:3307-3318.
94. Ioannou XP, Griebel P, Hecker R et al. The immunogenicity and protective efficacy of bovine herpesvirus 1 glycoprotein D plus Emulsigen are increased by formulation with CpG oligodeoxynucleotides. J Virol 2002; 76:9002-9010.
95. Van der Stede Y, Verdonck F, Vancaeneghem S et al. CpG-oligodinucleotides as an effective adjuvant in pigs for intramuscular immunizations. Vet Immunol Immunopathol 2002; 86:31-41.
96. Van der Stede Y, Cox E, Verdonck F et al. Reduced faecal excretion of F4+-E. coli by the intramuscular immunisation of suckling piglets by the addition of 1-a,25-dihydroxy vitamin D3 or CpG oligodeoxynucleotides. Vaccine 2003; 21:1023-1032.

97. Dalpke A, Zimmermann A, Heeg K. CpG DNA in the prevention and treatment of infections. BioDrugs 2002; 16:419-431.
98. Dittmer U, Olbrich AR. Treatment of infectious diseases with immunostimulatory oligodeoxynucleotides containing CpG motifs. Curr Opin Microbiol 2003; 6:472-477.
99. Temperton NJ, Quenelle DC, Lawson KM et al. Enhancement of humoral immune responses to a human cytomegalovirus DNA vaccine: adjuvant effects of aluminum phosphate and CpG oligodeoxynucleotides. J Med Virol 2003; 70:86-90.
100. Porter KR, Kochel TJ, Wu SJ et al. Protective efficacy of a dengue 2 DNA vaccine in mice and the effect of CpG immuno-stimulatory motifs on antibody responses. Arch Virol 1998; 143:997-1003.
101. Leutenegger CM, Boretti FS, Mislin CN et al. Immunization of cats against feline immunodeficiency virus (FIV) infection by using minimalistic immunogenic defined gene expression vector vaccines expressing FIV gp140 alone or with feline interleukin-12 (IL-12), IL-16, or a CpG motif. J Virol 2000; 74,10447-10457.
102. Shieh JJ, Liang CM, Chen CY et al. Enhancement of the immunity to foot-and-mouth disease virus by DNA priming and protein boosting immunization. Vaccine 2001; 19,4002-4010.
103. Olbrich AR, Schimmer S, Dittmer U. Preinfection treatment of resistant mice with CpG oligodeoxynucleotides renders them susceptible to friend retrovirus-induced leukemia. J Virol 2003; 77:10658-10662.
104. Fiedler M, Lu M, Siegel F et al. Immunization of woodchucks (Marmota monax) with hepatitis delta virus DNA vaccine. Vaccine 2001; 19,4618-4626.
105. Halperin SA, Van Nest G, Smith B et al. A phase I study of the safety and immunogenicity of recombinant hepatitis B surface antigen co-administered with an immunostimulatory phosphorothioate oligonucleotide adjuvant. Vaccine 2003; 21:2461-2467.
106. Davis HL, Suparto II, Weeratna RR et al. CpG DNA overcomes hyporesponsiveness to hepatitis B vaccine in orangutans. Vaccine 2000; 18:1920-1924.
107. Malanchere-Bres E, Payette PJ, Mancini M et al. CpG oligodeoxynucleotides with hepatitis B surface antigen (HBsAg) for vaccination in HBsAg-transgenic mice. J Virol 2001; 75:6482-6491.
108. Encke J, zu Putlitz J, Stremmel W et al. CpG immuno-stimulatory motifs enhance humoral immune responses against hepatitis C virus core protein after DNA-based immunization. Arch Virol 2003; 148:435-448.
109. Qiao M, Murata K, Davis AR et al. Hepatitis C virus-like particles combined with novel adjuvant systems enhance virus-specific immune responses. Hepatology 2003; 37:52-59.
110. Tritel M, Stoddard AM, Flynn BJ et al. Prime-boost vaccination with HIV-1 Gag protein and cytosine phosphate guanosine oligodeoxynucleotide, followed by adenovirus, induces sustained and robust humoral and cellular immune responses. J Immunol 2003; 171:2538-2547.
111. Ayash-Rashkovsky M, Weisman Z, Diveley J et al. Generation of Th1 immune responses to inactivated, gp120-depleted HIV-1 in mice with a dominant Th2 biased immune profile via immunostimulatory oligonucleotides—relevance to AIDS vaccines in developing countries. Vaccine 2002; 20:2684-2692.
112. yles RB, Higgins D, Chalk C et al. Use of immunostimulatory sequence-containing oligonucleotides as topical therapy for genital herpes simplex virus type 2 infection. J Virol 2002; 76:11387-11396.
113. Harandi AM, Eriksson K, Holmgren J. A protective role of locally administered immunostimulatory CpG oligodeoxynucleotide in a mouse model of genital herpes infection. J Virol 2003; 77:953-962.
114. Dong L, Mori I, Hossain MJ et al. An immunostimulatory oligodeoxynucleotide containing a cytidine-guanosine motif protects senescence-accelerated mice from lethal influenza virus by augmenting the T helper type 1 response. J Gen Virol 2003; 84:1623-1628.
115. Kovarik J, Bozzotti P, Love-Homan L et al. CpG oligodeoxynucleotides can circumvent the Th2 polarization of neonatal responses to vaccines but may fail to fully redirect Th2 responses established by neonatal priming. J Immunol 1999; 162:1611-1617.
116. Chu RS, McCool T, Greenspan NS et al. CpG oligodeoxynucleotides act as adjuvants for pneumococcal polysaccharide-protein conjugate vaccines and enhance antipolysaccharide immunoglobulin G2a (IgG2a) and IgG3 antibodies. Infect Immun 2000; 68:1450-1456.

117. Kovarik J, Bozzotti P, Tougne C et al. Adjuvant effects of CpG oligodeoxynucleotides on responses against T-independent type 2 antigens. Immunology 2001; 102:67-76.
118. Hancock GE, Heers KM, Smith JD et al. CpG containing oligodeoxynucleotides are potent adjuvants for parenteral vaccination with the fusion (F) protein of respiratory syncytial virus (RSV). Vaccine 2001; 19:4874-4882.
119. Cafaro A, Titti F, Fracasso C et al. Vaccination with DNA containing tat coding sequences and unmethylated CpG motifs protects cynomolgus monkeys upon infection with simian/human immunodeficiency virus (SHIV89.6P). Vaccine 2001; 19:2862-2877.
120. Boyaka PN, Tafaro A, Fischer R et al. Effective mucosal immunity to anthrax: neutralizing antibodies and Th cell responses following nasal immunization with protective antigen. J Immunol 2003; 170:5636-5643.
121. Al-Mariri A, Tibor A, Mertens P et al. Protection of BALB/c mice against Brucella abortus 544 challenge by vaccination with bacterioferritin or P39 recombinant proteins with CpG oligodeoxynucleotides as adjuvant. Infect Immun 2001; 69:4816-4822.
122. Raghavan S, Nystrom J, Fredriksson M et al. Orally administered CpG oligodeoxynucleotide induces production of CXC and CC chemokines in the gastric mucosa and suppresses bacterial colonization in a mouse model of Helicobacter pylori infection. Infect Immun 2003; 71:7014-7022.
123. Elkins KL, Rhinehart-Jones TR, Stibitz S et al. Bacterial DNA containing CpG motifs stimulates lymphocyte-dependent protection of mice against lethal infection with intracellular bacteria. J Immunol 1999; 162:2291-2298.
124. Krieg AM, Love-Homan L, Yi AK et al. CpG DNA induces sustained IL-12 expression in vivo and resistance to Listeria monocytogenes challenge. J Immunol 1998; 161:2428-2434.
125. Klinman DM, Conover J, Coban C. Repeated administration of synthetic oligodeoxynucleotides expressing CpG motifs provides long-term protection against bacterial infection. Infect Immun 1999; 67:5658-5663.
126. Oxenius A, Martinic MM, Hengartner H et al. CpG-containing oligonucleotides are efficient adjuvants for induction of protective antiviral immune responses with T-cell peptide vaccines. J Virol 1999; 73,4120-4126.
127. Freidag BL, Melton GB, Collins F et al. CpG oligodeoxynucleotides and interleukin-12 improve the efficacy of Mycobacterium bovis BCG vaccination in mice challenged with M. tuberculosis. Infect Immun 2000; 68,2948-2953.
128. Hogarth PJ, Jahans KJ, Hecker R et al. Evaluation of adjuvants for protein vaccines against tuberculosis in guinea pigs. Vaccine 2003; 21:977-982.
129. Cervi L, Borgonovo J, Egea M et al. Immunization of rats against Fasciola hepatica using crude antigens conjugated with Freund's adjuvant or oligodeoxynucleotides. Vet Immunol Immunopathol 2004; 97:97-104.
130. Zimmermann S, Egeter O, Hausmann S et al. CpG oligodeoxynucleotides trigger protective and curative Th1 responses in lethal murine leishmaniasis. J Immunol 1998; 160:3627-3630.
131. Stacey KJ, Blackwell JM. Immunostimulatory DNA as an adjuvant in vaccination against Leishmania major. Infect Immun 1999; 67:3719-3726.
132. Verthelyi D, Gursel M, Kenney RT et al. CpG oligodeoxynucleotides protect normal and SIV-infected macaques from Leishmania infection. J Immunol 2003; 170:4717-4723.
133. Gramzinski RA, Doolan DL, Sedegah M et al. Interleukin-12- and gamma interferon-dependent protection against malaria conferred by CpG oligodeoxynucleotide in mice. Infect Immun 2001; 69,1643-1649.
134. Su Z, Tam MF, Jankovic D et al. Vaccination with novel immunostimulatory adjuvants against blood-stage malaria in mice. Infect Immun 2003; 71:5178-5187.
135. Hirunpetcharat C, Wipasa J, Sakkhachornphop S et al. CpG oligodeoxynucleotide enhances immunity against blood-stage malaria infection in mice parenterally immunized with a yeast-expressed 19 kDa carboxyl-terminal fragment of Plasmodium yoelii merozoite surface protein-1 (MSP1(19)) formulated in oil-based Montanides. Vaccine 2003; 21:2923-2932.
136. Near KA, Stowers AW, Jankovic D et al. Improved immunogenicity and efficacy of the recombinant 19-kilodalton merozoite surface protein 1 by the addition of oligodeoxynucleotide and aluminum hydroxide gel in a murine malaria vaccine model. Infect Immun 2002; 70:692-701.

137. Coban C, Ishii KJ, Stowers AW et al. Effect of CpG Oligodeoxynucleotides on the Immunogenicity of Pfs25, a Plasmodium falciparum Transmission-Blocking Vaccine Antigen. Infect Immun 2004; 72:584-588.
138. Corral RS, Petray PB. CpG DNA as a Th1-promoting adjuvant in immunization against Trypanosoma cruzi. Vaccine 2000; 19:234-242.
139. Frank FM, Petray PB, Cazorla SI et al. Use of a purified Trypanosoma cruzi antigen and CpG oligodeoxynucleotides for immunoprotection against a lethal challenge with trypomastigotes. Vaccine 2003; 22:77-86.
140. Bozza S, Gaziano R, Lipford GB et al. Vaccination of mice against invasive aspergillosis with recombinant Aspergillus proteins and CpG oligodeoxynucleotides as adjuvants. Microbes Infect 2002; 4:1281-1290.
141. Svirshchevskaya EV, Kurup VP. Immunotherapy of allergic bronchopulmonary aspergillosis: a clinical and experimental approach. Front Biosci 2003; 8:s92-s101.
142. Maitta RW, Datta K, Lees A et al. Immunogenicity and Efficacy of Cryptococcus neoformans Capsular Polysaccharide Glucuronoxylomannan Peptide Mimotope-Protein Conjugates in Human Immunoglobulin Transgenic Mice. Infect Immun 2004; 72:196-208.
143. Bafica A, Scanga CA, Equils O et al. The induction of Toll-like receptor tolerance enhances rather than suppresses HIV-1 gene expression in transgenic mice. J Leukoc Biol 2004; 75:460-466.
144. Equils O, Schito ML, Karahashi H et al. Toll-like receptor 2 (TLR2) and TLR9 signaling results in HIV-long terminal repeat trans-activation and HIV replication in HIV-1 transgenic mouse spleen cells: implications of simultaneous activation of TLRs on HIV replication. J Immunol 2003; 170:5159-5164.
145. Agrawal S, Martin RR. Was induction of HIV-1 through TLR9? J Immunol 2003; 171:1621.
146. Lisziewicz J, Sun D, Weichold FF et al. Antisense oligonucleotide phosphorothioate complementary to gag mRNA blocks replication of human immunodeficiency virus in human peripheral blood cells. Proc Natl Acad Sci USA 1994; 91:7942-7946.
147. Carpentier AF, Auf G, Delattre JY. CpG-oligonucleotides for cancer immunotherapy: Review of the literature and potential applications in malignant giloma. Front Biosci 2003; 8:E115-E127.
148. Sfondrini L, Balsari A, Menard S. Innate immunity in breast carcinoma. Endocrine-Related Cancer 2003; 10:301-308.
149. Okamoto M, Sato M. Toll-like receptor signaling in anti-cancer immunity. J Med Invest 2003; 50:9-24.
150. Warren TL, Weiner GJ. Synergism between cytosine-guanine oligodeoxynucleotides and monoclonal antibody in the treatment of lymphoma. Semin Oncol 2002; 29(1 Suppl 2):93-97.
151. Wang D, Li Y, Yu D et al. Immunopharmacological and antitumor effects of second-generation immunomodulatory oligonucleotides containing synthetic CpR motifs. Int J Oncol 2004; 24:901-908.
152. Trinchieri G. Interleukin-12: a proinflammatory cytokine with immunoregulatory functions that bridge innate resistance and antigen-specific adaptive immunity. Annu Rev Immunol 1995; 13:251-276.
153. Silverman ES, Drazen JM. Immunostimulatory DNA for asthma: better than eating dirt. Am J Respir Cell Mol Biol 2003; 28:645-647.
154. Jain VV, Kitagaki K, Kline JN. CpG DNA and immunotherapy of allergic airway diseases. Clin Exp Allergy 2003; 33:1330-1335.
155. Spiegelberg HL, Takabayashi K, Beck L et al. DNA-based vaccines for allergic disease. Expert Rev Vaccines 2002; 1:169-177.
156. Bohle B. CpG motifs as possible adjuvants for the treatment of allergic diseases. Int Arch Allergy Immunol 2002; 129:198-203.
157. Agrawal DK, Edwin J, Kandimalla ER et al. Novel immunomodulatory oligonucleotides (IMOs) prevent development of allergic airway inflammation and airway hyperresponsiveness in asthma. Int Immunopharmacol 2004; 4:127-138.
158. Shirota H, Sano K, Hirasawa N et al. Novel roles of CpG oligodeoxynucleotides as a leader for the sampling and presentation of CpG-tagged antigen by dendritic cells. J Immunol 2001; 167:66-74.

159. Shirota H, Sano K, Kikuchi T et al. Regulation of murine airway eosinophilia and Th2 cells by antigen-conjugated CpG oligodeoxynucleotides as a novel antigen-specific immunomodulator. J Immunol 2000; 164:5575-5582
160. Kim SK, Raghupathi G, Cappello S et al. Effects of immunological adjuvant combinations on the antibody and T cell response to vaccination with MUCI-KLH and GD3-KLH conjugates. Vaccine 2000; 19:530-537
161. Sato Y, Roman M, Tighe H et al. Immunostimulatory DNA sequences necessary for effective intradermal gene immunization. Science 1996; 273:352-354.
162. Sasaki S, Takeshita F, Xin KQ et al. Adjuvant formulations and delivery systems for DNA vaccines. Methods 2003; 31:243-254.
163. Gurunathan S, Klinman DM, Seder RA. DNA vaccines: immunology, application, and optimization. Annu Rev Immunol 2000; 18:927-974.
164. Nahori MA, Lagranderie M, Lefort J et al. Effects of Mycobacterium bovis BCG on the development of allergic inflammation and bronchial hyperresponsiveness in hyper-IgE BP2 mice vaccinated as newborns. Vaccine 2001; 19:1484-1495.
165. Hopfenspirger MT, Parr SK, Hopp RJ et al. Mycobacterial antigens attenuate late allergic response, airway hyperresponsiveness, and bronchoalveolar lavage eosinophilia in a mouse model of bronchial asthma. Int Immunopharmacol 2001; 1:1743-1751.
166. Hopfenspirger MT, Parr SK, Townley RG et al. Attenuation of the late allergic response by mycobacterial antigens is independent of IgE in a mouse model of asthma. Allergology Int 2002; 51:21-32.
167. Hopfenspirger MT, Agrawal DK. Airway hyperresponsiveness, late allergic response, and eosinophilia are reversed with mycobacterial antigens in ovalbumin-presensitized mice. J Immunol 2002; 168:2516-2522.
168. Shirakawa T, Enomoto T, Shimazu S et al. The inverse association between tuberculin responses and atopic disorder. Science 1997; 275:77-79.
169. Aaby P, Shaheen SO, Heyes CB et al. Early BCG vaccination and reduction in atopy in Guinea-Bissau. Clin Exp Allergy 2000; 30:644-650.
170. Hersh EM, Gutterman JU, Mavligit GM. BCG as adjuvant immunotherapy for neoplasia. Ann Rev Med 1977; 28:489-515.
171. Bohle A, Brandau S. Immune mechanisms in bacillus Calmette-Guerin immunotherapy for superficial bladder cancer. J Urol 2003; 170:964-969.
172. Chabalgoity JA, Dougan G, Mastroeni P et al. Live bacteria as the basis for immunotherapies against cancer. Expert Rev Vaccines 2002; 1:495-505.
173. Skinner MA, Buddle BM, Wedlock DN et al. A DNA prime-Mycobacterium bovis BCG boost vaccination strategy for cattle induces protection against bovine tuberculosis. Infect Immun 2003; 71:4901-4907.
174. Horner AA, Raz E. Immunostimulatory sequence oligodeoxynucleotide-based vaccination and immunomodulation: two unique but complementary strategies for the treatment of allergic diseases. J Allergy Clin Immunol 2002; 110:706-712.
175. Martin R, Sullivan T, Bhagat L et al. A phase I placebo-controlled study in volunteers of escalating doses of HYB2055, a second-generation immunomodulatory agent based on CpG DNA. AACR-NCI-EORTC International Conference on Molecular Targets and Therapeutics, 2003, Boston, MA, Abstract # C100.
176. Hwang JJ, Martin R, Fox-Sinclair ES et al. A phase I study of a second-generation immunomodulatory oligonucleotide (HYB2055) in patients with advanced solid malignancies. AACR-NCI-EORTC International Conference on Molecular Targets and Therapeutics, 2003, Boston, MA, Abstract # C111.
177. Agrawal, S. Importance of nucleotide sequence and chemical modification of antisense oligonucleotides. Biochim Biophys Acta 1999; 1489:53-68.
178. Hackett CJ. Innate immune activation as a broad-spectrum biodefense strategy: prospects and research challenges. J Allergy Clin Immunol 2003; 112:686-694.

INDEX

D

E

F

G

H

I

S

T

U

V

W

Y

Z